Urban Air Pollution and Plant Responses

Urban Air Pollution and Plant Responses

The Critical Connection Between Urban Air Quality and Plants

Edited by

Ansuman Sahoo
Department of Botany, D.A.V. Degree College, Muzaffarnagar, Uttar Pradesh, India

Supriya Tiwari
Department of Botany, Laboratory of Ecotoxicology, Institute of Science, Banaras Hindu University, Varanasi, Uttar Pradesh, India

ELSEVIER

Elsevier
Radarweg 29, PO Box 211, 1000 AE Amsterdam, Netherlands
125 London Wall, London EC2Y 5AS, United Kingdom
50 Hampshire Street, 5th Floor, Cambridge, MA 02139, United States

ISBN: 978-0-443-40320-0

For Information on all Elsevier publications
visit our website at https://www.elsevier.com/books-and-journals

Publisher: Peter Llewellyn
Acquisitions Editor: Jennette McClain
Editorial Project Manager: Abhilasha Agarwal
Production Project Manager: Gomathi Sugumar
Cover Designer: Mark Rogers

Typeset by MPS Limited, Chennai, India

Contents

List of contributors

Naushad Ahamad
Department of Botany, University of Lucknow, Lucknow, Uttar Pradesh, India

Syed Zaki Ahmed
Department of Ecology and Environmental Sciences, School of Life Sciences, Pondicherry University, Puducherry, India

Carmen Arena
Department of Biology, University of Naples Federico II, Naples, Italy

Daniela Baldantoni
Department of Chemistry and Biology "Adolfo Zambelli", University of Salerno, Salerno, Italy

Krishna Kumar Choudhary
Department of Botany, MMV, Banaras Hindu University, Varanasi, Uttar Pradesh, India

Hamed Dadkhah-Aghdash
Department of Plant Biology, Faculty of Biological Sciences, Tarbiat Modares University, Tehran, Iran

Paul Fatoba
Department of Plant Biology, University of Ilorin, Ilorin, Kwara, Nigeria

Shivani Gupta
Department of Botany, Laboratory of Ecotoxicology, Institute of Science, Banaras Hindu University, Varanasi, Uttar Pradesh, India; Department of Chemical Engineering & Technology, Green Processing & Synthesis Lab, Indian Institute of Technology, Banaras Hindu University, Varanasi, Uttar Pradesh, India

Priyanka Khilar
Research and Development, Revelation's Biotech, Hyderabad, Telangana, India

Umesh Kumar
Department of Botany, Banaras Hindu University, Varanasi, Uttar Pradesh, India

Pradeep Kumar Mishra
Department of Chemical Engineering & Technology, Green Processing & Synthesis Lab, Indian Institute of Technology, Banaras Hindu University, Varanasi, Uttar Pradesh, India

Saumya Mishra
Department of Botany, Banaras Hindu University, Varanasi, Uttar Pradesh, India

Pradipta Kumar Mohapatra
Department of Botany, Ravenshaw University, Cuttack, Odisha, India

Giuseppe Muscari Tomajoli
Department of Biology, University of Naples Federico II, Naples, Italy

Mattia Napoletano
Department of Chemistry and Biology "Adolfo Zambelli", University of Salerno, Salerno, Italy

Clement Ogunkunle
Department of Plant Biology, University of Ilorin, Ilorin, Kwara, Nigeria

Ayodele Oluwatobi
Department of Biological Sciences, Ojaja University, Eiyenkorin, Kwara, Nigeria

Stephen Oyedeji
Department of Plant Biology, University of Ilorin, Ilorin, Kwara, Nigeria

Nikita Patel
Kishorbhai Institute of Agriculture Sciences and Research Centre, Uka Tarsadia University, Bardoli, Gujarat, India

Avanish Prajapati
Department of Botany, Banaras Hindu University, Varanasi, Uttar Pradesh, India

Richa Rai
Plant Ecology and Climate Change Science Division, CSIR-National Botanical Research Institute, Lucknow, Uttar Pradesh, India

Abdulfatai Rufai
Department of Plant Biology, Osun State University, Osogbo, Osun, Nigeria

Ansuman Sahoo
Department of Botany, D.A.V. Degree College, Muzaffarnagar, Uttar Pradesh, India

Rajesh Kumar Sharma
Department of Botany, Banaras Hindu University, Varanasi, Uttar Pradesh, India

Shasmita
Department of Botany, Dhenkanal Autonomous College, Dhenkanal, Odisha, India

Aditya Abha Singh
Department of Botany, University of Lucknow, Lucknow, Uttar Pradesh, India

Harshita Singh
Department of Botany, Institute of Science, Banaras Hindu University, Varanasi, Uttar Pradesh, India

Nidhi Singh
Centre of Experimental Medicine & Surgery, Institute of Medical Sciences, Banaras Hindu University, Varanasi, Uttar Pradesh, India

Prince Kumar Singh
Department of Botany, Banaras Hindu University, Varanasi, Uttar Pradesh, India

Priyanka Singh
Department of Botany, Institute of Science, Banaras Hindu University, Varanasi, Uttar Pradesh, India

Santosh Kumar Singh
Centre of Experimental Medicine & Surgery, Institute of Medical Sciences, Banaras Hindu University, Varanasi, Uttar Pradesh, India

Tejaswini Singh
Department of Ecology and Environmental Sciences, School of Life Sciences, Pondicherry University, Puducherry, India

Veer Singh
Department of Biosciences, School of Physical and Biological Sciences, FoSTA, Manipal University Jaipur, Jaipur, Rajasthan, India

Supriya Tiwari
Department of Botany, Laboratory of Ecotoxicology, Institute of Science, Banaras Hindu University, Varanasi, Uttar Pradesh, India

Rachna Tyagi
Department of Botany, D.A.V. College, Muzaffarnagar, Uttar Pradesh, India

Vineet Upadhyay
Department of Botany, MMV, Banaras Hindu University, Varanasi, Uttar Pradesh, India

About the editors

Ansuman Sahoo is an assistant professor in the Department of Botany at D.A.V. Degree College, Muzaffarnagar, Uttar Pradesh, India. His research focuses on exploring the impact of tropospheric ozone on leguminous crop plants and defining the cultivar-specific sensitivity towards elevated ozone stress. He has implored the strategic application of soil nitrogen amendments as a mitigative measure towards the management of ozone stress. His work also studies the temporal and spatial variations in tropospheric ozone concentrations at urban and rural areas, scouting the role of different ozone precursors in ozone formation. His credentials are well reflected in his published research articles in reputed international publications.

Supriya Tiwari is an associate professor in the Department of Botany at Banaras Hindu University, Varanasi, Uttar Pradesh, India. Her research focuses on the impact assessment and mechanistic evaluation of ozone phytotoxicity on agroecosystem productivity and its sustainable management. Her research team has been fortifying the global food security issues, as threatened by ozone stress, aligning with Sustainable Development Goals (SDGs), such as SDG 2 (zero hunger) and SDG 13 (climate change). Her research steers towards strategies for the management of phytotoxic effects of ozone, so as to sustain the qualitative and quantitative outputs of the agroecosystems, under a changing climate. With a strong academic background and a commitment to advancing scientific understanding, she has contributed to several research publications in reputed national and international journals, addressing the issues pertaining to ozone stress and its management in ozone-stressed agroecosystems.

Preface

Urban air pollution has emerged as one of the most urgent environmental concerns of the 21st century, fuelled by fast industrialization, urbanization, and an ever-increasing reliance on fossil fuels. The quality of the air we breathe continues to deteriorate as cities grow and human activity increases, posing major hazards to both human health and the integrity of natural ecosystems. Plants are among the most impacted elements of these ecosystems, acting as both active responders and silent victims of air pollution.

Urban Air Pollution and Plant Responses: The Critical Connection Between Urban Air Quality and Plants aims to provide a thorough understanding of the intricate interactions between airborne contaminants and plant systems. Plants, as solitary organisms, are constantly exposed to environmental stressors such as particulate matter, sulfur dioxide, nitrogen oxides, ozone, and heavy metals. These contaminants can disrupt plant physiology, morphology, and biochemical processes, affecting growth, production, and survival. At the same time, plants have exceptional adaptation systems that enable them to endure, resist, or even alleviate the negative impacts of pollution.

The core objective of this book is to bridge the gap between atmospheric studies and plant biology by offering a comprehensive view of how urban air pollution affects vegetation. It investigates the sources and types of urban air pollution, their distribution and transformation in the atmosphere, and their direct and indirect effects on plant life. Special attention is placed on physiological and biochemical responses, such as changes in photosynthesis, respiration, antioxidant activity, and secondary metabolite formation. These responses not only represent plant health but also provide useful metrics of environmental quality.

In addition to analyzing the adverse consequences of pollution, this book emphasizes the critical role of plants in reducing urban air pollution. Urban vegetation, which includes trees, shrubs, and green spaces, serves as a natural filter, absorbing pollutants, confining particulate matter, and improving overall air quality. The concept of phytoremediation and the selection of pollution-tolerant plant species are explored as long-term methods for urban environmental management.

This work is aimed at students, researchers, environmental scientists, and policymakers who want to understand the dynamics of urban ecosystems and the role of plants within them. It also seeks to promote awareness about the need to incorporate green infrastructure into urban development as a strategy for combating air pollution and increasing environmental resilience.

The information in this book is based on current scientific knowledge and research findings, with the goal of making complicated concepts understandable without sacrificing academic standards. It is intended that this book will inspire additional study and encourage the use of plant-based solutions to improve urban air quality.

Understanding the relationship between air pollution and plant responses is not only scientifically profound but also necessary for developing sustainable and habitable communities in an era of increasingly interrelated environmental concerns. This book represents a step toward that knowledge.

Ansuman Sahoo
Supriya Tiwari

Acknowledgments

The compilation of the edited volume of this book, entitled *Urban Air Pollution and Plant Responses: The Critical Connection Between Urban Air Quality and Plants,* has been a collaborative effort of all the contributors, and we would like to extend our sincere thanks and gratitude to them.

First and foremost, we extend our profound thanks to all the authors for their valuable insights and timely submissions. Their dedication, patience, and commitment have been instrumental in shaping the quality and scope of this book.

We are cordially grateful to the reviewers for their constructive feedback and thoughtful suggestions, which helped in refining the chapters and enhancing their overall quality.

We would also like to sincerely appreciate the support and guidance provided by the publishing team, whose professionalism and constant assistance ensured the smooth progress of this project from conception to publication.

We earnestly acknowledge our respective institutions for providing us with the infrastructure and resources required for the compilation of this volume.

CHAPTER ONE

Introduction to urban air pollution: nature and sources of urban air pollutants

Ansuman Sahoo[1], Priyanka Khilar[2], and Supriya Tiwari[3]
[1]Department of Botany, D.A.V. Degree College, Muzaffarnagar, Uttar Pradesh, India,
[2]Research and Development, Revelation's Biotech, Hyderabad, Telangana, India,
[3]Department of Botany, Laboratory of Ecotoxicology, Institute of Science, Banaras Hindu University, Varanasi, Uttar Pradesh, India

1.1 Introduction

The term "air pollution" describes the contamination of indoor or outdoor regions brought on by physical, chemical, or biological elements that change the atmosphere's inherent properties (Duan et al. 2024). Unchecked urbanization puts ecosystems and urban health at risk and accelerates environmental change across the globe (Dong et al. 2021). In 2019 alone, outdoor air pollution caused 4.14 million unintentional premature deaths in cities and the countryside globally, making it a serious public health concern (Sicard et al. 2023). Air pollution has grown to be a major barrier to the sustainable development of the economy and society due to the rapid advancement of global urbanization and industrialization (Li et al., 2025). At present, roughly 90% of the world's people breathe air that is more polluted than what the World Health Organization (WHO) says is safe. This causes about 8 million deaths per year (Li et al., 2025).

The Paris Climate Agreement, the WHO Air Quality Guidelines, and the United Nations Sustainable Development Goals are just a few of the international cooperation activities or projects that have been started to address this issue. This suggests that people's understanding of the negative and spatial spillover effects of air pollution is progressively growing, and that controlling air pollution has emerged as a major concern in environmental governance and worldwide public health (Li et al., 2025). The issue of air pollution in large cities—those with a population of more than 10 million—has multifaceted characteristics when compared to air pollution in other areas. On one hand, due to the distinct climate of cities,

Urban Air Pollution and Plant Responses
DOI: https://doi.org/10.1016/B978-0-443-40320-0.00012-7

the urban heat island effect can greatly increase air pollution (Piracha & Chaudhary, 2022). However, air pollution is caused by several factors, including population density, transportation, energy consumption, and industrial clustering. As a result, the impacts of urban air pollution are strongly correlated with the size of the city and show features of high intensity, complexity, and diverse sources (Han et al., 2023). There is an urgent need for focused research on air pollution in large cities due to the ongoing rapid growth of global cities in the coming decades.

Since air pollution affects public health and general quality of life, it is an important metric for assessing environmental sustainability. The growth of metropolitan areas is accompanied by an increase in the usage of energy, industrial pollutants, and vehicle traffic, all of which worsen the quality of air (Kahyaoğlu-Koračin et al., 2009). According to Abdolahnejad et al. (2018), the large concentration of automobiles and industry in metropolitan areas can lead to serious air pollution, endangering public health and plants. Since more than half of the world's population now resides in cities, it is crucial to comprehend how urbanization affects air quality. Abdolahnejad et al. (2018) claim that the high concentration of motor vehicles and industry in urban areas can cause severe air pollution, which puts plants and public health at risk. Understanding how urbanization impacts air quality is critical because over half of the world's population currently lives in cities.

1.2 Types and sources of air pollutants: a detailed overview

Air pollution is one of the most pressing environmental challenges of the modern world. As urbanization, industrialization, and population growth accelerate, the quality of the air we breathe continues to deteriorate. Polluted air not only affects the environment but also poses significant health risks, including respiratory diseases, cardiovascular disorders, and long-term ecological damage. Understanding the types of air pollutants and their sources is crucial for developing strategies to control pollution and protect public health.

Air pollutants can be broadly classified into primary and secondary pollutants, and further categorized based on their chemical composition, origin, and physical state (Table 1.1). This article explores these categories in detail. This chapter provides a detailed overview of various air pollutants,

Table 1.1 Classification of air pollutants based on their origin, chemical composition, and physical state.

Basis of classification	Category	Description	Air pollutants
Based on Origin (Type)	Primary Pollutants	Emitted directly from sources into the atmosphere	CO, SO_2, NO, NH_3, PM
	Secondary Pollutants	Formed by a chemical reaction between primary pollutants in the presence of sunlight/air	O_3, PAN, H_2SO_4, HNO_3, secondary PM
Based on Chemical Composition	Inorganic Pollutants	Composed of inorganic substances (non-carbon-based)	SO_2, NOx, CO, CO_2, NH_3, heavy metals (Pb, Hg)
	Organic Pollutants	Carbon-containing compounds	Volatile organic compounds (VOCs) (benzene, toluene), methane, and formaldehyde
	Particulate Pollutant	Solid/liquid particles suspended in air	PM2.5, PM10, soot, dust, smoke
Based on Physical State	Gaseous Pollutants Particulate Pollutants (Aerosols)	Exist in gaseous form in the atmosphere; sSolid or liquid particles suspended in air	CO, SO_2, NO_2, O_3, VOCs; dDust, smoke, mist, fumes, fly ash

categorizing them systematically to elucidate their diverse characteristics and environmental impacts. Specifically, pollutants are differentiated based on their chemical makeup, point of emission into the atmosphere, and physical state as gases, liquids, or solids. This comprehensive classification scheme allows for a nuanced understanding of their formation mechanisms, atmospheric residence times, and subsequent ecological and health implications (Karri et al., 2024). Understanding these distinctions is crucial for developing effective mitigation strategies and regulatory frameworks tailored to specific pollutant types and their unique environmental behaviors (Khan, 2020; Odubo & Kosoe, 2024). Air pollutants can originate from a wide array of sources, encompassing both natural phenomena like volcanic eruptions and dust storms, as well as anthropogenic activities, such as industrial processes, vehicular emissions, and agricultural practices (Temiz, 2024). The detailed classification of natural and anthropogenic sources of urban air pollutants is discussed in Table 1.2. These can be broadly classified as primary pollutants, directly emitted from their sources, or secondary pollutants, which form through subsequent chemical reactions in the atmosphere (Manisalidis et al., 2020). For instance, primary particulate matter (PM) is directly discharged, whereas secondary PM, along with photochemical smog and acid precipitation, arises from atmospheric transformations of precursors (Korkmaz & Ceylan, 2021; Nasir et al., 2024). Further classification includes re-emission sources, where primary or secondary pollutants accumulate on surfaces before being re-emitted into the atmosphere (Ite & Ite, 2024). These classifications are fundamental to comprehending the complex interplay between emission sources and atmospheric chemistry, ultimately influencing air quality and its broader environmental

Table 1.2 Natural and anthropogenic sources of urban air pollutants.

Category	Source type	Pollutants	Reference
Natural	Dust	$PM_{2.5}$, PM_{10}	Madheshiya et al. (2022)
	Volcanic eruptions	SO_2, CO_2, ash, $PM_{2.5}$, PM_{10}	Liu, et al. (2022)
	Biogenic emissions	Volatile organic compounds (VOCs) (isoprene, terpenes)	Liu, et al. (2022)
	Wildfires	CO, CO_2, $PM_{2.5}$, PM_{10}, NOx	Liu, et al. (2022)
Anthropogenic	Vehicular emissions	$PM_{2.5}$, PM_{10}, NO_2, CO, and SO_2	Buele and Criollo-Casignia (2026)
	Industrial units	SO_2, NOx, PM_{10}, toxic metals	Selami et al. (2025)
	Diesel generators, biomass burning	CO, NOx, $PM_{2.5}$, PM_{10}	Akomolafe et al. (2024)
	Construction and road dust	$PM_{2.5}$, PM_{10}	Akomolafe et al. (2024)
	Waste burning	CO, CO2, CH4, VOCs, $PM_{2.5}$, PM_{10}	Khan (2022)

ramifications (Hutton, 2018; Tiwari et al., 2024). This detailed categorization, therefore, serves as a foundational framework for analyzng the atmospheric fate and transport of these substances, thereby informing targeted intervention strategies (Pénard-Morand & Annesi-Maesano, 2004). A more granular classification further distinguishes between criteria air pollutants, which are regulated due to their widespread presence and known health effects, and hazardous air pollutants, which pose significant risks even at lower concentrations (Bhardwaj et al., 2023). Beyond these, pollutants can also be classified based on their emission characteristics and chemical composition, such as volatile organic compounds (VOCs), persistent organic pollutants, and heavy metals (Galán-Madruga, 2021). Further differentiation can be made between primary pollutants, emitted directly into the atmosphere, and secondary pollutants, which arise from chemical reactions among primary pollutants and other atmospheric constituents (Khallaf, 2011). For example, sulfur dioxide and nitrogen oxides are primary pollutants, whereas sulfuric acid (H_2SO_4) and nitric acid, major components of acid rain, are secondary pollutants formed from their atmospheric reactions (Santillán-Lima et al., 2023). Such transformations highlight the dynamic nature of atmospheric chemistry, where precursor emissions can lead to a cascade of secondary pollutant formations (Govande et al., 2024; Letsoalo et al., 2024).

Primary pollutants, exemplified by sulfur dioxide, nitrogen oxides, carbon monoxide, methane, VOCs, polycyclic aromatic hydrocarbons, and PM, are directly released from emission sources (Souza et al., 2024). Conversely, secondary pollutants, such as tropospheric ozone and certain components of PM (e.g., sulfates and nitrates), are not directly emitted but are formed through complex photochemical and chemical reactions involving primary precursors in the ambient atmosphere (Gupta et al., 2021; Meng, 2024). This distinction is critical for developing effective air quality management strategies, as controlling secondary pollutants often necessitates regulating their primary precursors rather than the secondary species directly (Kim et al., 2021). The diverse groups of air pollutants include gaseous pollutants, such as nitrogen oxides and sulfur dioxide, VOCs, and greenhouse gases like carbon dioxide (CO_2) and methane, in addition to toxic air pollutants such as lead and mercury (Bhat et al., 2024; Onivefu & Imarhiagbe, 2024). These pollutants, varying in their physical state as solid particles, liquid droplets, or gases, can further be categorized by their origin as either natural or anthropogenic, and by their formation pathway as primary or secondary (Chmielewski, 2011). The complexity of atmospheric reactions can further result in some pollutants exhibiting characteristics of both primary and secondary origins (Chmielewski, 2011). For instance, some PM can be directly emitted (primary) but can also form in the atmosphere from gaseous precursors (secondary), blurring this categorization (Bhat et al., 2024). For example, while primary $PM_{2.5}$ is directly emitted, a significant fraction of ambient $PM_{2.5}$ is secondary, forming from precursor gases such as sulfur dioxide, nitrogen oxides, and VOCs (Ketjalan et al., 2023). Ground-level ozone, a prominent secondary pollutant, exemplifies this as it forms from the photochemical reactions of nitrogen oxides and VOCs in the presence of sunlight and stagnant air (Partap et al., 2023). Similarly, secondary inorganic aerosols, like sulfates and nitrates, are formed through atmospheric oxidation of SO_2 and NO_x, respectively, highlighting the interconnectedness of various pollutant formation pathways (Ojha et al., 2022). These intricate formation pathways underscore the challenge of mitigating air pollution, as interventions often require a holistic approach targeting multiple precursor compounds across various emission sources. An example of this duality is NO_2, which, although primarily formed via chemical reactions, has an average of 10% of its urban street concentration originating directly from primary traffic exhaust (Wang, 2023). This demonstrates the nuanced classification challenges within air quality management, where a single pollutant can exhibit multiple formation mechanisms and origins (Asilevi et al., 2025). The chemical composition of air pollutants varies significantly, ranging from simple inorganic gases like carbon monoxide and sulfur dioxide to complex organic compounds such as polycyclic aromatic hydrocarbons and dioxins (Huang et al., 2024). This diversity necessitates an understanding of their individual physiochemical properties, as these directly influence their atmospheric residence time, transport mechanisms, and ultimate environmental and health impacts. For instance, while oxides of nitrogen, carbon monoxide, sulfur dioxide, VOCs, and carbonaceous and noncarbonaceous primary particles are direct emissions, secondary pollutants like ozone and certain PM fractions are products of atmospheric chemical reactions (Bălă et al., 2021). The transformation of primary gaseous pollutants, such as nitrogen monoxide and sulfur dioxide, into secondary particle pollutants like nitric acid and sulfate aerosols is mediated by atmospheric oxidants, including hydroxyl radicals, ozone, and nitrate radicals, alongside heterogeneous reactions (Miao et al., 2023). These reactive processes, often driven by solar radiation and temperature, significantly contribute to the formation of secondary organic aerosols and modify the composition of existing PM, thereby influencing the overall air pollution mixture (Lelieveld & Pöschl, 2017; Rider et al., 2014). Moreover, the interplay between various atmospheric constituents, including nitrogen oxides and CO_2, can lead to the formation of secondary PM, such as nitrate aerosols, further intensifying the complexity of the atmospheric pollutant burden (Taha et al., 2025).

This dynamic system involves intricate feedback loops where meteorological conditions, such as wind speed, temperature, and solar radiation, modulate pollutant dispersion, reaction rates, and overall atmospheric concentrations (Apte & Manchanda, 2024; Melki, 2017). Furthermore, variability in these atmospheric conditions influences the physical state and chemical transformation of pollutants, with different types of PM having varying toxicological properties due to their complex and dynamic composition (Costabile et al., 2022; Garcia et al., 2019; Levy et al., 2014). This complexity in composition is often tied to their origin, with a significant proportion of fine $PM_{2.5}$ deriving from anthropogenic sources like vehicular emissions and biomass combustion, which further contribute to their hazardous characteristics (Vega et al., 2025). For instance, precursor substances like sulfur oxides, nitrogen oxides, and ammonia can transform into secondary aerosols, such as sulfates, nitrates, and ammonium salts, which constitute $PM_{2.5}$ through various chemical reactions in the atmosphere (Pang et al., 2025). The formation of these secondary species is often initiated by radical-driven oxidation of primary pollutants, a process fundamentally governed by atmospheric oxidizing capacity (Li et al., 2026). Such atmospheric chemical reactions, specifically the oxidation of sulfur dioxide and nitrogen dioxide into particulate sulfate and nitrate, respectively, are critical pathways for secondary pollutant formation in urban photochemistry (Khoder, 2002). These reactions are profoundly influenced by factors such as relative humidity, the ratio of VOCs to NO_x, and light availability (Schneider et al., 2024). The partitioning of nitrate to the particle phase is additionally modulated by acid–base chemistry involving ammonia, contingent on ambient temperature, relative humidity, and particle pH (Kroll et al., 2020). This highlights the critical role of atmospheric conditions and precursor availability in determining the ultimate speciation and phase distribution of secondary aerosols, which directly impacts their deposition rates and subsequent environmental effects. Moreover, the interaction between different $PM_{2.5}$ components and meteorological variables, like temperature and humidity, can lead to complex dependencies, where higher temperatures might increase sulfate concentrations but decrease nitrates and organic components due to volatility (Tai et al., 2012).

The complex composition of fine PM, consisting of a dynamic mixture of organic and inorganic compounds, results from precursors emitted from both natural and anthropogenic sources (Duarte & Duarte, 2020; Zaccardi et al., 2024). Primary PM, such as dust and sea salts, originates directly from sources like agricultural activities and natural phenomena, whereas secondary PM, including sulfates and nitrates, forms through complex atmospheric reactions involving gaseous precursors (Chyzhykov & Mathews, 2024; Wang et al., 2018). These secondary aerosols often arise from the oxidation of sulfur dioxide, largely from fossil fuel combustion, which forms sulfuric acid that can then be neutralized by ammonia into ammonium sulfate and bisulfate, contributing to $PM_{2.5}$ acidity (Advanced Topics in Environmental Health and Air Pollution Case Studies, 2011). Similarly, nitrogen oxides from combustion sources are oxidized to nitric acid, which subsequently reacts with ammonia to form ammonium nitrate, further contributing to the secondary inorganic aerosol burden (Garcia et al., 2019; Myriokefalitakis et al., 2024). Beyond these inorganic species, secondary organic aerosol formation from VOCs through oxidation processes constitutes a significant fraction of $PM_{2.5}$, exhibiting varied chemical structures and physical properties depending on their precursors and atmospheric conditions. Furthermore, the intricate interplay between nitrogen oxides and ammonia is a central driver of global PM concentrations, with their complex thermodynamic interactions influencing the sensitivity of aerosol PM to these precursors (Nenes et al., 2020). In dry environments, surface temperature strongly influences gas-phase reactions for fine particle formation, whereas in humid conditions, aqueous-phase reactions involving secondary inorganic aerosols become more activated, with relative humidity being as crucial as temperature in this process (Choi et al., 2023). Additionally, aluminum and iron, pervasive components of ambient dust, undergo transformations in the presence of water, thereby increasing their reactivity and solubility within the aerosol phase (Yang et al., 2025). This transformation process can significantly alter their environmental mobility and biological availability, impacting their potential to cause adverse health effects. This further underscores the multifaceted nature of $PM_{2.5}$, where even trace elemental constituents undergo dynamic transformations that dictate their environmental fate and toxicological potency, particularly concerning respiratory and cardiovascular health impacts.

Given the heterogeneous composition and varying origins, a comprehensive classification of air pollutants becomes essential for understanding their atmospheric behavior and designing effective mitigation strategies. This classification can be systematically approached by considering their chemical composition, predominant physical state, and primary source categories, which collectively dictate their atmospheric residence time and potential for long-range transport (Crippa et al., 2019). For instance, classifying pollutants based on their chemical makeup allows for detailed study of reaction pathways, such as the multiphase and heterogeneous chemical processes that significantly influence aerosol loading and growth, particularly in polluted atmospheres (Acharja et al., 2023). Moreover, understanding the varying physical states of pollutants, ranging from gaseous to particulate forms, is crucial for assessing their dispersion characteristics and deposition mechanisms (Khare, 2023). Such a classification framework enables more precise identification of specific emission sources, allowing for targeted regulatory interventions to improve air quality and mitigate the adverse health impacts associated with diverse pollutant exposures (Fuzzi et al., 2015; Mastromatteo et al., 2024). For example, the inhalation of fine $PM_{2.5}$, characterized by its small aerodynamic diameter, has been unequivocally linked to respiratory and cardiovascular diseases, and increased mortality, a phenomenon partly mediated by the oxidative potential of PM-bound chemicals (Jouannais & Pizzol, 2021; Li et al., 2022). This deleterious effect is often exacerbated by the presence of trace metals in aerosol particles, as their physicochemical properties and speciation profoundly influence their bioavailability and capacity to induce oxidative stress in biological systems (Giorio et al., 2022). The diverse nature of PM, categorized by aerodynamic equivalent diameter into fractions, like PM_{10}, $PM_{2.5}$, and $PM_{1.0}$, necessitates detailed characterization for effective environmental pollution mitigation, given its interactions with other atmospheric components and contributions to secondary pollutant formation (Bilo et al., 2023). PM is further categorized into coarse and fine ($PM_{2.5}$) fractions based on their aerodynamic diameter, with PM10 encompassing particles with diameters up to 10 μm and $PM_{2.5}$, including those up to 2.5 μm, along with ultrafine particles below 0.1 μm (Coll et al., 2018; Ilaghi et al., 2024; Lang et al., 2025). These classifications are critical because the size of the particles dictates their atmospheric residence time and their ability to penetrate the human respiratory system, with finer particles ($PM_{2.5}$ and $PM_{1.0}$) posing greater health risks due to deeper pulmonary penetration and systemic translocation (Brook et al., 2010; Majewski et al., 2023).

1.3 Factors affecting air pollution in urban areas

Urbanization, while driving economic growth, concurrently exacerbates air-quality degradation within metropolitan centers due to heightened anthropogenic emissions. This phenomenon necessitates a comprehensive understanding of the multifaceted factors contributing to urban air pollution, including geographical, meteorological, demographic, and economic elements (Özcan & Çubukçu, 2018). Specifically, the interplay between urban spatial structures, population density, and prevailing meteorological conditions significantly influences the dispersion and concentration of atmospheric pollutants. Such complexities underscore the importance of examining both natural factors, like topography and wind patterns, and human-induced factors, such as industrial development and energy consumption, in shaping urban air quality dynamics. This paper aims to synthesize the current understanding of these contributing factors, categorizing them into urban geography and meteorological influences, population density and economic activity, traffic patterns and urban sprawl, and the overarching impact of climate change and the urban heat island effect. The escalating reliance on fossil fuels for industrial processes and transportation further intensifies the emission of PM, nitrogen oxides, and VOCs, all of which critically compromise urban atmospheric health (Akomolafe et al., 2024). The combined effects of diverse emission sources and atmospheric parameters, such as wind patterns, are directly responsible for the concentrations of pollutants in urban environments. These pollutants, originating from diverse sources, exhibit varying spatial distributions depending on city morphology and regional economic activities.

1.3.1 Urban geography and meteorological factors

Meteorological conditions, encompassing wind fields, relative humidity, and temperature inversions, play a crucial role in the dispersion, transformation, and accumulation of air pollutants (Huang et al., 2025). For instance, calm winds can lead to the stagnation of air masses, concentrating pollutants, while strong winds can facilitate their dispersion over wider areas. Temperature inversions, characterized by an atmospheric layer in which temperature increases with altitude, trap pollutants near the ground, leading to severe air quality degradation, especially in urban basins (Advanced Topics in Environmental Health and Air Pollution Case Studies, 2011). Furthermore, such geographical configurations, particularly urban canyons formed by dense building layouts, impede natural ventilation and contribute to pollutant entrapment, thereby exacerbating localized air quality issues. This is particularly evident in cities situated in mountain valleys, where the constrained topography, combined with industrial activity and traffic congestion, significantly elevates PM2.5 concentrations. Moreover, the seasonal variability of meteorological factors, such as increased surface ozone during warmer periods, further compounds the complexity of urban air pollution dynamics. Concurrently, human activities, including industrial emissions and vehicular exhaust, directly influence the generation of pollutants, demonstrating a strong spatial heterogeneity in $PM_{2.5}$ concentrations (Chen et al., 2023).

1.3.2 Industrial emissions

Industrial activities, particularly in regions with rapid economic development, are significant contributors to atmospheric pollutant loads, including PM and gaseous emissions (Luo et al., 2023). This is particularly evident in urban, agricultural, and industrial areas where a substantial proportion of $PM_{2.5}$ mass originates from sources, such as vehicular emissions, biomass combustion, inorganic ions, and mineral dust (Vega et al., 2025). This intricate relationship highlights how population density and economic indicators, such as the value added of secondary production and local fiscal expenditures, are profoundly linked to industrial operations and transportation activities, thereby becoming notable sources of $PM_{2.5}$ emissions (Han et al., 2023). For example, specific industrial sectors, like manufacturing and power generation, release substantial quantities of sulfur dioxide, nitrogen oxides, and VOCs, which are precursors to the formation of secondary PM. The intensification of industrial growth and human activity inevitably correlates with elevated emission levels, directly contributing to increased $PM_{2.5}$ concentrations, especially in regions characterized by rapid industrialization, high population density, and significant heavy-industry contributions. For instance, studies in developing countries have shown a direct correlation among industrialization, urbanization, and heightened $PM_{2.5}$ concentrations, with industrial output often serving as a primary socioeconomic determinant of air quality degradation. This connection between economic output and environmental impact is further complicated by the fact that certain indicators, such as night light and gross domestic Product, reflect regional development levels but simultaneously contribute to environmental pollution through ongoing industrialization and modernization processes (Lü et al., 2015).

1.3.3 Vehicle emissions and congestion

Traffic patterns significantly influence urban air quality, with congested roadways leading to higher pollutant concentrations from vehicle emissions, including nitrogen oxides, carbon monoxide, and PM. These emissions are particularly problematic in densely populated urban cores where slow-moving or idling vehicles generate substantial quantities of harmful aerosols and gases (Khan, 2020). The spatial distribution of vehicle emissions is further exacerbated by urban sprawl, which necessitates longer commutes and increased reliance on private transportation, thereby amplifying regional air pollution burdens. Moreover, the efficiency of pollutant dispersion is often compromised in such expansive urban landscapes, as the increased travel distances contribute to a more pervasive distribution of vehicle-related pollutants across broader geographic areas. Conversely, compact urban forms, by shortening commuting distances, can mitigate vehicular emissions, though highly dense areas may still experience localized pollution hotspots due to concentrated traffic and reduced air circulation. Additionally, the morphological characteristics of cities, such as the presence of valleys or coastal proximity, can create bowl-like effects that trap pollutants, impeding their dispersal and leading to elevated concentrations, especially during temperature inversions. This phenomenon is particularly pronounced in cities with high population density and substantial economic activity, as these areas often exhibit increased traffic volumes and industrial emissions, further contributing to air quality degradation (Fig. 1.1). The total number of vehicles in an area is a significant predictor of $PM_{2.5}$ concentrations, confirming the direct relationship between traffic volume and PM levels (Pope et al., 2024). Studies have consistently shown a significant correlation between vehicle miles traveled and air pollution levels in urban counties, indicating that efforts to reduce vehicular impact are more effective in metropolitan areas (Ngo et al., 2024). Furthermore, urban sprawl has been directly correlated with increased vehicle tailpipe emissions and higher concentrations of air pollution, largely due to more vehicles per household and decreased utilization of public transportation (Liu et al., 2016). While compact development can reduce overall transportation emissions per capita, it may

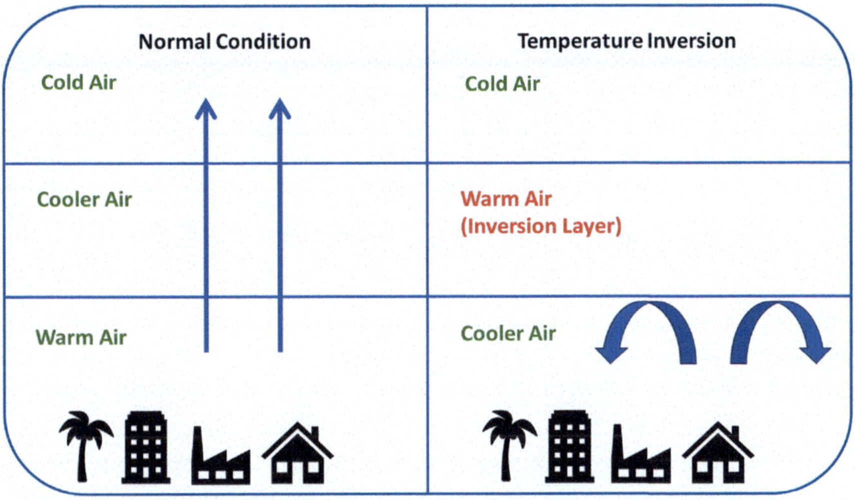

Figure 1.1 A comparison between normal and temperature inversion conditions illustrates the trapping of pollutants and smoke beneath a warm air cap, creating stagnant air conditions.

also lead to higher localized $PM_{2.5}$ concentrations within urban hotspots, potentially increasing population exposure (Mansfield et al., 2014). This dynamic interplay highlights the complex relationship among urban planning, transportation infrastructure, and air quality, emphasizing the necessity of integrated strategies to mitigate the adverse effects of vehicular emissions.

1.3.4 Impact of urban planning on air quality

Effective urban planning, incorporating principles of smart growth and mixed-use development, can significantly influence air quality outcomes by minimizing travel distances and promoting sustainable transportation modes. Such strategies often include the integration of public transportation networks, pedestrian-friendly infrastructure, and cycling pathways, which collectively reduce reliance on private vehicles and, subsequently, lower vehicular emissions (Wen, 2024). Moreover, the strategic placement of green spaces and the implementation of stringent emission standards for both industrial and vehicular sources are critical components in enhancing urban air quality and promoting public health. Indeed, the detrimental effects of urban sprawl on air quality are well-documented, with studies indicating that low-density, vehicle-oriented development exacerbates air pollution (Talkhabi et al., 2024). Conversely, maintaining a rational development intensity, avoiding both excessively high and low densities, has been shown to be a more effective approach to mitigating air pollution from on-road transportation, particularly concerning $PM_{2.5}$ concentrations.

1.3.5 Climate change and Urban Heat Island effect

The interplay between climate change and the Urban Heat Island effect creates a feedback loop that intensifies air pollution in urban areas, as elevated temperatures can accelerate photochemical reactions leading to increased ground-level ozone formation and reduce the efficiency of pollutant dispersion. This exacerbation of pollution is further compounded by the urban heat island effect, which traps pollutants near the surface, hindering their vertical dispersion and prolonging their atmospheric residence time. This phenomenon, often described as an Urban Pollution Island, frequently coexists with and is intrinsically linked to the Urban Heat Island effect, collectively impairing urban meteorological conditions and public health. The intensified heat within urban environments not only alters atmospheric stability but also accelerates the formation of secondary pollutants, thereby creating a more persistent and concentrated dome of atmospheric contamination over cities. Moreover, the removal of vegetation during urban expansion diminishes the natural air purification capacity, contributing to heightened pollutant concentrations and exacerbating the urban heat island effect. Polluted urban areas, characterized by the urban heat island effect, absorb and trap heat, leading to higher local temperatures that disrupt climate patterns and amplify heatwaves and energy demands . The complex interplay between climate change and air pollution is further highlighted by how climate-induced meteorological shifts, such as altered temperature, rainfall, and wind patterns, directly influence the concentration and distribution of atmospheric pollutants (Lai et al., 2025).

1.4 Conclusion

Air pollution is a complex and evolving problem that has a big effect on the health of people, the environment, and the livability of cities. This manuscript emphasizes the intricacy of air pollutants by categorizing them according to their origin, composition, physical state, and formation processes. The difference between primary and secondary pollutants shows how atmospheric chemistry changes over time. For example, emissions from primary pollutants can lead to the creation of hazardous secondary pollutants.

The review shows that both natural and man-made sources cause air pollution. Urban areas are more at risk because they have a lot of people, factories, and transportation systems. Fine $PM_{2.5}$ is one of the most dangerous pollutants because it can have serious effects on health. Furthermore, weather circumstances, urban morphology, traffic congestion, and industrial emissions all have a significant impact on pollution concentration and dispersion. The interplay between climate change with the Urban Heat Island effect exacerbates air pollution, resulting in complicated feedback loops that heighten environmental and health concerns. Addressing urban air pollution requires a comprehensive and integrated approach that includes sustainable urban design, cleaner industrial technology, efficient transportation systems, and stringent regulatory frameworks. Future research should prioritize region-specific tactics and enhanced monitoring techniques in order to better understand pollution dynamics and design tailored responses. Finally, increasing air quality is critical to maintaining a healthy environment and meeting long-term sustainable development goals.

References

Abdolahnejad, A., Jafari, N., Mohammadi, A., Miri, M., & Hajizadeh, Y. (2018). Mortality and morbidity due to exposure to ambient NO_2, SO_2, and O_3 in Isfahan in 2013–2014. *International journal of preventive medicine, 9*(1), 11.

Acharja, P., Ghude, S. D., Sinha, B., Barth, M. C., Govardhan, G., Kulkarni, R., Sinha, V., Kumar, R., Ali, K., Gültepe, I., Petit, J., & Rajeevan, M. (2023). Thermodynamical framework for effective mitigation of high aerosol loading in the indo-gangetic plain during winter. *Scientific Reports, 13*(1). https://doi.org/10.1038/s41598-023-40657-w.

Akomolafe, O. O., Olorunsogo, T., Anyanwu, E. C., Osasona, F., Ogugua, J. O., & Daraojimba, O. H. (2024). AIR quality and public health: A review of urban pollution sources and mitigation measures [Review of AIR quality and public health: A review of urban pollution sources and mitigation measures]. *Engineering Science & Technology Journal, 5*(2), 259. https://doi.org/10.51594/estj.v5i2.751 Fair East Publishers.

Apte, J. S., & Manchanda, C. (2024). High-resolution urban air pollution mapping [Review of High-resolution urban air pollution mapping]. *Science (New York, N.Y.), 385*(6707), 380. https://doi.org/10.1126/science.adq3678.

Asilevi, P. J., Dzidzorm, E. N., Boakye, P., & Quansah, E. (2025). Nitrogen dioxide (NO_2) meteorology and predictability for air quality management using TROPOMI. *Npj Clean Air, 1*(1). https://doi.org/10.1038/s44407-024-00003-4.

Bălă, G.-P., Râjnoveanu, R., Tudorache, E., Motișan, R., & Oancea, C. (2021). Air pollution exposure—the (in)visible risk factor for respiratory diseases [Review of Air pollution exposure—the (in)visible risk factor for respiratory diseases]. *Environmental*

Science and Pollution Research, 28(16), 19615. https://doi.org/10.1007/s11356-021-13208-x Springer Science+Business Media.

Bhardwaj, L. K., Singh, V. V., Dwivedi, K., & Rai, S. (2023). A comprehensive review on sources, types, impact and challenges of air pollution. *Preprints.Org.*. https://doi.org/10.20944/preprints202311.1245.v1.

Bhat, R., Jain, A., Rakesh, C., Khan, I., Pahwa, S., & Abood, A. S. (2024). Improvement of the quality of air and the implementation of sustainable technologies in green technological innovation. *E3S Web of Conferences, 505*, 3011. https://doi.org/10.1051/e3sconf/202450503011.

Bilo, F., Cirelli, P., & Borgese, L. (2023). Elemental analysis of particulate matter by X-ray fluorescence methods: A green approach to air quality monitoring. *TrAC Trends in Analytical Chemistry, 170*, 117427. https://doi.org/10.1016/j.trac.2023.117427.

Brook, R. D., Rajagopalan, S., Pope, C. A., Brook, J. R., Bhatnagar, A., Diez–Roux, A. V., Holguín, F., Hong, Y., Luepker, R. V., Mittleman, M. A., Peters, A., Siscovick, D. S., Smith, S. C., Whitsel, L. P., & Kaufman, J. D. (2010). Particulate matter air pollution and cardiovascular disease [Review of *particulate matter air pollution and cardiovascular disease*]. *Circulation, 121*(21), 2331. https://doi.org/10.1161/cir.0b013e3181dbece1 Lippincott Williams & Wilkins.

Buele, J., & Criollo-Casignia, D. (2026). Effects of vehicular emissions on urban air quality in ecuador and implications for respiratory health. *Sustainability, 18*(3), 1262.

Chen, T., Zhou, P., Xu, W., Liu, Y., & Sun, S. (2023). Spatial-temporal evolution characteristics and influencing factors of PM2.5 in the yangtze river basin. *Polish Journal of Environmental Studies*. https://doi.org/10.15244/pjoes/173105.

Chmielewski, A. G. (2011). *Monitoring, control and effects of air pollution*. InTech. https://doi.org/10.5772/998.

Choi, W., Ho, C., & Kim, K.-Y. (2023). Critical contribution of moisture to the air quality deterioration in a warm and humid weather. *Scientific Reports, 13*(1), 4260. https://doi.org/10.1038/s41598-023-31316-1.

Chyzhykov, D., & Mathews, B. (2024). Measurements of water-soluble ions in particulate matter 2.5 in polish rural areas: Identifying possible sources. *Water Air & Soil Pollution, 235*(7). https://doi.org/10.1007/s11270-024-07265-4.

Coll, P., Cazaunau, M., Boczkowski, J., Zysman, M., Doussin, J., Gratien, A., Dérumeaux, G., Pini, M., Biagio, C. D., Pangui, É., Gaimoz, C., Hüe, S., Relaix, F., Vatanian, A. D., Coll, I., Michoud, V., Formenti, P., Forêt, G., Thavaratnasingam, L., ... Lanone, S. (2018). Pollurisk: An innovative experimental platform to investigate health impacts of air quality. *WIT Transactions on Ecology and the Environment, 1*, 557. https://doi.org/10.2495/air180521.

Costabile, F., Decesari, S., Vecchi, R., Lucarelli, F., Curci, G., Massabò, D., Rinaldi, M., Gualtieri, M., Corsini, E., Menegola, E., Canepari, S., Massimi, L., Argentini, S., Busetto, M., Iulio, G. D., Liberto, L. D., Paglione, M., Petenko, I., Russo, M., ... Facchini, M. C. (2022). On the redox-activity and health-effects of atmospheric primary and secondary aerosol: Phenomenology. *Atmosphere, 13*(5), 704. https://doi.org/10.3390/atmos13050704.

Crippa, M., Janssens-Maenhout, G., Guizzardi, D., Dingenen, R. V., & Dentener, F. (2019). Contribution and uncertainty of sectorial and regional emissions to regional and global PM_{2.5} health impacts. *Atmospheric Chemistry and Physics, 19*(7), 5165 https://doi.org/10.5194/acp-19-5165-2019.

Dong, L., Chen, B., Huang, Y., Song, Z., & Yang, T. (2021). Analysis on the characteristics of air pollution in China during the COVID-19 outbreak. *Atmosphere, 12*(2), 205.

Duan, X., Chen, X., Li, C., Witherspoon, E., Burcar, E., Wang, Z., & Li, H. (2024). Progress of MOFs and their derivatives for mitigating air pollution. *Advanced Composites and Hybrid Materials*, 7(4), 114.

Duarte, R. M. B. O., & Duarte, A. C. (2020). Urban atmospheric aerosols: Sources, analysis, and effects. *Atmosphere, 11*(11), 1221. https://doi.org/10.3390/atmos11111221.

Fuzzi, S., Baltensperger, U., Carslaw, K. S., Decesari, S., Gon, H. D., van der, Facchini, M. C., Fowler, D., Koren, I., Langford, B., Lohmann, U., Nemitz, E., Pandis, S.N, Riipinen, I., Rudich, Y., Schaap, M., Slowik, J. G., Spracklen, D. V., Vignati, E., Wild, M., ... Gilardoni, S. (2015). Particulate matter, air quality and climate: Lessons learned and future needs. *Atmospheric Chemistry and Physics, 15*(14), 8217. https://doi.org/10.5194/acp-15-8217-2015.

Galán-Madruga, D. (2021). Implication of secondary atmospheric pollutants in the air quality: A case-study for ozone. In S. Abdul Rehman Khan (Ed.). *Environmental sustainability - Preparing for tomorrow*IntechOpenhttps://doi.org/10.5772/intechopen.95481.

Garcia, J., Borrega, J., & Coelho, L. (2019). Airborne PM impact on health, overview of variables, and key factors to decision making in air quality. In J. Del Real Olvera (Ed.). *Air Pollution - Monitoring, Quantification and Removal of Gases and Particles*IntechOpenhttps://doi.org/10.5772/intechopen.79695.

Giorio, C., DaRonco, S., Marco, V. D., Badocco, D., Battaglia, F. P., Soldà, L., Pastore, P., & Tapparo, A. (2022). Emerging investigator series: Aqueous-phase processing of atmospheric aerosol influences dissolution kinetics of metal ions in an urban background site in the Po Valley. *Environmental Science Processes & Impacts, 24*(6), 884. https://doi.org/10.1039/d2em00023g.

Govande, A., Attada, R., Shukla, K. K., Muralidharan, S., Kunchala, R. K., Chilukoti, N., & Kaushik, G. (2024). Analyzing and predicting ventilation coefficient over india using long-term reanalysis datasets and hybrid machine learning approach. *Research Square (Research Square)* https://doi.org/10.21203/rs.3.rs-4551619/v1.

Gupta, A., Farjad, B., Wang, G., Eum, H., & Dubé, M. G. (2021). *Integrated environmental modelling framework for cumulative effects assessment.* Calgary: University of Calgary Press. https://doi.org/10.1515/9781773852003.

Han, J., Yang, Y., Yang, X., Wang, D., Wang, X., & Sun, P. (2023). Exploring air pollution characteristics from spatio-temporal perspective: A case study of the top 10 urban agglomerations in China. *Environmental Research, 224*, 115512.

Huang, H. N., et al. (2025). Impact of agricultural straw open-field burning on concentrations of six criteria air pollutants in China. *Environmental Pollution, 373.* https://doi.org/10.1016/j.envpol.2025.126109.

Huang, W., Hong-bing, X., Wu, J., Ren, M., Yang, K., & Qiao, J. (2024). Toward cleaner air and better health: Current state, challenges, and priorities [Review of Toward cleaner air and better health: Current state, challenges, and priorities]. *Science (New York, N.Y.), 385*(6707), 386. https://doi.org/10.1126/science.adp7832.

Hutton, G. (2018). Appendix A: Selected solutions. In *Programming in Haskell.* Cambridge: Cambridge University Press, 263–279. https://doi.org/10.1017/cbo9781316784099.020.

Ilaghi, M., Kafi, F., Shafiei, M., Zangiabadian, M., & Nasiri, M. J. (2024). Dietary supplementations to mitigate the cardiopulmonary effects of air pollution toxicity: A systematic review of clinical trials [Review of Dietary supplementations to mitigate the cardiopulmonary effects of air pollution toxicity: A systematic review of clinical trials]. *PLoS One, 19*(6). https://doi.org/10.1371/journal.pone.0304402 Public Library of Science.

Ite, A. E., & Ite, I. A. (2024). Organic micropollutants in the air: Origin and occurence, environmental and health effects, and mitigation strategies. *Environment & Ecosystem Science, 8*(1), 36. https://doi.org/10.26480/ees.01.2024.36.52.

Jouannais, P., & Pizzol, M. (2021). Ex-ante LCA for microalgae-based veterinary molecules in finfish aquaculture: How to assess the environmental performance of unknown molecules? Abstract from Setac Europe 31st annual meeting, Sevilla, Spain https://vbn.aau.dk/ws/portalfiles/portal/462154350/Europe2021_Abstract_Bookv2.pdf

Kahyaoğlu-Koračin, J., Bassett, S. D., Mouat, D. A., & Gertler, A. W. (2009). Application of a scenario-based modeling system to evaluate the air quality impacts of future growth. *Atmospheric Environment, 43*(5), 1021–1028.

Karri, R. R., Vera, T., Hassan, S. K., Khoder, M. I., Dehghani, M. H., Mubarak, N. M., & Ravindran, G. (2024). Chapter 1 - Classification, sources, and occurrence of outdoor air pollutants: A comprehensive overview. In M. H. Dehghani, R. R. Karri, T. Vera, & S. K. M. Hassan (Eds.). *Health and Environmental Effects of Ambient Air Pollution: Volume 1: Air Pollution, Human Health, and the Environment* (pp. 1–34). Academic Press. https://doi.org/10.1016/b978-0-443-16088-2.00002-8.

Ketjalan, A., Humphries, U. W., & Suadee, W. (2023). A study on PM2.5 concentration in Bangkok, Thailand: A case study of Bang Na Station. *International Journal of ADVANCED AND APPLIED SCIENCES, 10*(10), 55. https://doi.org/10.21833/ijaas.2023.10.006.

Khallaf, M. K. (2011). The impact of air pollution on health, economy, environment and agricultural sources. *In InTech eBooks*. https://doi.org/10.5772/1000.

Khan, S. A. R. (2020). *Environmental sustainability - Preparing for tomorrow*. IntechOpen. https://doi.org/10.5772/intechopen.91568.

Khan, Y. A. (2022). Risk of mortality due to COVID-19 and air pollution in Pakistan. *Environmental Science and Pollution Research, 29*(2), 2063–2072.

Khare, R. (2023). Chemistry and bonding between air quality and climate change. *International Journal for Research in Applied Science and Engineering Technology, 11*(6), 1310. https://doi.org/10.22214/ijraset.2023.53823.

Khoder, M. I. (2002). Atmospheric conversion of sulfur dioxide to particulate sulfate and nitrogen dioxide to particulate nitrate and gaseous nitric acid in an urban area. *Chemosphere, 49*(6), 675. https://doi.org/10.1016/s0045-6535(02)00391-0.

Kim, J., Lee, J., Han, J., Choi, J., Kim, D.-G., Park, J., & Lee, G. (2021). Long-term assessment of ozone nonattainment changes in south Korea compared to US, and EU Ozone guidelines. *Asian Journal of Atmospheric Environment, 15*(4). https://doi.org/10.5572/ajae.2021.098.

Korkmaz, Ş., & Ceylan, Z. (2021). An overview: The effects of particulate matters, an important atmospheric pollutant, on the spread of Covid19. *JURNAL KESEHATAN LINGKUNGAN, 13*(3), 159. https://doi.org/10.20473/jkl.v13i3.2021.159-165.

Kroll, J. H., Heald, C. L., Cappa, C. D., Farmer, D. K., Fry, J. L., Murphy, J. G., & Steiner, A. L. (2020). The complex chemical effects of COVID-19 shutdowns on air quality. *Nature Chemistry, 12*(9), 777. https://doi.org/10.1038/s41557-020-0535-z.

Lai, G, et al. (2025). High-Quality Development and Decoupling Economic Growth from Air Pollution: Evidence from Daily Electricity Consumption in Fujian. *Sustainability, 17*(4). https://doi.org/10.3390/su17041489.

Lang, T., Lipp, A.-M., & Wechselberger, C. (2025). Xenobiotic toxicants and particulate matter: Effects, mechanisms, impacts on human health, and mitigation strategies. *PubMed, 15*(4). https://doi.org/10.3390/jox15040131.

Lelieveld, J., & Pöschl, U. (2017). Chemists can help to solve the air-pollution health crisis. *Nature, 551*(7680), 291. https://doi.org/10.1038/d41586-017-05906-9.

Letsoalo, M. R., Mogashane, T. M., Mashale, K., Ntsasa, N., Mkhohlakali, A., & Tshilongo, J. (2024). Sustainable approaches to monitoring urban particulate matter monitoring: Challenges and innovations. In *IntechOpen eBooks*. IntechOpen https://doi.org/10.5772/intechopen.1007884.

Levy, I., Mihele, C., Lu, G., Narayan, J., Hilker, N., & Brook, J. R. (2014). Elucidating multipollutant exposure across a complex metropolitan area by systematic deployment of a mobile laboratory. *Atmospheric Chemistry and Physics, 14*(14), 7173. https://doi.org/10.5194/acp-14-7173-2014.

Li, P., Ren, Y., Bi, F., Long, F., Li, J., Zhang, H., Wu, Z., & Li, H. (2026). Advances and challenges in understanding atmospheric oxidizing capacity in China: Insights from chemical mechanisms and model applications. *Toxics, 14*(2), 159. https://doi.org/10.3390/toxics14020159.

Li, R., Luo, Y., Zhu, X., Zhang, J., Hua, P., Wang, Z., & Li, H. (2025). Increasing health burdens driven by global trade induced air pollution. *Earth's Future, 13*(1) e2024EF004814.

Li, T., Yu, Y., Sun, Z., & Duan, J. (2022). A comprehensive understanding of ambient particulate matter and its components on the adverse health effects based from epidemiological and laboratory evidence. *Particle and Fibre Toxicology, 19*(1), 67. https://doi.org/10.1186/s12989-022-00507-5.

Liu, J, et al. (2016). Air pollutant emissions from Chinese households: A major and underappreciated ambient pollution source. *The Proceedings of the National Academy of Sciences, 113*(28), 7756–7761. https://doi.org/10.1073/pnas.1604537113.

Liu, C., Cai, J., Chen, R., Sera, F., Guo, Y., Tong, S., & Kan, H. (2022). Coarse particulate air pollution and daily mortality: A global study in 205 cities. *American Journal of Respiratory and Critical Care Medicine, 206*(8), 999–1007.

Lü, J, et al. (2015). Air Pollution Exposure and Physical Activity in China: Current Knowledge, Public Health Implications, and Future Research Needs. *Int. J. Environ. Res. Public Health, 12*(11), 14887–14897. https://doi.org/10.3390/ijerph121114887.

Madheshiya, P., Gupta, G. S., Sahoo, A., & Tiwari, S. (2022). Biomonitoring tools and bioprogramming: An overview. *New Paradigms in Environmental Biomonitoring Using Plants,* 341–366.

Majewski, G., Klik, B., Rogula-Kozłowska, W., Rogula-Kopiec, P., Rybak, J., Radziemska, M., & Liniauskienė, E. (2023). Assessment of heavy metal inhalation risks in urban environments in Poland: A case study. *Journal of Ecological Engineering, 24*(11), 330. https://doi.org/10.12911/22998993/171591.

Manisalidis, I., Stavropoulou, E., Stavropoulos, A., & Bezirtzoglou, E. (2020). Environmental and health impacts of air pollution: A review [Review of *Environmental and Health Impacts of Air Pollution: A Review*]. *Frontiers in Public Health, 8*. https://doi.org/10.3389/fpubh.2020.00014 Frontiers Media.

Mansfield, T. J., et al. (2014). The Effects of Urban Form on Ambient Air Pollution and Public Health Risk: A Case Study in Raleigh, North Carolina. *Risk Analysis, 35*(5).

Mastromatteo, N., Drudi, L., Gallione, D., Bellopede, R., & Clerico, M. (2024). Assessment of environmental parameters in natural coastal scenery and compositional by means of an innovative approach. *Research Square (Research Square)*. https://doi.org/10.21203/rs.3.rs-4790082/v1.

Melki, P. (2017). Health impact of airborne particulate matter in Northern Lebanon: From a pilot epidemiological study to physico-chemical characterization and toxicological effects assessment. *HAL (Le Centre Pour La Communication Scientifique Directe)*. https://tel.archives-ouvertes.fr/tel-02088915.

Meng, Y. (2024). Effect of air pollution on the colour of the sky. *E3S Web of Conferences, 553*, 3010. https://doi.org/10.1051/e3sconf/202455303010.

Miao, Y., Zhang, G., Li, X., & Xue, C. (2023). Editorial: Physical and chemical processes within the planetary boundary layer and their impacts on air pollution. *Frontiers in Environmental Science, 11*. https://doi.org/10.3389/fenvs.2023.1221546.

Moldoveanu, A. (2011). *Advanced topics in environmental health and air pollution case studies*. InTech. https://doi.org/10.5772/997.

Myriokefalitakis, S., Karl, M., Weiss, K. A., Karagiannis, D., Athanasopoulou, E., Kakouri, A., Bougiatioti, A., Liakakou, E., Stavroulas, I., Papangelis, G., Grivas, G., Paraskevopoulou, D., Speyer, O., Mihalopoulos, N., & Gerasopoulos, E. (2024). Analysis of secondary inorganic aerosols over the greater Athens area using the EPISODE–CityChem source dispersion and photochemistry model. *Atmospheric Chemistry and Physics, 24*(13), 7815. https://doi.org/10.5194/acp-24-7815-2024.

Nasir, A., Aslam, R. A., Ali, F., & Nasır, A. (2024). Air pollution from industrial emissions and its control in Pakistan: Current situation, challenges, and way forward. In M. Saleh, A. I. Hassan, & R. F. Aglan (Eds.). *Advances and Challenges in Hazardous Waste Management*IntechOpenhttps://doi.org/10.5772/intechopen.1004052.

Nenes, A., Pandis, S.N, Weber, R. J., & Russell, A. G. (2020). Aerosol pH and liquid water content determine when particulate matter is sensitive to ammonia and nitrate availability. *Atmospheric Chemistry and Physics, 20*(5), 3249. https://doi.org/10.5194/acp-20-3249-2020.

Ngo, N. S., et al. (2024). The impact of urban form on the relationship between vehicle miles traveled and air pollution. *The impact of urban form on the relationship between vehicle miles traveled and air pollution, 28*. https://doi.org/10.1016/j.trip.2024.101288.

Odubo, T. C., & Kosoe, E. A. (2024). Sources of air pollutants: Impacts and solutions. In S. C. Izah, M. C. Ogwu, & A. Shahsavani (Eds.). *Air Pollutants in the Context of One Health. The Handbook of Environmental Chemistry, vol 134* (pp. 75–121). Cham: Springer. https://doi.org/10.1007/698_2024_1127.

Ojha, N., Soni, M., Kumar, M., Gunthe, S. S., Chen, Y., & Ansari, T. (2022). Mechanisms and pathways for coordinated control of fine particulate matter and Ozone [Review of *Mechanisms and pathways for coordinated control of fine particulate matter and ozone*]. *Current Pollution Reports, 8*(4), 594. https://doi.org/10.1007/s40726-022-00229-4.

Onivefu, A. P., & Imarhiagbe, O. (2024). Types of air pollutants. In S. C. Izah, M. C. Ogwu, & A. Shahsavani (Eds.). *Air Pollutants in the Context of One Health . The Handbook of Environmental Chemistry, vol 134* (pp. 123–160). Cham: Springer Nature. https://doi.org/10.1007/698_2024_1128.

Ozcan, N. S., & Cubukcu, N. M. (2018). The relationship between urban air pollution and urban planning decisions. *Asian Journal of Quality of Life, 3*(11), 181–192.

Pang, M., Jin, J., Segers, A., Lin, H.X., Wang, G., Liao, H.B., & Han, W. (2025). Zeeman: A Deep Learning Regional Atmospheric Chemistry Transport Model. *arXiv (Cornell University)*. https://doi.org/10.48550/arxiv.2510.06140

Partap, M., Sharma, D., HN, D., Chandel, A., Thakur, M., VipashaVerma, & Bhargava, B. (2023). Phytoremediation toward air pollutants: Latest status and current developments. In M. Eyvaz, A. Albahnasawi, & M. Y. D. Alazaiza (Eds.). *Air Pollution - Latest Status and Current Developments. In Environmental Sciences*IntechOpenhttps://doi.org/10.5772/intechopen.111559.

Pénard-Morand, C., & Annesi-Maesano, I. (2004). Air pollution: From sources of emissions to health effects. *Breathe, 1*(2), 108. https://doi.org/10.1183/18106838.0102.108.

Piracha, A, & Chaudhary, M. T. (2022). Urban air pollution, urban heat island and human health: a review of the literature. *Sustainability, 14*(15). https://doi.org/10.3390/su14159234.

Pope, CA, et al. (2024). Atherosclerotic Cardiovascular Disease Mortality and PM2.5 Air Pollution — Role of Pollution Sources? *The New England Journal of Medicine Evidence, 3*(12). 10.1056/EVIDe2400371.

Rider, C. V., Boekelheide, K., Catlin, N. R., Gordon, C. J., Morata, T. C., Selgrade, M. K., Sexton, K. G., & Simmons, J. E. (2014). Cumulative risk: Toxicity and interactions of physical and chemical stressors. In *Carolina digital repository (University of North Carolina at Chapel Hill)* https://doi.org/10.17615/k2n3-7d16.

Santillán-Lima, P., Llerena, M. R., Lima, J. C. S., Molina-Granja, F., Rivera, D. M. C., & Lozada-Yánez, R. (2023). Assessment of the concentration of settleable particulate matter using geographic information systems in the central ecuadorian highlands. *ICST Transactions on Scalable Information Systems*. https://doi.org/10.4108/eetsis.4156.

Schneider, E., Czech, H., Hartikainen, A., Hansen, H. J., Gawlitta, N., Ihalainen, M., Yli-Pirilä, P., Somero, M., Kortelainen, M., Louhisalmi, J., Orasche, J., Fang, Z., Rudich, Y., Sippula, O., Rüger, C. P., & Zimmermann, R. (2024). Molecular composition of fresh and aged aerosols from residential wood combustion and gasoline car with modern emission mitigation technology. *Environmental Science Processes & Impacts, 26*(8), 1295. https://doi.org/10.1039/d4em00106k.

Selami, N., Sèze, G., Khatir, R., Tahri, F., Tahri, A., & Benabadji, N. (2025). Impact of weather regimes on solar power variability in Algeria. *Energies, 19*(1), 167.

Sicard, P., Agathokleous, E., Anenberg, S. C., De Marco, A., Paoletti, E., & Calatayud, V. (2023). Trends in urban air pollution over the last two decades: A global perspective. *Science of the Total Environment, 858*, 160064.

Souza, B. S.de, Severo, L. G. R., Berlese, D. B., & Barros, M. P.de (2024). Análise ambiental integrada da qualidade do ar, variáveis meteorológicas e comunidade de aves em área urbana subtropical. *Revista de Gestão Ambiental e Sustentabilidade, 13*(1). https://doi.org/10.5585/2024.24055.

Taha, S., Idoudi, S., Alhamdan, N., Ibrahim, R., Surkatti, R., Amhamed, A. I., & Alrebei, O. F. (2025). Comprehensive review of health impacts of the exposure to nitrogen oxides (NO_x), carbon dioxide (CO_2), and particulate matter (PM). *Journal of Hazardous Materials Advances, 19*, 100771. https://doi.org/10.1016/j.hazadv.2025.100771.

Tai, A. P. K., Mickley, L. J., Jacob, D. J., Leibensperger, E. M., Zhang, L., Fisher, J. A., & Pye, H. O. T. (2012). Meteorological modes of variability for fine particulate matter (PM_{2.5}) air quality in the United States: Implications for PM_{2.5} sensitivity to climate change. *Atmospheric Chemistry and Physics, 12*(6), 3131. https://doi.org/10.5194/acp-12-3131-2012.

Talkhabi, H, et al. (2024). Integrating Tehran metropolitan air pollution into the current transport system and sprawl growth: an emphasis on urban performance and accessibility. *Discover Cities, 1*. https://doi.org/10.1007/s44327-024-00008-4.

Temiz, C. (2024). Analyzing Air Pollution Trends in Eastern Türkiye Over the Last Decade Using a Multivariate Statistical Approach. *Research Square (Research Square)*. https://doi.org/10.21203/rs.3.rs-3987349/v1

Tiwari, A., Pandey, M., Tirkey, A., Tiwari, A., Dubey, R., & Pandey, S. K. (2024). Image based analytical approaches for study of particulate matter (PM) in air. *Frontiers in Environmental Science, 12*. https://doi.org/10.3389/fenvs.2024.1362422.

Vega, E., Ruíz-Olivares, A., Miquelajauregui, Y., Rentería-Campos, R., Bramwell, L., Feytmans, R. B. M., Namdeo, A., Núñez-Enríquez, J. C., Entwistle, J., Jaimes-Palomera, M., Torres-Rojas, N. M., Rascón-Pacheco, R. A., Duarte-Rodríguez, D. A., McNally, R., García-Burgos, J., Resendiz-Martinez, C. G., Fragoso-Chino, Á., Hayes, L., Portas, A., & Mejía-Arangué, J. M. (2025). Association between daily variations in the levels of atmospheric O_3, PM2.5, and NO_2 and the frequency of hospital visits due to respiratory diseases and hypertension in Mexico city using generalized additive mixed models. *Frontiers in Public Health, 13*. https://doi.org/10.3389/fpubh.2025.1593285.

Wang, J., Chen, S., Zhu, M., Miao, C., Song, Y., & He, H. (2018). Particulate matter and respiratory diseases: How far have we gone? *Journal of Pulmonary & Respiratory Medicine, 8*(4). https://doi.org/10.4172/2161-105x.1000465.

Wang, Y. (2023). Air quality estimations at local scale accounting for indoor and outdoor pollutants emissions. *HAL (Le Centre Pour La Communication Scientifique Directe)*. https://pastel.hal.science/tel-04368505.

Wen, Z (2024). Combined short-term and long-term emission controls improve air quality sustainably in China. *Nature Communications, 15*, 5169.

Yang, J., Hasheminassab, S., Franklin, M., Zhang, A., Diner, D. J., Pinto, J. P., & Yung, Y. L. (2025). Prediction of Ambient PM2.5 chemical components in southern california using machine learning. . *Carolina Digital Repository (University of North Carolina at Chapel Hill)*. https://doi.org/10.17615/rn91-wx32.

Zaccardi, C., Valentini, P., Ippoliti, L., & Schmidt, A.M. (2024). On the PM2.5 – Mortality Relationship: A Bayesian Dynamic Model for Spatio-Temporal Confounding. *arXiv (Cornell University)*. https://doi.org/10.48550/arxiv.2405.16106

CHAPTER TWO

Chlorophyll *a* fluorescence as a measure of plant responses to air pollutants

Shasmita[1] **and Pradipta Kumar Mohapatra**[2]
[1]Department of Botany, Dhenkanal Autonomous College, Dhenkanal, Odisha, India,
[2]Department of Botany, Ravenshaw University, Cuttack, Odisha, India

2.1 Introduction

Plants are exposed to different biotic and abiotic stresses, and all of them more or less influence plant growth and metabolism. Anthropogenic activities, such as lifestyle habits, mechanized cultivation practices, chemical intervention, consumerism, urbanization, and industrialization, are releasing a variety of pollutants, such as carbon dioxide (CO_2), carbon monoxide (CO), oxides of sulfur (SO_x) and nitrogen (NO_X), particulate matter (PM), volatile organic compounds (VOCs), heavy metals (HMs), acids, ozone (O_3), and polycyclic aromatic hydrocarbons, to the atmosphere. While some of these gases are utilized by plants for growth and metabolism, and some others support plant growth at low concentrations, most of them adversely affect plant physiological and biochemical processes. Air pollution associated with climate change, warming of the atmosphere, and loss of seasonality also create associated and cumulative adverse conditions for plants, compelling them to adapt, mitigate, or perish (Moores, 2009; Manisalidis et al., 2020). Adaptations of plants to air pollutants occur at morphological, biochemical, physiological, and molecular levels (Chatterjee et al., 2024); however, most air pollutants inflict visible and invisible damage to plants.

Photosynthesis is an important physiological process that determines growth, development, and yield attributes of plants. Air pollutants often cause a reduction in carbon dioxide fixation, leading to alteration of plant growth and biochemical composition. The photosynthetic machinery from photosystems (PS) to chloroplasts is affected by air pollutants. There

Urban Air Pollution and Plant Responses
DOI: https://doi.org/10.1016/B978-0-443-40320-0.00007-3

are consistent efforts to understand how air pollutants affect the structural and functional characteristics of photosynthetic pigment–protein complexes, intersystem components of photoelectron transport, and the thylakoid and chloroplast membranes, and thereby modify the photosynthetic performance. The effects of air pollutants on the leaf architecture and stomatal functions, which lead to a reduction in gas exchange and rate of photosynthesis, have also been studied (Flowers et al., 2007; Oguntimehin et al., 2010; Singh et al., 2017; Hu et al., 2021; Kohan et al., 2021).

Chlorophyll *a* fluorescence (ChlF) has been extensively used over the last few decades to understand the impact of different pollutants on the photochemistry of plants (Strasser et al., 2004; Ismail et al., 2014; Gao et al., 2022; Swain et al., 2024; Mishra et al., 2024). This phenomenon is closely related to the state of photosystem II (PSII), which plays a leading role in the photosynthetic process, and therefore it has become a powerful tool to investigate the light reaction of photosynthesis under various physical or chemical and biotic stresses (Mohapatra et al., 1997; Lazar, 2003; Stirbet & Govindjee, 2011; Jena et al., 2012; Duan et al., 2019; Shasmita et al., 2018, 2019, 2021; Swain et al., 2024). The possibility of performing easy, fast, and direct measurements of the fluorescence, even under light conditions, has opened new frontiers for in situ analysis of air pollutants. The activity of PSII in response to various abiotic (Flowers et al., 2007; Mohapatra et al., 2010; Jena et al., 2012; Sarkar & Ray, 2016; Singh et al., 2017; Hu et al., 2021; Swain et al., 2024; Mishra et al., 2024) and biotic (Shasmita et al., 2018, 2019, 2021) stresses has been determined by using ChlF. The air pollutants may react with PSII and modify ChlF emission, which varies qualitatively and quantitatively with the type and concentrations of pollutants. Thus, ChlF has unlocked new frontiers in the analysis of the effect of in situ air pollutants on the structure and function of PSII, as well as in the photosynthetic performance of plants. This chapter presents an overview of different types of air pollutants and their effects on photosynthesis and ChlF with special emphasis on the use of fast fluorescence kinetics and the pattern changes of the related bioenergetics parameters. It also throws light on how different gas exchange parameters help in monitoring plant health status in response to air pollutants. The research achievements on gas exchange and prompt fluorescence have been discussed to present a wholesome picture of the effects of air pollutants on photosynthesis.

2.2 Air pollutants affecting plant health

Depending upon the mode of discharge into the environment, the air pollutants that impact air quality and cause health issues are categorized into two types, i.e., (1) primary air pollutants and (2) secondary air pollutants. The primary pollutants include gaseous molecules, such as CO_2, CO, SO_x, NO_x, PM, VOCs, and HMs, which are discharged directly into the environment. HMs derived from traffic emissions, such as Cd (cadmium), Cu (copper), Cr (chromium), Pb (lead), As (arsenic), Hg (mercury), and Sb (antimony), remain in the air as volatile and particulate elements/compounds. Other air pollutants, such as sulfates, nitrates, ammonia, black carbon, organic carbon, and mineral dust, are released into the atmosphere by both natural and anthropogenic means. In addition, due to longer atmospheric duration, PM is the most widespread air pollutant. Depending upon particle size, PM is categorized into (1) coarse PM (particle size $\leq$ 10 μm), (2) fine PM (particle size $\leq$ 2.5 μm), and (3) ultrafine PM (particle size $\leq$ 0.1 μm) (WHO, 2021; Chatterjee et al., 2024; Antenozio et al., 2024). PM is released from construction, forest fires, volcanic eruptions, fly ash, smoke, dust, burning of waste, household, vehicular exhaust, and industrial emissions (Mahilang et al., 2021; Antenozio et al., 2024). Secondary pollutants include O_3, secondary PM and VOCs, sulfuric acid and nitric acid, which are derived from photochemical interaction (triggered by sunlight) between primary pollutants, such as CO, SO_X, and NO_X, present in the air, leading to the formation of acid rain and photochemical smog (Chatterjee et al., 2024; Antenozio et al., 2024).

The vegetation acts as a natural sponge, which absorbs air pollutants. The extended leaf area and microstructures present on the leaf surface increase the deposition of PM, thereby altering optical properties of the leaf, leading to reduced light intensity vis-à-vis increased light attenuation. SO_X, NO_X, and O_3 pollutants cause foliar injury, loss of pigment, premature senescence, and decreases rate of photosynthesis and growth. Deposition of PM on leaf surface decreases plant height, root and shoot length, branching, root/shoot ratio, biomass, growth, number of leaves, tillers, panicles, flowers, trichomes, and increases leaf temperature, wax content, thickness, leaf area damage, and vegetative bud damage (Chatterjee et al., 2024; Jumrani et al., 2024). Deposition of PM also increases stomatal clogging, thereby affecting gaseous exchange, leading to reduced photosynthesis and respiration. The distribution of beneficial bacteria and fungi in leaf endosphere and phyllosphere is also disturbed by pollutants (Chatterjee et al., 2024; Manisalidis et al., 2020). Air

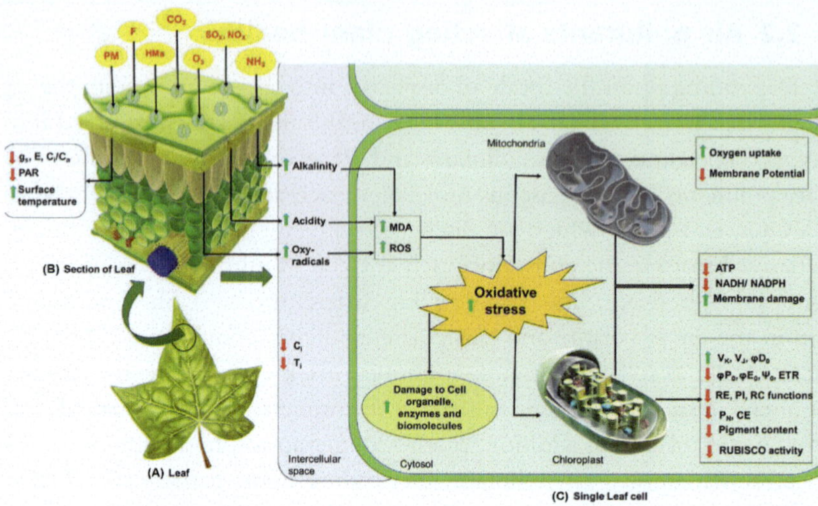

Figure 2.1 Effect of common air pollutants on photosynthetic and respiratory activities in leaves. The parameters related to photosynthesis and stomatal functions are mentioned in 'b and c'. The change in level of oxyradicals, acidity, and alkalinity is induced in the intercellular environment primarily by O_3, SO_X, and NO_X, and NH_3, respectively. Such a change triggers ROS, which causes oxidative stress and minimizes chloroplast and mitochondrial functions. '↑' indicates an increase and '↓' indicates a decrease in the magnitude of parameters taken. Abbreviations are explained in Tables 2.1 and 2.2.

pollution affects plant health by inducing different stress responses, such as increasing reactive oxygen species (ROS) burst, deregulating production of ROS scavenging enzymes such as catalase (CAT), superoxide dismutase (SOD), and peroxidase, and enhancing the synthesis of secondary metabolites (e.g., flavonoids) involved in defense pathways (Manisalidis et al., 2020). Oxidative damage affects the photoelectron transport, primary and secondary photochemistry, and photosystem efficiency, which in turn affect the rate of photosynthesis and other biochemical processes (Fig. 2.1). Tables 2.1 and 2.2.

2.3 Chlorophyll *a* fluorescence: understanding the basic concepts

ChlF kinetics is a simple, nondestructive, inexpensive, fast, and powerful tool for assessment of architecture and behavior of photosynthetic machinery (Govindjee, 1995; Strasser et al., 2004; Stirbet & Govindjee, 2011; Kalaji et al., 2016). The ChlF signal comes mainly from PSII and is

Table 2.1 The equations and explanations for commonly used bioenergetic parameters of OJIP fluorescence transients. (The equations are based on Force et al., 2003; Strasser et al., 2004; Stirbet & Govindjee, 2011; Shasmita et al., 2018; 2024.)

Bioenergetic parameters	Equations	Explanation of parameters
OJIP inflections		
F_0	-	Minimal fluorescence measured at 20 μs
F_L	-	Fluorescence recorded at 'L' inflection (or 150 μs) of OJIP transient
F_K	-	Fluorescence recorded at 'K' inflection (or 300 μs) of OJIP transient
F_J	-	Fluorescence recorded at 'J' inflection (or 2 ms) of OJIP transient
F_I	-	Fluorescence recorded at 'I' inflection (or 5 ms) of OJIP transient
F_H	-	Fluorescence recorded at 'H' inflection (or 20 ms) of OJIP transient
F_G	-	Fluorescence recorded at 'G' inflection (or 100 ms) of OJIP transient
F_M	-	Maximum fluorescence measured when all the PSII reaction centres (RCs) are closed (tF_M)
Area	-	Area between the OJIP transient and the line $F = F_M$
Variable fluorescence parameters		
F_V	$F_t - F_0$	Variable fluorescence recorded at time 't'
W_L	$(F_L - F_0)/(F_J - F_0)$	Relative variable fluorescence at L inflection of OJIP transient to the amplitude F_J
W_K	$(F_K - F_0)/(F_J - F_0)$	Relative variable fluorescence at K inflection of OJIP transient to the amplitude F_J
V_L	$(F_L - F_0)/F_V$	Relative variable fluorescence at the L inflection of OJIP transient to the amplitude F_M
V_K	$(F_K - F_0)/F_V$	Relative variable fluorescence at the K inflection of OJIP transient to the amplitude F_M
V_J	$(F_J - F_0)/F_V$	Relative variable fluorescence at the J inflection of OJIP transient to the amplitude F_M
V_I	$(F_I - F_0)/F_V$	Relative variable fluorescence at the I inflection of OJIP transient to the amplitude F_M
M_0	$4 \times [(F_K - F_0)/F_V]$	Net rate of PSII RCs closure
OEC(F)	$(1-V_K)/V_J$	Fraction of functional oxygen evolving complexes (OECs)

(Continued)

Table 2.1 (Continued)

Bioenergetic parameters	Equations	Explanation of parameters
Quantum yield and efficiency parameters		
φP_0	F_V/F_M	Maximum quantum yield of primary PSII photochemistry
ψ_0	$1 - V_J$	Efficiency of PSII exciton transfer beyond Q_A
φE_0	$\varphi P_0 \times \psi_0$	Quantum yield of electron transport from Q_A to Q_B
φD_0	$1 - \varphi P_0$	Quantum yield of energy dissipation
ψRE	$1 - V_I$	The efficiency with which a PSII trapped electron is transferred to PSI acceptors
φRE	$\varphi P_0 \times (1 - V_J)$	Quantum yield of reduction of end electron acceptors of PSI per photon absorbed
Energy flux parameters		
$ABS/\ RC$	$(M_0/V_J) \times (1/\varphi P_0)$	Absorption flux per active RC of PSII
$TR_0/\ RC$	M_0/V_J	Trapped energy flux per active RC of PSII
$ET_0/\ RC$	$(M_0/V_J) \times \psi_0$	Electron transport flux per active RC of PSII
$DI_0/\ RC$	$(ABS/RC) - (TR_0/RC)$	Dissipated energy flux per active RC of PSII
Performance indexes		
PI_φ	F_V/F_0	Performance index of primary photochemistry
PI_ψ	ψ_0/V_J	Performance index of electron transport
PI_{ABS}	$[\psi_0/(1 - \psi_0)] \times [\varphi P_0/(1 - \varphi P_0)] \times (RC/ABS)$	Performance index on absorption basis
PI_{TOTAL}	$PI_{ABS} \times [\delta RE/(1 - \delta RE)]$	Performance index up to the reduction of the end electron acceptor of PSI

Table 2.2 Bioenergetic and photosynthetic parameters were taken to analyze the effect of different air pollutants on photosynthesis and photoelectron transport of plants.

Air pollutant	Plant	Bioenergetic parameters studied	Photosynthetic parameters studied	Remark	References
O_3	*Phaseolus vulgaris*	F_0, φP_0, qP, qN	P_N, g_s, $V_{c,max}$	– Increase in F_0; – Decrease in yield, P_N, φP_0, qP and $V_{c,max}$ (in S156 – sensitive variety) and qN; No change in g_s	Flowers et al. (2007)
	Pisum sativum	φP_0, qP, qN	P_N, g_s, C_i	– Increase in qN; – Decrease in Chl *a* content, P_N, g_s, C_i, φP_0, and qP	Ismail et al. (2014)
	Koelreuteria paniculata, Ginkgo biloba, Pinus bungeana, Platycladus orientalis	F_0, F_M, F_V, electron transport rate (ETR), φP_0		– Increase in relative conductivity; – Decrease in Chl content, φP_0, ETR; – *K. paniculata* and *G. biloba* were more sensitive to O_3	Gao et al. (2022)
O_3 + Fluoranthene	*Impatiens* spp.	F_0, F_M, φP_0	P_N, g_s, C_i	– Increase in Ci, F_0; – Decrease in Chl content, P_N, g_s, CE, F_M, and φP_0	Oguntimehin et al. (2010)
SO_X	*Pinus massoniana, Schima superba, Castanopsis fissa, Acmena acuminatissima, Cryptocarya concinna*	F_V; PSII functions	-	– Increase in damage to PSII; – Decrease in F_V	Liu et al. (2007)
	Euonymus Kiautschovicus, Ligustrum vicaryi, Syringa oblata	F_0, F_0', F_M, F_M', F_V', F', φP_0, F_V'/F_M'	P_N, C_i, C_a, g_s, L_S	– Increase in oxidative damage, stomatal resistance, photoinhibition of PSII, sulfur accumulation, C_i; – Decrease in Chl *a*, *b*, and carotenoids content, P_N, g_s, φP_0, F_V'/F_M'	Duan et al. (2019)

(Continued)

Table 2.2 (Continued)

Air pollutant	Plant	Bioenergetic parameters studied	Photosynthetic parameters studied	Remark	References
NO_X	*Xanthoria parietina, Flavoparmelia caperata, Parmotrema perlatum*	F_0, F_M, φP_0, qN, rETR	-	– Increase in qN; – Decrease in F_0, F_M, φP_0 and rETR	Piccotto et al. (2011)
NO_X, NO_2, SO_2, CH_4, CO	*Lactuca sativa*	-	P_N, g_s, C_i	– Increase in ion leakage, proline, catalase (CAT), APX, and (GPX activity; – Decrease in growth attributes, Chl content, P_N, g_s, C_i	Kohan et al. (2021)
PM	*Lagerstroemia speciosa*	-	P_N, VPD, E, g_s, g_r, WUE, T_i	– Increase in g_r, WUE, VPD of air; – Decrease in P_N, T_i, g_s, E	Singh et al. (2017)
Airborne HMs	*Lactuca sativa*	F_0, F_M, F_V, φP_0	P_N, g_s	– Increase in lipid peroxidation, superoxide dismutase (SOD) activity; – Decrease in Chl *a*, *b*, *a*/*b*, carotenoids content, shoot length, F_M, F_V, φP_0, P_N, g_s	Hassan et al. (2013)
Acid rain	Barley	F_0, F_M, F_V, Area, PI_{ABS}, V_J, V_I, φP_0, PI_φ, ABS/reaction centre (RC), DI_0/RC,	P_N, P_{max}, E, g_s, C_i, LCP, R_d	– Increase in LCP, R_d, ABS/RC, DI_0/RC – Decrease in Chl *a*, *b*, *a*/*b*, carotenoids content, P_N, P_{max}, E, g_s, F_V, F_M, φP_0, PI_φ – No significant effect on Ci and F_V/F_M	Hu et al. (2021)

Abbreviations: *APX*, ascorbate peroxidase; *CAT*, catalase; C_a, ambient CO_2 concentration; C_i, intercellular CO_2; *CE*, carboxylation efficiency; *E*, rate of transpiration; *ETR*, electron transfer rate; *F'*, steady state fluorescence; *F0'*, minimal fluorescence from a light-adapted leaf; *FM'*, maximum chlorophyll fluorescence in light-adapted state; *FV'*, variable fluorescence level of leaves in light; *FV'/FM'*, maximum quantum efficiency of PSII photochemistry in light adapted state; *GPX*, guaiacol peroxidase; g_r, stomatal resistance; g_s, stomatal conductance; *LCP*, light compensation point; L_S, relative decrease in photosynthesis imposed by stomatal closure; *PM*, particulate matter; P_{max}, maximum rate of photosynthesis; P_N, net rate of photosynthesis; R_d, dark respiration; *rETR*, relative electron transport rate; *SOD*, superoxide dismutase; T_i, leaf intercellular temperature; *VPD*, vapour pressure deficit; $V_{c,max}$, RuBP saturated rate of carboxylation.

the reemitted excess energy that is neither involved in photochemical processes, quenching, nor dissipated as heat, as photochemistry, heat dissipation, and fluorescence are competing processes. Fluorescence measurement is used to quantify the proportion of nonphotochemical dissipation of absorbed light that is increased in a pattern, which is universal from algae to higher plants (Maxwell & Johnson, 2000). ChlF thus provides numerous parametric data that allow deeper analysis of physiological processes associated with the light phase of photosynthesis. During the past decades, equipment with different operating principles and numerous mathematical tools has been developed for the purpose of making it possible to measure the emission from plants with different complexity, and growing in varied habitats (Baker & Rosenqvist, 2004; Stirbet & Govindjee, 2011; Kalaji et al., 2016; Banks, 2017).

Signals of ChlF can be detected from samples well adapted to darkness (when all photosystems are functionally open), as well as from samples under ambient light (where a fraction of photosystems are at work). The fast (or prompt) fluorescence analysis is made from the sample preconditioned with dark adaptation (typically up to 20 min) for bringing the photosystems to a relaxed state, followed by a saturating pulse of actinic light between 2000 and 5000 $\mu E/m^2s$ (Strasser et al., 2004; Stirbet & Govindjee, 2011; Mohapatra et al., 2014; Kalaji et al., 2016). When algae or higher plants are kept in darkness up to about 20 min and then illuminated, a polyphasic ChlF is observed in ≤ 1 s with a so-called OJIP transient curve reflecting different phenomena occurring in and around PSII, as well as in H_2O-$NADP^+$ electron transport pattern (Strasser et al., 2004; Strasser et al., 2010; Stirbet & Govindjee, 2011; Kalaji et al., 2016). In fast rise, the 'O' denotes origin or first measured minimal fluorescence (20 μs), 'J' and 'I' (currently taken as 'H' peak) refer to intermediate fluorescence at 2 and 20 ms (which are visible in the OJIP transient), and 'P' is the maximal fluorescence at about 300 to 500 ms. There are also hidden peaks between O and J at 150 and 300 μs known as L and K inflections, respectively, which are not visible in the native OJIP rise but can be made visible by differential fluorescence spectra. However, leaves exposed to high temperatures (about 47°C) show these peaks because of the effect of the treatment on the donor side of PSII (Lazar, 2003). Additional invisible peaks at 5 ms (currently taken as 'I' peak) and 100 ms (G peak) are seen when the noncyclic electron transport is disturbed between PQ-Cyt b_6f complex, and between PSII and ferredoxin reducing system, respectively (Strasser et al., 1995; Strasser et al., 2010; Stirbet & Govindjee, 2011; Kalaji et al., 2016; Shasmita et al., 2019). Analysis of shape of fast or OJIP phase of fluorescence induction curve helps to monitor the photosynthetic efficiency of different components of electron transport chain (ETC) of chloroplast, such as reaction centre (RC), antenna complexes, and components of donor and acceptor side of PSII (e.g., water splitting, plastoquinone (PQ), etc.) in response to different biotic and abiotic stresses (Stirbet & Govindjee, 2011; Kalaji et al., 2016; Yan et al., 2024; Shasmita et al., 2019, 2021; Swain et al., 2024; Mishra et al., 2024).

The rise in O–J part of OJIP transient corresponds to closure of some PSII RCs in response to reduction of the primary electron acceptor of PSII, which is a tightly bound PQ, i.e., Q_A. The L and K inflections correspond to disruption in the activity of oxygen-evolving complex (OEC) present on the PSII donor side (Kalaji et al., 2016; Swain et al., 2024; Shasmita et al., 2024). J–I rise of OJIP transient is related to the reduction of secondary electron acceptors of PSII, which include a weakly bound PQ, i.e., Q_B, PQ, cytochrome b_6f complex (cyt b_6f), and plastocyanin (PC) (Guha et al., 2013; Kalaji et al., 2016). The increase in emission in the I–P phase of the OJIP transient is attributed to a reduction in components present on the acceptor side of PSI, such as ferredoxin (Fd) and oxidized nicotinamide adenine dinucleotide phosphate (NADP+), as well as the effects on the end reactions of photoelectron transport (Kalaji et al., 2016). I, H, and G inflections represent the status of electron flow from PQ to cyt b_6f, cyt b_6f to PSI, and PSI to FeS centers, respectively (Strasser et al., 2004; Stirbet & Govindjee, 2011). When algal samples and plants are dark-adapted, the PSII acceptor side is in an oxidized state (PSII RCs are in the open state), and fluorescence is minimal, denoted by F_0. Different ChlF or bioenergetic parameters have been used in the analysis of plant health under abiotic and biotic stresses (Table 2.1; Force et al., 2003; Strasser et al., 2004; Stirbet & Govindjee, 2011; Shasmita et al., 2019, 2021, 2024; Swain et al., 2024). Prompt fluorescence analysis also allows the assessment of the probability of the movement of a trapped exciton into the electron transport chain beyond Q_A until it reaches PSI acceptors (Strasser et al., 2004; Strasser et al., 2010). Additionally, specific energy fluxes expressed per active RC, the phenomenological energy fluxes per cross-section (CS), absorption (ABS), trapping (TR_0), thermal dissipation (DI_0), and electron transport rate (ETR) beyond RC can be calculated.

2.4 Gas exchange parameters: role in analysis of the effect of air pollutants on plants

The plant photosynthetic machinery and the related physiological processes are among the most critical components affected by air pollutants. Most air pollutants, notably CO, SO_x, NO_x, and O_3, interact with plants at the level of the leaf surface, diffuse through stomata, and move into the intracellular spaces, thereby becoming available to the mesophylls (Fig. 2.1; Castagna and Ranieri, 2009; Räsänen et al., 2017). Once absorbed, these toxic compounds can interfere with stomatal conductance (g_s), leaf transpiration (E), photoelectron transport, and CO_2 fixation through photosynthesis (P_N) (Rai et al., 2011). Altered function of stomata results in an increase in leaf intercellular temperature (T_i) and a decrease in C_i of the leaf interior. At the cellular level, altered cytosolic pH and oxyradical concentrations induce the synthesis and accumulation of malondialdehyde and ROS, which impose oxidative stress. The functional efficiency of the chloroplast is negatively affected, leading to poor energy conservation and carbon fixation. A number of studies have been conducted where analyzing the changes in P_N, g_s, and E helps to monitor the physiological status of plants exposed to various air pollutants, such as O_3, PM, SO_x, NO_x, NH_3, and acids (Table 2.2).

The decrease in photosynthetic efficiency due to the interference of pollutants affects plant central carbon metabolism, resulting in the generation of ROS, which play a major role in plant responses to air pollutants (Yendrek et al., 2015). In strawberry (*Fragaria* x *ananassa*), treatment with different low, medium, and high doses of CO (133–533 ppb), NOx and SO_2 (0.05–0.2 ppm), caused continuous ROS production in leaves with a negative effect on g_s and P_N (Muneer et al., 2014). Lower chlorophyll content, reduced P_N and g_s, and an overall decrease in plant biomass and seed yield were also observed in wheat (*Triticum aestivum*), mustard (*Brassica campestris*), beet (*Beta vulgaris*), and mung bean (*Vigna radiata*) (Agrawal et al., 2006) on solo and combined exposure to NO_2 (29–80 ppb), SO_2 (30–40 ppb), and O_3 (9–58 ppb) (Papazian & Blande, 2020). In *Pisum sativum*, low SO_2 concentrations (0.2 and 0.5 ppm) caused a reduction in g_s (Ismail et al., 2014), and in other plants, SO_2 exposure caused a decrease in P_N both due to stomatal and nonstomatal regulations (Duan et al., 2019). When plants are exposed to low concentrations of SO_2, the absorbed SO_2 is oxidized and used to synthesize proteins (Robinson et al., 1998). However, at high concentrations, it disrupts thylakoid function and interferes with the electron transport chain (Heber & Hueve, 1997; Mostowska, 1997). Guderian (2012) reported that high SO_2 concentration causes stomatal closure by affecting the membrane potential of guard cells, but opined that at low concentrations, SO_2 enhances the rate of protein synthesis. Contrastingly, in *Arabidopsis*, adaptation to SO_2 by controlling stomatal closure via $Ca^{2+}$and H_2O_2 signaling, and by regulation of gene expression in guard cells has been observed (Yi et al., 2014). However, combined CO_2 and SO_2 exposure caused increased toxicity. Similarly, combined exposure to SO_x and NO_x leads to the production of oxyradicals, which disrupt cellular membranes and cause a decrease in pigment content, $P_{N,}$ and g_s (Saini et al., 2019; Duan et al., 2019). Reduction of P_N has been reported due to oxidative damage of chlorophyll molecules caused by superoxide radicals generated by H_2SO_3 during SO_2 metabolism due to a lack of carotenoid protection (Rai, 2016; Chatterjee et al., 2024).

O_3 stimulates stomatal opening and causes damage to the chloroplast membrane, leading to ion leakage, thereby causing a delay in stomatal responses and making plants more sensitive to ozone (Ismail et al., 2014). O_3-induced ROS cause injury to the plasma membrane via lipid peroxidation and consequent loss of membrane integrity. The fluid membrane becomes passively permeable to K^+ and Ca^{2+}, resulting in ion exclusion and reduced membrane potential (Ainsworth et al., 2014; Pellegrini et al., 2015). Swelling of thylakoids, an increase of plastoglobuli per chloroplast, and disruption of membranes lead to leakage of ions and a consequent change in photosynthetic capacity (Pellegrini, 2014).

A negative correlation was observed between P_N and C_i in two pea plant cultivars (e.g., Little Marvel and Perfection) affected with O_3 and suggested the involvement of nonstomatal factors, leading to inhibition of the electron transport chain in the chloroplast (Ismail et al., 2014). However, the positive impact of low levels of O_3 or short exposure has also been reported (Mikkelsen et al., 2015; Oksanen & Kontunen-Soppela, 2021). Combination treatment of *Impatience* with O_3 and fluoranthene decreased photosynthetic efficiency and g_s, even at high C_i, thereby showing the nonstomatal regulation (Oguntimehin et al., 2010). Nonstomatal factors include: (1) a decrease in CO_2 diffusion to the mesophylls, (2) a reduction in PSII activity, (3) a decrease in the rate of photoelectron transport, and (4) reduced ribulose-1,5- bisphosphate carboxylase/ oxygenase (RUBISCO) activity. A strong and significant negative correlation was observed between P_N and C_i, or between g_s and C_i, but this differed by plant species. Ismail et al. (2014) observed that the P_N vs. C_i curve was fitted to a nonrectangular hyperbolic model, confirming the nonstomatal limitations of P_N under O_3 treatment.

Long-term treatments to low-concentration mixtures of O_3, NO_2, and SO_2 caused an increase in the content of D1 protein of PSII, enhancing P_N, whereas exposure to high concentrations reduced it (Dhir, 2016). Inhibition of photosynthesis also occurred due to loss of carboxylation efficiency (CE) and RUBISCO activity (Farage et al., 1991). Stimulated acid rain mimicking the pollution by NOx and SOx caused a decrease of P_N, E, g_s, and C_i in barley plants (Hu et al., 2021). High (4 μL/L) concentrations of NO_2 could enhance nitrogen metabolism and photosynthetic efficiency in mulberry (*Morus alba*) with improved light absorption and P_N (Wang et al., 2019), but negative effects have been reported by other authors (Dhir, 2016). Exposure to low levels of NO_2 increases RUBISCO, as well as glycolate oxidase activities, thereby causing increased production of photosynthates. On the other hand, it is observed that damage to the photosynthetic apparatus, upon prolonged exposure to high NO_2 concentration (>1 μL/L), occurs by swelling of chloroplast membranes. NO_2 inhibits P_N by uncoupling electron transport and inducing structural alterations. Photosynthetic inhibition could also be caused by NOx due to competition for nicotinamide adenine dinucleotide phosphate (NADPH) for nitrite reduction and carbon assimilation in chloroplasts.

Deposition of PM leads to a reduction in effective leaf area and g_s due to closure of stomata. Decrease in g_s causes a decline in P_N and E rate, leading to a decrease in water use efficiency (WUE), CE, leaf longevity, and conductance (Singh et al., 2017). Deposition of SO_x and PM damages the epicuticular wax and also leads to increase in leaf T_i, thereby making the plant vulnerable to drought and temperature stress (Singh et al., 2017; Saini et al., 2019; Popek et al., 2018a, 2018b). It was observed that PM reduces the fixation of CO_2 in the chloroplast, and is coupled with the reduction of the photosynthetic ETC resulting in the production of ROS (Popek et al., 2018a, 2018b). In *Azadirachta indica,* it has been observed that PM reduced the concentration of RUBISCO, which could also be a strong reason behind the reduced P_N (Govindaraju et al., 2018; Chatterjee et al., 2024).

Air pollutants from vehicle exhaust affect *Lactuca sativa* by causing an increase in ion leakage, proline content, and ROS scavenging enzymes, such as CAT, ascorbate peroxidase (APX), and guaiacol peroxidase (GPX) activity, and a decrease in height, fresh and dry shoot weight, leaf area, pigment content, P_N, g_s, and C_i (Kohan et al., 2021). Deposition of airborne HMs caused an increase in lipid peroxidation and SOD activity, and a decrease in pigment content (Hassan et al., 2013). Acid rain has a strong negative effect on the photosynthesis of the plant. The acidic components decreased P_N, E, and g_s due to nonstomatal factors, such as a decline in chlorophyll and carotenoid content in barley (Hu et al., 2021).

2.5 Chlorophyll *a* fluorescence: role in analysis of the effect of air pollutants on plants

ChlF measurement is used as a primary indicator of physiological stress faced by plants. Different bioenergetic parameters of primary photochemistry, relative variable fluorescence rates, RC functions, phenomenological energy fluxes, etc., have been used to study the fluorescence kinetics associated with air pollutants deposition on leaf surfaces and their entry into the leaf mesophylls (Table 2.2). Depending upon the type and level of stress, the shape of the OJIP fluorescence transient, and the values of various derived ChlF parameters vary. A general indication of the toxic effects of air pollutants is an increase in the level of F_0 caused by an increase in the proportion of nonfunctional photosystems (Flowers et al., 2007). Higher F_0 suggested low excitation capture and disruption of electron transport in the chloroplast, leading to poor photosynthetic efficiency in sensitive plant species. Almost all air pollutants, at toxic concentrations, cause a decrease in the intensity of fluorescence at the intermediate inflections and at the F_M level, primarily due to a decrease in the pigment content. However, prompt fluorescence measurements have shown that all parameters from photosystem functions to the end reaction are more or less affected by air pollutants (Fig. 2.2).

O_3 exposure affects bioenergetic parameters, such as F_0, F_V, F_M, F_V/F_M, φE_0, φP_0, ETR, qP, and qN (Table 2.2). Popovic et al. (2003) reported that when *Avena sativa* leaves were exposed to O_3 (46 ppb for 7 h) during the growing season, the qN increased by 35% while φP_0 and qP decreased by 37% and 20%, respectively. Leaves of *Bromus erectus, Rumex obtusifolius,* and *Solanum lycopersicum* showed higher qN when exposed to O_3 at higher concentrations than the ambient levels (Nussbaum et al., 2001; Calatayud & Barreno, 2001). This showed a decrease in the electron transport efficiency induced by O_3, leading to nonphotosynthetic energy dissipation as a protection mechanism (Thwe et al., 2014; Ismail et al., 2014). On the other hand, studies with spruce, watermelon (*Citrullus lamatus*), several species of *Sphagnum,* and lichens found little or no effect of O_3 on φP_0 and energy quenching (Gimeno et al., 1999; Calatayud & Barreno, 2000; Nussbaum et al., 2001; Calatayud et al.,

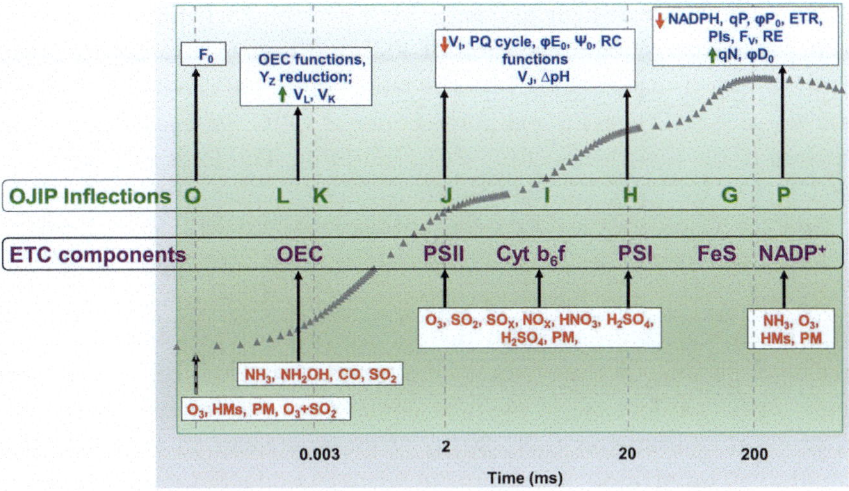

Figure 2.2 The illustrative scheme of the effect of air pollutants on various components of the photoelectron transport chain (ETC) as measured through OJIP fluorescence transients' analysis. A typical OJIP prompt fluorescence transient measured for a period of 1s has been given. The inflections of the transient that are visible in (and calculated from) the transient are presented against the electron transport intersystem components. The boxes below the ETC components are the pollutants affecting at various sites of the ETC. The boxes above the OJIP inflections are the parameters that are affected by the pollutants and can be calculated from the transient at mentioned inflections. '↑' indicates increase and '↓' indicates decrease in magnitude of parameters taken. Abbreviations are mentioned in Tables 2.1 and 2.2.

2002), thus showing a high variability among plant species with regard to the sensitivity to O_3. Flowers et al. (2007) observed a significant increase in F_0 by the impairment of the transport of excitation energy from light-harvesting complexes to RCs. There was also a decrease in φP_0, qP, and the maximum rate of carboxylation of ribulose-1,5-bisphosphate via RUBISCO ($V_{c,max}$) with an increase in O_3 concentration (15 to 60 nmol/ mol) in the sensitive variety of *Phaseolus vulgaris,* thereby making them more susceptible to O_3 injury, whereas no such effect was seen in the tolerant variety. The decrease in quantum yield of electron transport is responsible for downregulation of photosynthetic electron transport and reduction of photochemical efficiency due to the damage to PSII RCs (Ainsworth et al., 2014; Thwe et al., 2014). Similarly, in *Pisum sativum,* a significant reduction in φP_0, qP, and φE_0 and an increase in qN has been observed by the exposure to O_3 (71–90 nL/L) for five successive years. Reduction in φP_0 and φE_0 suggested inhibition of the ability to use photon energy by the plants in the presence of ambient O_3

concentration whereas high qN suggested lower photochemical quenching leading to greater photoinactivation of PSII and nonradiative dissipation of energy as heat (Ismail et al., 2014). The reduction in the density of thylakoid proteins, such as D1, cyt f, and polypeptides of OEC has been reported on exposure to the toxic levels of O_3, SO_x, and NO_x. This leads to a decline in the CE and oxidative damage caused by indirect heat-related injuries to RUBISCO (Castagna et al., 2001; Volgusheva et al., 2011; Ismail et al., 2014).

Exposure of leaves of *Impatiens* spp. to O_3 (130 ppb) or O_3 in combination with fluoranthene (10 μM) caused a decrease in F_0, F_M, and φP_0, which was consistent with the pattern of variation observed in gas exchange parameters, suggesting photoinhibition due to nonstomatal factors such as a decrease in pigment content and photosystem density in the leaves (Oguntimehin et al., 2010). With an increase in O_3 concentration (0 to 80 nmol/mol), an upsurge in relative conductivity and a decrease in φP_0, φE_0, ETR, and pigment content were observed in *Koelreuteria paniculata, Ginkgo biloba, Pinus bungeana*, and *Platycladus orientalis* (Gao et al., 2022). The authors also observed that as compared to *P. bungeana* and *P. orientalis*, there was approximately twofold decline in the rate of φP_0 and 1.2 to 2 times increase in electrical conductivity in *K. paniculata* and *G. biloba* suggesting that the two later species were more sensitive to O_3 than the former plant species. An increase in O_3 concentration decreases the light-capturing ability and pigment content by promoting chlorophyll degradation of the plants, resulting in reduced energy utilization efficiency for photochemical reactions and increased heat dissipation. As absorbed quantum energy is dissipated as heat energy, the ETR also decreases, suggesting damage to the photosynthetic structure and obstruction in electron transport (Thwe et al., 2014; Gao et al., 2022).

Exposure of plants to SO_x and NO_x pollutants affects bioenergetic parameters, such as F_0, F', F_M, F_M', φP_0, qP, qN, and ETR (Table 2.2). Piccotto et al. (2011) observed a decrease in F_0, F_M, φP_0, and qN in three foliose lichens, i.e., *Xanthoria parietina, Flavoparmelia caperata*, and *Parmotrema perlatum*, exposed to vehicular traffic containing NO_x for different durations. Short exposure (6 weeks) did not cause a significant change in the fluorescence transients and the associated bioenergetics, whereas prolonged exposure (12 weeks) forced a significant increase in F_0 and a decrease in φP_0 and qN in *F. caperata* and *P. perlatum*; however, *X. parietina* did not show any significant change in the fluorescence pattern. Exposure of *Pinus massoniana, Schima superba, Castanopsis fissa, Acmena acuminatissima*, and *Cryptocarya concinna* to different concentrations of $NaHSO_3$ (e.g., 20,

50, and 100 mM) resulted in a decrease in thylakoid membrane fluidity and an increase in the number of damaged PSII (Liu et al., 2007). A decrease in the magnitude of OJIP rise has been observed at all concentrations of $NaHSO_3$ in all species except *C. concinna*, which suggests that this plant is less sensitive to SO_2 (Liu et al., 2007). Exposure of *Euonymus kiautschovicus, Ligustrum vicaryi*, and *Syringa oblata* to SO_2 (25, 50, 100, 200 mg/m^3) caused a significant increase in oxidative damage to membranes, photoinhibition of PSII, and a decrease in φP_0 in a dose-dependent manner because of photoinhibition of PSII RCs (Wang et al., 2014; Duan et al., 2019). A reduction in the quantum yield of primary photochemistry of the dark- and light-adapted plants exposed to SO_2 indicated a decrease in quantum efficiency of operating PSII, and altered photosynthetic efficiency due to impaired ability of plants to use photon energy. Downregulation of PSII, reduced RUBISCO activity, and damage to PSII RCs effected the cumulative damage to the photosynthetic efficiency of the exposed plants (Baker & Oxborough, 2004; Duan et al., 2019).

The photosystem RCs are more vulnerable to SO_2 treatment than the light-harvesting antenna system. Decreases in quantum yield (φP_0; φPSII) result from inactivation of a part of the PSII, thereby reducing the photochemical energy conversion in PSII. The inhibition of electron transfer occurs at the site close to the RC of PSII. Studies conducted using isolated chloroplasts indicate that the oxidizing side of PSII is more affected by SO_2 than the reducing side (Liu et al., 2007; Wang et al., 2014; Dhir, 2016; Lee et al., 2017; Duan et al., 2019). However, the xanthophyll cycle (de-epoxidation) and antioxidant system (1,1-diphenyl-2-picrylhydrazyl radical-scavenging capacity) are the two protective mechanisms active during SO_2 exposure for dissipation of excess absorbed energy as heat (Demmig-Adams & Adams, 1996). The accessory pigment, lutein, participates in the nonradiative energy dissipation through the quenching of ChlF, leading to high qN. The activation of the xanthophyll cycle under the influence of SO_2 is the result of a decrease in intrathylakoid pH caused by adenosine triphosphate (ATP) consumption or as a consequence of SO_2-inhibited carbon assimilation (Liu et al., 2006). Such acidification of the thylakoid lumen is known to slow down PQ cycle due to feedback inhibition (Mohapatra et al., 1996).

Ammonia toxicity causes saturation of membrane lipids, resulting in an increase in membrane permeability of leaf tissues (plasmolysis and necrosis). It uncouples the ΔpH across the photosynthetic membrane, as well as damages the Mn cluster of OEC of PSII (Navarro et al., 2013; Oyala et al., 2015;

Markou et al., 2016). Ammonia toxicity also increases the sensitivity of PSII to photodamage but does not affect the PSII repair rate (Drath et al., 2008). As the proportion of oxidized PSII increases with an increase in ammonia toxicity, elevated concentrations of ammonia in the atmosphere affects the prompt fluorescence rise making significant reduction in the magnitude of fluorescence rise at all inflections. However, V_J increase was observed due to a decrease in the capacity for reoxidation of Q_A (Markou et al., 2016). φP_0, Ψ_0, and φE_0 decreased while φD_0, ABS/RC, and DI_0/RC increased with increasing NH_3 concentration (Navarro et al., 2013; Oyala et al., 2015; Markou et al., 2016).

Acid rain has a very strong impact on the photosynthetic processes of the plants. It damages chloroplast and foliar structure, decreases seed germination, water content of leaf, yield, photosynthetic rate, and CAT activity whereas increases GPX and SOD activity (Fig. 2.1; Hu et al., 2021). Hu et al. (2021) observed that acidity (pH 3.5 to 6.5) altered the shape of the OJIP transient in both the selected varieties of barley. The highest ChlF was observed in plants exposed to acid rain, resulting a pH of 6.5. An increase in acidity induced a small change in O step (F_0) but the magnitude of fluorescence at J, I, and P inflections of the OJIP transient significantly decreased. Acidity slowed down the electron flow from PQ to cyt b_6f complex, consequently affecting the performance of the energy production, which could be marked from the reduction in φP_0, TR_0/RC, and ABS/RC, and increase in the value of DI_0/RC. The number of inactive PSII RC increased with an increase in acidity, resulting in decreased photosynthesis in barley, cucumber, rice, and soybean (Qiu et al., 2001; Yu et al., 2002; Sun et al., 2016; Hu et al., 2021). Lu et al. (2001) reported that inactivation of some PSII reaction centers can lead to an increase in ABS/RC.

Exposure to PM causes a decrease in φP_0, and φE_0 in *Clusia hilariana* (Pereira et al., 2009). Electron transport activity, photosynthetic efficiency of PSII, and transpiration rate drastically decreased in *Eugenia uniflora* (Neves et al., 2009; Rai, 2016) and urban side plants (Popek et al., 2018a, 2018b) due to high amounts of PM. In *Quercus coccifera* and *Zygophyllum prismatocarpum*, PM resulted in a reduced primary photochemistry and increased dust deposition on leaves. When the stress increased, the pool size of electron acceptor decreases on the reducing side of PSII (Dhir, 2016). Łukowski et al. (2020) also observed higher quantum efficiency (1.7%–2.4%) and photosynthetic performance (32%–49%) of PSII in *Betula pendula*, *Quercus robur*, and *Tilia cordata* under control conditions than in

PM-polluted conditions. On the other hand, enhancement of the performance was observed in *Schinus terebinthifolius* (Kuki et al., 2008), thus confirming that the effect of PM on plants is species specific. HMs, like Hg, Pb, and Cd, present in the PM from vehicles, mining activities, and industry sources, gets deposited on the plants surface and cause necrosis, chlorosis, a decrease in pigment content, an increase in membrane damage leading to leakage of ions and lipid peroxidation, as well as alter leaf gas exchange properties and activities of ROS scavenging enzymes, such as SOD, CAT, and POX (Fig. 2.1; Hassan et al., 2013). HMs pollution has a negative impact on photosynthesis by affecting OEC, PC, the electron transport system of PSII, and enzymes of carbon dioxide fixation cycles (Popovic et al., 2003; Calatayud, 2007). Hassan et al. (2013) observed a decrease of F_M, F_V, and φP_0 and an increase in F_0 due to an increased proportion of closed and/or nonfunctional photosystems. Qin et al. (2023) reported that Pb-loaded PM reduced φP_0 and potential activity of PSII reaction centers (F_v/F_0) in the two populations of *M. floridulus*. The qP and ETR decreased while qN increased under Pb stress.

2.6 Regulation of respiration

Air pollutants not only cause visible damage to leaves but also interfere with respiratory processes. If the rate of photosynthesis in plants is high, a small change in the rate of respiration does not affect the carbon balance of the plant. On the other hand, if photosynthesis is affected by the pollutant exposure, a change in respiration can lead to a change in growth and yield of the plant. This is especially observed when the environmental conditions, like light and temperature, limit photosynthesis in the presence of inhibitory concentrations of pollutants. However, the effects of air pollutants on the process of respiration (dark respiration and photorespiration), which is an important component of the carbon budget, have been less studied than of photosynthesis (Kozioł & Whatley, 2013). Exposure of plants to toxic concentrations of air pollutants for a prolonged period inflicts irreversible damage to the chloroplast and mitochondria, thereby affecting oxygen uptake and use. Respiration is also influenced by the pollutant-induced changes of the stomatal behavior, as well as pollutant accumulation in the leaf interior. In *Triticum aestivum*, it is observed that SO_2, alone or in combination with elevated levels of CO_2, increases dark respiration, whereas CO_2 alone causes reduced O_2 consumption, compared to the combination (Agrawal & Deepak, 2003). Further, it was observed that elevated levels of CO_2 do not significantly enhance respiration as long as light intensity and temperature are not limiting. Ballantyne (1973) observed that sodium sulfite inhibited respiratory ATP formation, which could be reversed by the addition of oxidized glutathione, in both bean and corn mitochondria.

SO_2 concentration caused an initial decrease in photosynthetic rate and an increased rate of respiration in plants (Gheorghe & Ion, 2011). Markou et al. (2016) reported that in *Arthospira plantensis* and *Chlorella vulgaris*, respiration was also negatively affected by free NH_3, causing reduced ATP production, whereas the respiratory oxygen consumption increased to compensate for the energy demand. The degree of inhibition of respiration increased with the concentration of free NH_3. However, the authors observed that *A. plantensis* species was more tolerant of ammonia than *C. vulgaris* due to the higher intrathylakoid pH and the relatively high internal pH that prevents intracellular accumulation of ammonia (Belkin & Boussiba, 1991a, 1991b). The effect of ammonia on mitochondria may result in uncoupling of electron transport as observed with other respiratory inhibitors and toxicants (Yamano & Morita, 1993; Mohapatra et al., 1997; Mohapatra & Schiewer, 1996, 2000).

Depending on the degree of injury inflicted on plants by air pollutants, respiration is either inhibited or stimulated. As the repair processes utilize energy, the rate of respiration in the nondamaged tissues adjacent to these necrotic areas increases to subsidize the adjacent infected cells with energy for adaptation and stress mitigation. This causes wasteful loss of carbohydrates and energy, which are normally used in growth and development. On the other hand, the respiration in the affected tissues decreases because of the distorted physical and biochemical nature of the cell organelle. If exposures are not extreme, physiological processes are altered without any visible damage, whereas with prolonged exposure, irreversible reduction of respiration leads to reductions in growth and metabolism (Kozioł & Whatley, 2013; Gupta, 2016). On exposure to a low concentration of pollutants, respiration is usually stimulated due to the stimulation of detoxification and repair mechanisms. In response to many pollutants, a shift from the glycolytic pathway to the pentose phosphate pathway is often observed in order to produce the resources for repair. The NADPH produced by such a metabolic shift is used as the electron source for the anabolic reactions and for the maintenance of the antioxidant defense system. If the pollutant exposure periods are short, this enhanced use of energy benefits the plant for stress alleviation by preventing the pollutants from reaching the sensitive metabolic sites and/or scavenging the toxic pollutant species (Kozioł & Whatley, 2013; Gupta, 2016).

2.7 Conclusion and future prospects

Continuous increase in the magnitude of air pollution and addition of new pollutants to the atmosphere have created adversity for the efficient growth of plants. Physiological modifications affect photosynthetic performance, which is quantitatively reflected in the rates of carbon fixation and gain of biomass. ChlF and gas exchange parameters are found useful to monitor the plant health in order to adopt protective strategies. Measurements of the fluorescence in plants are done quantitatively and are related easily to others dependent processes. Therefore, ChlF is a useful tool to analyze the health of the plant even when visual injuries are not observed. However, many derived bioenergetics parameters from the fast fluorescence transient are relative values, which makes it difficult for comparison across the experimental setups. Therefore, contradictory results in in vivo systems are possible with variation in the experimental plants, pollutant types, and concentrations. Reproducibility of the fluorescence measurement has been improved by combining it with the studies of the photosynthetic process by gas exchange or oxygen evolution. The ratio of fluorescence yields, energy quenching, and PS performance has been proven effective to provide a signature on the status of the plant health. Other fluorescence parameters, such as energy dissipation, ETRs, and end reaction performances, provide sufficient insight into the effect of the pollutant on plant physiological efficiency.

While these parameters are used as indicators of the functional state of the photosynthetic apparatus, they are not quantitative at the molecular level. This becomes more complicated when pollutants are present in toxic combinations targeting multiple sites in the photoelectron transport chain and the enzymes of carbon fixation cycles. Therefore, the choice of fluorescence parameter, that is to be used, should be considered on the basis of the site of inhibitory action at the molecular level. The change in PSII functional states induced by pollutants will require extensive work to analyze various plant species in order to have a better understanding on the mechanisms of toxicity at the molecular level. Simultaneously, molecular evidence can be provided to support the fluorescence data, and the correlation will help to recommend fluorescence parameters that are reliable and quick analytical tools for plant health analysis. This will open the way for further advancement of the usefulness of environmental toxicology bioassays involving the chlorophyll fluorescence technique. Furthermore, it will also provide support to the efforts made to understand and remediate the effects of climate change on crop productivity and global food security.

References

Agrawal, M., & Deepak, S. S. (2003). Physiological and biochemical responses of two cultivars of wheat to elevated levels of CO_2 and SO_2, singly and in combination. *Environmental Pollution (Barking, Essex: 1987), 121*, 189–197.

Agrawal, M., Singh, B., Agrawal, S. B., Bell, J. N. B., & Marshall, F. (2006). The effect of air pollution on yield and quality of mungbean grown in peri-urban areas of Varanasi. *Water, Air, and Soil Pollution, 169*, 239–254.

Ainsworth, E. A., Serbin, S. P., Skoneczka, J. A., & Philip, A. (2014). Using leaf optical properties to detect ozone effects on foliar biochemistry. *Photosynthesis Research, 119*, 65–76.

Antenozio, M. L., Caissutti, C., Caporusso, F. M., Marzi, D., & Brunetti, P. (2024). Urban air pollution and plant tolerance: Omics responses to ozone, nitrogen oxides, and particulate matter. *Plants, 13*, 2027.

Baker, N. R., & Oxborough, K. (2004). Chlorophyll fluorescence as a probe of photosynthetic productivity. In G. C. Papageorgiou, & Govindjee (Eds.). *Chlorophyll a fluorescence: A signature of photosynthesis* (pp. 65–82). London, UK: Kluwer Academic.

Baker, N. R., & Rosenqvist, E. (2004). Applications of chlorophyll fluorescence can improve crop production strategies: An examination of future possibilities. *Journal of Experimental Botany, 55*, 1607–1621.

Ballantyne, D. J. (1973). Sulphite inhibition of ATP formation in plant mitochondria. *Phytochemistry, 12*(6), 1207–1209.

Banks, J. M. (2017). Continuous excitation chlorophyll fluorescence parameters: A review for practitioners. *Tree Physiology, 37*, 1128–1136.

Belkin, S., & Boussiba, S. (1991a). High internal pH conveys ammonia resistance in *Spirulina platensis*. *Bioresource Technology, 38*, 167–169.

Belkin, S., & Boussiba, S. (1991b). Resistance of *Spirulina platensis* to ammonia at high pH values. *Plant & Cell Physiology, 32*, 953–958.

Calatayud, A. (2007). Chlorophyll a fluorescence as indicator of atmospheric pollutant effects. *Toxicology and Environmental Chemistry, 89*(4), 627–639.

Calatayud, A., & Barreno, E. (2000). Foliar spraying with Zineb increases fruit productivity and alleviates oxidative stress in two tomato cultivars. *Photosynthetica, 38*, 149–154.

Calatayud, A., & Barreno, E. (2001). Chlorophyll *a* fluorescence, antioxidant enzymes and lipid peroxidation in tomato in response to ozone and benomyl. *Environmental Pollution (Barking, Essex: 1987), 115*, 283–289.

Calatayud, A., Ramirez, J. W., Iglesias, D. J., & Barreno, E. (2002). Effects of ozone on photosynthetic CO_2 exchange, chlorophyll *a* fluorescence and antioxidant systems in lettuce leaves. *Physiologia Plantarum, 116*, 308–316.

Castagna, A., Nali, C., Ciompi, S., Lorenzini, G., Soldatini, G. F., & Ranieri, A. (2001). Ozone exposure affects photosynthesis of pumpkin (*Cucurbita pepo*) plants. *The New Phytologist, 152*, 223–229.

Castagna, A., & Ranieri, A. (2009). Detoxification and repair process of ozone injury: From O3 uptake to gene expression adjustment. *Environmental Pollution, 157*(5), 1461–1469. 10.1016/j.envpol.2008.09.029.

Chatterjee, S., Mandal, M., Kakkar, M., Basak, G., Khan, N. B., Chakraborty, R., Popek, R., Barman, C., & Sarkar, A. (2024). An insight to strategical responses of particulate pollution in plants: From phenome to genome. *Plant Stress, 14*, 100671.

Demmig-Adams, B., & Adams, W. W. (1996). The role of xanthophyll cycle carotenoids in the protection of photosynthesis. *Trends in Plant Science, 1*, 21–26.

Dhir, B. (2016). Air pollutants and photosynthetic efficiency of plants. In U. Kulshrestha, & P. Saxena (Eds.). *Plant responses to air pollution* (pp. 71–84). Singapore: Springer.

Drath, M., Kloft, N., Batschauer, A., Marin, K., Novak, J., & Forchhammer, K. (2008). Ammonia triggers photodamage of photosystem II in the cyanobacterium *Synechocystis* sp. strain PCC 6803. *Plant Physiology, 147*, 206–215.

Duan, J., Fu, B., Kang, H., Song, Z., Jia, M., Cao, D., & Wei, A. (2019). Response of gas-exchange characteristics and chlorophyll fluorescence to acute sulfur dioxide exposure in landscape plants. *Ecotoxicology and Environmental Safety, 171*, 122–129.

Farage, P. K., Long, S. P., Lechner, E. G., & Baker, N. R. (1991). The sequence of the change within the photosynthetic apparatus of wheat following short term exposure to ozone. *Plant Physiology, 95*, 529–535.

Flowers, M. D., Fiscus, E. L., Burkey, K. O., Booker, F. L., & Dubois, J. J. B. (2007). Photosynthesis, chlorophyll fluorescence, and yield of snap bean (*Phaseolus vulgaris* L.) genotypes differing in sensitivity to ozone. *Environmental and Experimental Botany, 61*, 190–198.

Force, L., Critchley, C., & van Rensen, J. J. (2003). New fluorescence parameters for monitoring photosynthesis in plants. *Photosynthesis Research, 78*(1), 17–33.

Gao, Y., Liu, Y., Chen, B., Tao, Y., Cui, C., Wen, Y., Deng, W., Chen, Q., & Yuan, X. (2022). Response of fluorescence and chlorophyll physiological characteristics of typical urban trees to ozone stress. *Atmosphere, 13*, 1885.

Gheorghe, I. F., & Ion, B. (2011). The effects of air pollutants on vegetation and the role of vegetation in reducing atmospheric pollution. In M. K. Khallaf (Ed.). *The impact of air pollution on health, economy, environment and agricultural sources* (pp. 256–259). Shanghai, China: Intech Publisher.

Gimeno, B. S., Bermejo, V., Reinert, R. A., Zheng, Y., & Barnes, J. D. (1999). Adverse effects of ambient ozone on watermelon yield and physiology at a rural site in Eastern Spain. *The New Phytologist, 144*, 245–260.

Govindaraju, M., Ganeshkumar, R. S., Suganthi, K., & Suganthi, P. (2018). Functional analysis of RUBISCO protein in selected plants grown under air pollution stress environment. *Indian Journal of Plant Physiology, 23*, 201–208.

Govindjee (1995). Sixty-three years since Kautsky: Chlorophyll *a* fluorescence. *Australian Journal of Plant Physiology, 22*, 131–160.

Guderian, R. (2012). Air pollution: Phytotoxicity of acidic gases and its significance in air pollution control. In *Springer Science & Business Media, vol 92*, 11–40. Berlin.

Guha, A., Sengupta, D., & Reddy, A. R. (2013). Polyphasic chlorophyll *a* fluorescence kinetics and leaf protein analyses to track dynamics of photosynthetic performance in mulberry during progressive drought. *Journal of Photochemistry and Photobiology B: Biology, 119*, 71–83.

Gupta, A. (2016). Effect of air pollutants on plant gaseous exchange process: Effect on stomata and respiration. In U. Kulshrestha, & P. Saxena (Eds.). *Plant responses to air pollution* (pp. 85–92). Singapore: Springer.

Hassan, I. A., Basahi, J. M., & Ismail, I. M. (2013). Gas exchange, chlorophyll fluorescence and antioxidants as bioindicators of airborne heavy metal pollution in Jeddah, Saudi Arabia. *Current World Environment, 8*(2), 203–213.

Heber, U., & Hueve, K. (1997). Action of SO_2 on plants and metabolic detoxification of SO_2. *International Review of Cytology, 177*, 255–286.

Hu, H., Hua, W., Shen, A., Zhou, H., Sheng, L., Lou, W., & Zhang, G. (2021). Photosynthetic rate and chlorophyll fluorescence of barley exposed to simulated acid rain. *Environmental Science and Pollution Research International, 28*(31), 42776–42786.

Ismail, I. M., Basahi, J. M., & Hassan, I. A. (2014). Gas exchange and chlorophyll fluorescence of pea (*Pisum sativum* L.) plants in response to ambient ozone at a rural site in Egypt. *The Science of the Total Environment, 497-498*, 585–593.

Jena, S., Acharya, S., & Mohapatra, P. K. (2012). Variation in effects of four OP insecticides on photosynthetic pigment fluorescence of *Chlorella vulgaris* Beij. *Ecotoxicology and Environmental Safety, 80*, 111–117.

Jumrani, K., Bhatia, V. S., Hussain, S., Kataria, S., Yang, X., & Brestic, M. (2024). Effect of shading on leaf anatomical structure, photosynthesis characteristics and chlorophyll fluorescence of soybean (*Glycine max*). *Journal of Agronomy and Crop Science, 210*, e12783.

Kalaji, H. M., Jajoo, A., Oukarroum, A., Brestic, M., Zivcak, M., Samborska, I. A., Cetner, M. D., Łukasik, I., Goltsev, V., & Ladle, R. J. (2016). Chlorophyll *a* fluorescence as a tool to monitor physiological status of plants under abiotic stress conditions. *Acta Physiologiae Plantarum, 38*, 102.

Kohan, A., Haghighi, M., Mirghaffari, N., & Ehtemam, M. H. (2021). Effect of air pollution resulting from exhaust emission on the morphological, physiologic and biochemical responses of lettuce (*Lactuca sativa* var. *longifolia*). *Journal of Plant Process and Function, 9*(40), 43–53.

Kozioł, M. J., & Whatley, F. R. (2013). *Gaseous air pollutants and plant metabolism*. London: Butterworth-Heinemann.

Kuki, K. N., Oliva, M. A., Pereira, E. G., Costa, A. C., & Cambraia, J. (2008). Effects of simulated deposition of acid mist and iron ore particulate matter on photosynthesis and the generation of oxidative stress in *Schinus terebinthifolius* Radii and *Sophora tomentosa* L. *The Science of the Total Environment, 403*(1-3), 207–214.

Lazar, D. (2003). Chlorophyll *a* fluorescence rise induced by high light illumination of dark-adapted plant tissue studied by means of a model of photosystem II and considering photosystem II heterogeneity. *Journal of Theoretical Biology, 220*, 469–503.

Lee, H. K., Khaine, I., Kwak, M. J., Jang, J. H., Lee, T. Y., Lee, J. K., Kin, I. R., Kim, W. I., Oh, K. S., & Woo, S. Y. (2017). The relationship between SO_2 exposure and plant physiology: A mini review. *Horticulture, Environment, and Biotechnology, 58*(6), 523–529.

Liu, N., Peng, C. L., Lin, Z. F., Lin, G. Z., & Pan, X. P. (2007). Effects of simulated SO_2 pollution on subtropical forest succession: Toward chlorophyll fluorescence concept. *Pakistan Journal of Botany, 39*(6), 1921–1935.

Liu, N., Peng, C. L., Lin, Z. F., Zhang, L. L., & Pan, X. P. (2006). Changes in photosystem II activity and leaf reflectance features of several subtropical woody plants under simulated SO_2 treatment. *Journal of Integrative Plant Biology, 48*, 1274–1286.

Lu, C. M., Zhang, J. H., Zhang, Q. D., Li, L. B., & Kuang, T. Y. (2001). Modification of photosystem II photochemistry in nitrogen deficient maize and wheat plants. *Journal of Plant Physiology, 158*, 1423–1430.

Łukowski, A., Popek, R., & Karolewski, P. (2020). Particulate matter on foliage of *Betula pendula, Quercus robur*, and *Tilia cordata*: Deposition and ecophysiology. *Environmental Science and Pollution Research, 27*, 10296–10307.

Mahilang, M., Deb, M. K., & Pervez, S. (2021). Biogenic secondary organic aerosols: A review on formation mechanism, analytical challenges and environmental impacts. *Chemosphere, 262*, 127771.

Manisalidis, I., Stavropoulou, E., Stavropoulos, A., & Bezirtzoglou, E. (2020). Environmental and health impacts of air pollution: A review. *Frontiers in Public Health, 8*, 14.

Markou, G., Depraetere, O., & Muylaert, K. (2016). Effect of ammonia on the photosynthetic activity of *Arthrospira* and *Chlorella*: A study on chlorophyll fluorescence and electron transport. *Algal Research, 16*, 449–457.

Maxwell, K., & Johnson, G. N. (2000). Chlorophyll fluorescence – A practical guide. *Journal of Experimental Botany, 51*, 659–668.

Mikkelsen, B. L., Jorgensen, R. B., & Lyngkjaer, F. M. (2015). Complex interplay of future climate levels of CO_2, ozone, and temperature on susceptibility to fungal diseases in barley. *Plant Pathology, 64*, 319–327.

Mishra, S., Swain, B. B., & Mohapatra, P. K. (2024). Enrichment augments tolerance of *Synechocystis* sp. PCC 6803 to imidacloprid. *Journal of Environmental Toxicology Research, 1*(1), 1–13.

Mohapatra, L., Chandrakala, Y., Samantaray, P. K., & Mohapatra, P. K. (2014). OJIP fluorescence transient as a tool for analysis of plant responses to insecticides: A review. *Plant Sciences Research, 36*, 6–16.

Mohapatra, P. K., Schubert, H., & Schiewer, U. (1996). Short term toxicity effect of dimethoate on transthylakoid pH gradient of intacte *Synechocystis* sp. PCC 6803 cells. *Bulletin of Environmental Contamination and Toxicology, 57*, 722–728.

Mohapatra, P. K., & Schiewer, U. (1996). Influence of dimethoate on the structure and function of the natural phytoplankton assemblage of the Darss-Zingst Bodden chain reared in a laboratory. *Polish Journal of Environmental Studies, 5*, 31–36.

Mohapatra, P. K., & Schiewer, U. (2000). Dimethoate and quinalphos toxicity: Pattern of photosynthetic pigment degradation and recovery in *Synechocystis* sp. PCC 6803. *Archiv für Hydrobiologie, 99*, 79–94.

Mohapatra, P. K., Khillar, R., Hansdah, B., & Mohanty, R. C. (2010). Photosynthetic and fluorescence responses of *Solanum melangena* L. to field application of dimethoate. *Ecotoxicology and Environmental Safety, 73*, 78–83.

Mohapatra, P. K., Schubert, H., & Schiewer, U. (1997). Effect of dimethoate on photosynthesis and pigment fluorescence of *Synechocystis* sp. PCC 6803. *Ecotoxicology and Environmental Safety, 36*, 231–237.

Moores, F. C. (2009). Climate change and air pollution: Exploring the synergies and potential for mitigation in industrializing countries. *Sustainability, 1*, 43–54.

Mostowska, A. (1997). Environmental factors affecting chloroplasts. In M. Pessarakli (Ed.). *Handbook of photosynthesis* (pp. 407–426). New York: Marcel Dekker Inc.

Muneer, S., Kim, T. H., Choi, B. C., Lee, B. S., & Lee, J. H. (2014). Effect of CO, NOx and SO_2 on ROS production, photosynthesis and ascorbate–glutathione pathway to induce *Fragaria*× *annasa* as a hyperaccumulator. *Redox Biology, 2*, 91–98.

Navarro, M. P., Ames, W. M., Nilsson, H., Lohmiller, T., Pantazis, D. A., Rapatskiy, L., Nowaczyk, M. M., Neese, F., Boussac, A., & Messinger, J. (2013). Ammonia binding to the oxygen evolving complex of photosystem II identifies the solvent-exchangeable oxygen bridge (μ-oxo) of the manganese tetramer. *Proceedings of the National Academy of Sciences, 110*, 15561–15566.

Neves, N. R., Oliva, M. A., da, C., Centeno, D., Costa, A. C., Ribas, R. F., & Pereira, E. G. (2009). Photosynthesis and oxidative stress in the restinga plant species *Eugenia uniflora* L. exposed to simulated acid rain and iron ore dust deposition: Potential use in environmental risk assessment. *The Science of the Total Environment, 407*(12), 3740–3745.

Nussbaum, S., Geissmann, M., Eggenberg, P., Strasser, R. J., & Fuhrer, J. (2001). Ozone sensitivity in herbaceous species as assessed by direct and modulated chlorophyll fluorescence techniques. *Journal of Plant Physiology, 158*, 757–766.

Oguntimehin, I., Kondo, H., & Sakugawa, H. (2010). The use of Sunpatiens (*Impatiens* spp.) as a bioindicator of some simulated air pollutants- using an ornamental plant as bioindicator. *Chemosphere, 81*(2), 273–281.

Oksanen, E., & Kontunen-Soppela, S. (2021). Plants have different strategies to defend against air pollutants. *Current Opinion in Environmental Science & Health, 19*, 100222.
Oyala, P. H., Stich, T. A., Debus, R. J., & Britt, R. D. (2015). Ammonia binds to the dangler manganese of the photosystem II oxygen-evolving complex. *Journal of the American Chemical Society, 137*, 8829–8837.
Papazian, S., & Blande, J. D. (2020). Dynamics of plant responses to combinations of air pollutants. *Plant Biology (Stuttgart), 22*(1), 68–83.
Pellegrini, E. (2014). PSII photochemistry is the primary target of oxidative stress imposed by ozone in *Tiliaamericana. Urban Forestry & Urban Greening, 13*, 94–102.
Pellegrini, E., Campanella, A., Paolocci, M., Trivellini, A., Gennai, C., Muganu, M., Nali, C., & Lorenzini, G. (2015). Functional leaf traits and diurnal dynamics of photosynthetic parameters predict the behavior of grapevine varieties towards ozone. *PLoS One, 10*(8), e0135056.
Pereira, E. G., Oliva, M. A., Kuki, K. N., & Cambraia, J. (2009). Photosynthetic changes and oxidative stress caused by iron ore dust deposition in the tropical CAM tree *Clusia hilariana. Trees, 23*, 277–285.
Piccotto, M., Bidussi, M., & Tretiach, M. (2011). Effects of the urban environmental conditions on the chlorophyll *a* fluorescence emission in transplants of three ecologically distinct lichens. *Environmental and Experimental Botany, 73*, 102–107.
Popek, R., Łukowski, A., & Grabowski, M. (2018a). Influence of particulate matter accumulation on photosynthetic apparatus of *Physocarpus opulifolius* and *Sorbaria sorbifolia. Polish Journal of Environmental Studies, 27*(5), 2391–2396.
Popek, R., Przybysz, A., Gawrońska, H., Klamkowski, K., & Gawroński, S. W. (2018b). Impact of particulate matter accumulation on the photosynthetic apparatus of roadside woody plants growing in the urban conditions. *Ecotoxicology and Environmental Safety, 163*, 56–62.
Popovic, R., Dewez, D., & Juneau, P. (2003). Applications of chlorophyll fluorescence in ecotoxicology: Heavy metal, herbicides and air pollutants. In J. R. DeEll, & P. M. A. Toivonen (Eds.). *Practical applications of chlorophyll fluorescence in plant biology* (pp. 150–183). Massachusetts: Springer.
Qin, J., Jiang, X., Qin, J., Zhao, H., Dai, M., Liu, H., & Chen, X. (2023). Effects of lead pollution on photosynthetic characteristics and chlorophyll fluorescence parameters of different populations of *Miscanthus floridulus. Processes, 11*, 1562.
Qiu, D. L., Liu, X. H., & Guo, S. Z. (2001). Effects of simulated acid rain on longan photosynthesis. *Acta Horticulturae, 558*, 301–304.
Rai, P. K. (2016). Impacts of particulate matter pollution on plants: Implications for environmental biomonitoring. *Ecotoxicology and Environmental Safety, 129*, 120–136.
Rai, R., Rajput, M., Agrawal, M., & Agrawal, S. B. (2011). Gaseous air pollutants: A review on current and future trends of emissions and impact on agriculture. *Journal of Scientific Research, 55*, 77–102.
Räsänen, J. V., Leskinen, J. T. T., Holopainen, T., Joutsensaari, J., Pasanen, P., & Kivimäenpää, M. (2017). Titanium dioxide (TiO_2) fine particle capture and BVOC emissions of *Betula pendula* and *Betula pubescens* at different wind speeds. *Atmospheric Environment (Oxford, England: 1994), 152*, 345–353.
Robinson, M. F., Heath, J., & Mansfield, T. A. (1998). Disturbances in stomatal behaviour caused by air pollutants. *Journal of Experimental Botany, 49*, 461–469.
Saini, D. K., Garg, S. K., & Kumar, M. (2019). Major air pollutants and their effects on plant and human health: A review. *Plant Archives, 19*(2), 3273–3278.
Sarkar, R. K., & Ray, A. (2016). Submergence-tolerant rice withstands complete submergence even in saline water: Probing through chlorophyll *a* fluorescence induction O–J–I–P transients. *Photosynthetica, 54*, 275–287.

Shasmita, Mohapatra, D., Mohapatra, P. K., Naik, S. K., & Mukherjee, A. K. (2019). Priming with salicylic acid induces defense against bacterial blight disease by modulating rice plant photosystem II and antioxidant enzymes activity. *Physiology and Molecular Plant Pathology, 108*, 101427.

Shasmita, Mohapatra, P. K., Naik, S. K., & Mukherjee, A. K. (2021). Improved photosystem II and defense enzymes activity in rice (*Oryza sativa* L.) by biopriming against *Xanthomonas oryzae* pv. *oryzae*. *Functional Plant Biology: FPB, 48*(3), 298–311.

Shasmita, Swain, B. B., Mishra, S., Mohapatra, D., Naik, S. K., & Mohapatra, P. K. (2024). Functional variations in efficiency of PSII during leaf ontogeny in the tropical plant *Saraca asoca*. *Functional Plant Biology, 51*, 1–13.

Shasmita, Swain, H., Ray, A., Mohapatra, P. K., Sarkar, R. K., & Mukherjee, A. K. (2018). Riboflavin (Vitamin B2) mediated defence induction against bacterial leaf blight: Probing through chlorophyll *a* fluorescence induction O–J–I–P transients. *Functional Plant Biology, 45*(12), 1251–1261.

Singh, H., Savita, Sharma, R., Sinha, S., Kumar, M., Kumar, P., Verma, A., & Sharma, S. K. (2017). Physiological functioning of *Lagerstroemia speciosa* L. under heavy roadside traffic: An approach to screen potential species for abatement of urban air pollution. *3 Biotechnology*, 7(1), 61.

Stirbet, A., & Govindjee (2011). On the relation between the Kautsky effect (chlorophyll *a* fluorescence induction) and Photosystem II: Basics and applications of the OJIP fluorescence transient. *Journal of Photochemistry and Photobiology B: Biology, 104*, 236–257.

Strasser, R. J., Srivastava, A., & Govindjee (1995). Polyphasic chlorophyll *a* fluorescence transient in plants and cyanobacteria. *Photochemistry and Photobiology, l61*, 32–42.

Strasser, R. J., Tsimilli-Michael, M., & Srivastava, A. (2004). Analysis of the chlorophyll *a* fluorescence transient. In G. C. Papageorgiou, & Govindjee (Eds.). *Chlorophyll* a *fluorescence: A signature of photosynthesis* (pp. 321–362). Dordrecht, The Netherlands: Springer.

Strasser, R. J., Tsimilli-Michael, M., Qiang, S., & Goltsev, V. (2010). Simultaneous *in vivo* recording of prompt and delayed fluorescence and 820-nm reflection changes during drying and after rehydration of the resurrection plant *Haberlea rhodopensis*. *Biochimica et Biophysica Acta, 1797*, 1313–1326.

Sun, J. W., Hu, H. Q., Li, Y. L., Wang, L. H., Zhou, Q., & Huang, X. H. (2016). Effects and mechanism of acid rain on plant chloroplast ATP synthase. *Environmental Science and Pollution Research, 23*, 18296–18306.

Swain, B. B., Mishra, S., Samal, S., Adak, T., Mohapatra, P. K., & Ayyamperumal, R. (2024). Chlorpyrifos enrichment enhances tolerance of *Anabaena* sp. PCC 7119 to dimethoate. *Environmental Research, 249*, 118310.

Thwe, A. A., Vercambre, G., Gautier, G. H. F., Phattaralerphong, J., & Kasemsap, P. (2014). Response of photosynthesis and chlorophyll fluorescence to acute ozone stress in tomato (*Solanum lycopersicum* Mill.). *Photosynthetica, 52*(1), 105–116.

Volgusheva, A., Yakovleva, O. V., Kukarskikh, G. P., Riznichenko, G. Y., & Krendeleva, T. E. (2011). Performance index in assessing the physiological state of trees in urban ecosystems. *Biophysics, 56*, 90–95.

Wang, H., Maher, B. A., Ahmed, I. A., & Davison, B. (2019). Efficient removal of ultrafine particles from diesel exhaust by selected tree species: Implications for roadside planting for improving the quality of urban air. *Environmental Science & Technology, 53*(12), 6906–6916.

Wang, L., Deng, L. Y., & Li, X. (2014). Chlorophyll biosynthesis metabolism of golden-leaf plants to SO_2 stress. *Chinese Journal of Ecology, 33*(9), 2382–2387.

WHO. (2021). *Global air quality guidelines: Particulate matter (PM2.5 and PM10), ozone, nitrogen dioxide, sulfur dioxide and carbon monoxide*. Geneva, Switzerland: World Health Organization.

Yamano, T., & Morita, S. (1993). Effect of pesticides on isolated rat hepatocites, mitochondria and microsomes. *Archives of Environmental Contamination and Toxicology, 25*, 271–278.

Yan, W., Lu, Y., Guo, L., Liu, Y., Li, M., Zhang, B., Zhang, B., Zhang, L., Qin, D., & Huo, J. (2024). Effects of drought stress on photosynthesis and chlorophyll fluorescence in Blue Honeysuckle. *Plants, 13*, 2115.

Yendrek, C. R., Koester, R. P., & Ainsworth, E. A. (2015). A comparative analysis of transcriptomic, biochemical, and physiological responses to elevated ozone identifies species-specific mechanisms of resilience in legume crops. *Journal of Experimental Botany, 66*, 7101–7112.

Yi, H., Liu, X., Yi, M., & Chen, G. (2014). Dual role of hydrogen peroxide in *Arabidopsis* guard cells in response to sulfur dioxide. *Advances in Toxicology, 2014*, 407368.

Yu, J. Q., Ye, S., & Huang, L. (2002). Effects of simulated acid precipitation on photosynthesis, chlorophyll fluorescence, and antioxidative enzymes in *Cucumis sativus* L. *Photosynthetica, 40*, 331–335.

CHAPTER THREE

Morpho-anatomical responses of woody species to urban air pollution

Hamed Dadkhah-Aghdash[1], Carmen Arena[2], Giuseppe Muscari Tomajoli[2], Mattia Napoletano[3], and Daniela Baldantoni[3]
[1]Department of Plant Biology, Faculty of Biological Sciences, Tarbiat Modares University, Tehran, Iran,
[2]Department of Biology, University of Naples Federico II, Naples, Italy,
[3]Department of Chemistry and Biology "Adolfo Zambelli", University of Salerno, Salerno, Italy

3.1 Introduction

In urban areas, anthropogenic activities, such as traffic and industrial plants, alongside natural sources, negatively affect air quality (Ali et al., 2019). Air pollution, because of its harmful effects, is one of the main environmental challenges in cities (Gheshlaghpoor et al., 2023; Mousavi Javardi et al., 2023), impairing plant communities (Fiorentino et al., 2025). The effects of air pollution on woody species depend on different factors such as characteristics and concentration of air pollutants, exposure time, plant type, age, and interaction with other environmental factors (Abeyratne & Ileperuma, 2006a).

Mechanisms of plants for adaptation to pollution stress relate to two factors: (1) the avoidance of plants to absorb pollutants and (2) the entery of pollutants into leaves and increase their tolerance to air pollution (Taylor, 1978). Because of differences in ecophysiology and morphology (leaf surface characteristics such as cuticle, stomata, trichomes, and epicuticular wax), the responses of different plant species to air pollution and their ability to improve the air quality are different (Neinhuis & Barthlott, 1998). Woody species during the exposure to air pollution use different strategies such as allocation of resources to defensive structures, optimization of their defense tissues and organs, and improving their capacity for tolerance of air pollution (Yan et al., 2024). Different responses of five tree species to urban air pollution of Beijing (China) were investigated, and results showed that

Urban Air Pollution and Plant Responses
DOI: https://doi.org/10.1016/B978-0-443-40320-0.00009-7

the morphological and anatomical responses of plants were species-specific. *Fraxinus chinensis* Roxb. and *Koelreuteria paniculata* Laxm.were tolerant species and their tolerance suggested their employment in removing air pollution in urban green spaces of highly polluted regions (Yan et al., 2024).

During the exposure to air pollution, plants undergo physiological changes (Hasnain et al., 2024; Mehmood et al., 2024; Salimi & Dadkhah Aghdash, 2019) before the visible damages in their structures, for example, collapse of leaf tissue, yellowing, alterations in growth, or premature loss of foliage, and so on (Liu & Ding, 2008; Mousavi Javardi et al., 2023). Leaves are sensitive organs for environmental changes, being characterized by a large contact surface with the surrounding environment (Kuster et al., 2016). For this reason, and considering their fundamental role in plants, leaves exert a paramount role in responses of plants to air pollution, as well as in the accumulation and deposition of different air pollutants. Leaf morphological and anatomical responses are some of the most evident plant responses to air pollution. In this regard, a lot of research investigated the response of different plant types (conifers and deciduous species) to air pollution and the influence of different parameters on their responses to air pollution. The sensitivity of several woody species to pollutants, indeed, makes them good indicators, useful in air quality biomonitoring, and helping planners manage urban cities (Ghahremaninejad et al., 2021; Horaginamani & Ravichandran, 2010).

Several morphological traits, such as leaf area and leaf number, may modulate the level of exposure to airborne pollutants (Jochner et al., 2015). There are different functional traits that plants use for responding and adapting to urban environments (Yan et al., 2024). The anatomical traits of leaves can reflect plant adaptation to changing environmental conditions, thereby influencing ecosystem functions owing to their ability to regulate gas exchange and light absorption to some degree (He et al., 2018). Stomatal traits show the ability of plants to respond to environmental changes (Li et al., 2021), since stomatal density could be increased and decreased in highly polluted regions (Abbasi et al., 2018; Khosropour et al., 2019).

In this chapter, we explore the diverse responses of woody species to air pollution in urban ecosystems, focusing on their detrimental effects on plant growth, development, reproduction, and overall health. The primary aim is to provide an integrative overview of how air pollutants impact woody plants at multiple levels of biological organization, from whole-plant morphology to tissue and cellular ultrastructure. A special emphasis is placed on identifying key physiological and morpho-anatomical traits involved in these responses. By synthesizing current knowledge, we aim to highlight critical response mechanisms, discuss influencing parameters, and propose promising directions for future research.

Understanding how woody plants respond and adapt to air pollution is essential for safeguarding the ecological functionality of urban forests. An innovative perspective lies in examining the tight interplay between structural integrity and physiological function, recognizing that changes in morphology and anatomy often reflect, or even precede, disruptions in photosynthesis, water relations, and resource allocation. By investigating how woody species tolerate and survive in polluted environments, we can gain critical insights into the mechanisms that allow them to continue delivering key ecosystem services, such as carbon sequestration, air purification, and temperature regulation. This integrative approach not only improves our understanding of plant resilience but also informs urban planning and species selection strategies for more sustainable green infrastructure.

3.2 Whole-plant level responses

The response of plants to environmental pollution varies with species, type of pollutant, and the duration of exposure. Generally, plants exhibit responses to various air pollutants at different levels of organization, ranging from whole plant to cell. This section explores the common effects of air pollution on plants, with particular attention to woody species. Woody plants, characterized by slower growth compared to herbaceous plants, tend to experience prolonged exposure to pollutants, leading to cumulative damage that can impair organs, tissues, and overall physiological function. It remains challenging to determine whether structural damage precedes disruptions in functional processes such as photosynthesis, or vice versa (Roy et al., 2024). Nevertheless, there is often a strict coupling between physiological and morphological processess that together determine the plant health in a specific environment. Woody species also exhibit a strong resistance to pollutants primarily due to the synergistic action of detoxification mechanisms, which prevent the occurrence of irreversible damages (Antenozio et al., 2024).

3.2.1 Growth inhibition

Urban and periurban ecosystems are increasingly exposed to elevated concentrations of air pollutants, including ground-level ozone (O_3), sulfur dioxide (SO_2), nitrogen oxides (NO_x), and particulate matter (PM_{10} and $PM_{2.5}$). These airborne pollutants exert significant stress on the physiology and development of plants, particularly in woody species, which persist in the environment over a long time and are chronically exposed. One of the most visible outcomes of this exposure is the inhibition of growth, manifested through reductions in height, biomass accumulation, and shifts in biomass allocation between above and below-ground tissues. These structural symptoms are indicative of deeper physiological disturbances, including damage to the photosynthetic apparatus, water and nutrient uptake, and hormonal regulation (Ainsworth & Long, 2005; Xu, 2021). Pollutants enter plants mainly through stomatal apertures and can either exert direct cytotoxic effects or initiate oxidative stress cascades via the formation of reactive oxygen species (ROS). These processes affect both cellular integrity and whole-plant metabolism. The extent of the impact depends on the species, phenological stage, duration of exposure, and the type and concentration of pollutants involved (Ainsworth & Long, 2005). An overview of the pollutant impact on woody plants is summarized in Fig. 3.1.

3.2.2 Plant height and shoot growth

Plant height is a primary indicator of vegetative vigor and developmental success, integrating the net effect of genetics,resources availability, and competitive ability. Under polluted conditions, especially with high levels of ozone, a significant reduction in shoot elongation has been documented. Tropospheric ozone (O_3), being a highly reactive molecule, damages key cellular components, including DNA, lipids, and proteins, ultimately impairing cell division and expansion in growing tissues. This results in the shortening of internodes and reduced elongation of primary and secondary stems (Frei, 2015; Kangasjärvi et al., 2005).

Controlled exposure experiments using open-top chambers (OTCs) have revealed marked growth suppression in several European and North American tree species. For instance, *Fagus sylvatica* L. (European beech) and *Fraxinus excelsior* L. (common ash) showed a 20%–35% decrease in terminal shoot length when exposed to ambient ozone concentrations 1.5 times higher than the background level for one growing season (Karlsson et al.,

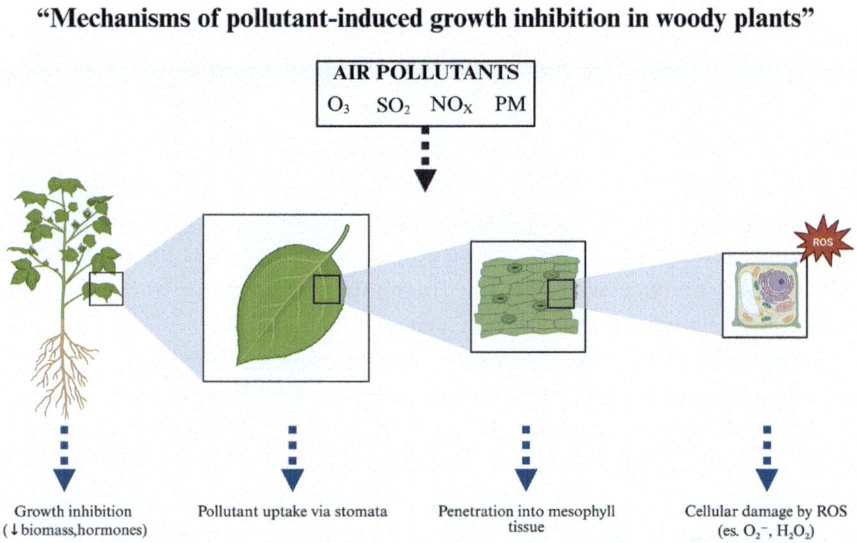

Figure 3.1 ***Schematic overview of processes impacting woody plants due to pollution, from the whole organism to organs, tissues, and cells. Note: Created in BioRender. Muscari Tomajoli, G. (2026) https://BioRender.com/tx6s1d8.*** Schematic overview of processes impacting woody plants due to pollution, from the whole organism to organs, tissues, and cells. Note: Created in BioRender. Muscari Tomajoli, G. (2026) https://BioRender.com/tx6s1d8. No permission required.

2004). Similarly, conifers such as *Picea abies* (L.) H. Karst. and *Pinus sylvestris* L. demonstrated shorter needles, smaller crown expansion, and significant reductions in branch number (Huttunen & Manninen, 2013; Karlsson et al., 2004).

Pollutants interfere with the hormonal regulation of shoot elongation, particularly auxins and gibberellins. Decreased gibberellic acid (GA) synthesis, induced by oxidative stress, has been implicated in reduced cell elongation in growing tissues (Wilkinson & Davies, 2010). Under chronic exposure, plants often exhibited variations in leaf functional traits, developing thinner laminae, reducing specific leaf area (SLA) and leaf area index (LAI), and increasing leaf dry matter content—all indicative of structural and metabolic shifts toward a more conservative growth strategy (Matyssek et al., 2012; Poorter et al., 2009). These changes ultimately impair the photosynthetic surface, reducing the carbon gain potential of the plant and initiating a feedback loop of further growth limitation (Arena et al., 2013).

3.2.3 Biomass reduction and alteration in the root: shoot ratio

Biomass accumulation is a direct measure of plant productivity and carbon fixation. Pollutant-induced growth inhibition translates into measurable reductions in both above-ground (stem, branches, leaves) and below-ground (roots) biomass. Above-ground tissues are most directly affected by foliar absorption of pollutants. Stomatal dysfunction is a central mechanism: pollutants such as O_3 and NO_x cause either overclosure (leading to reduced CO_2 uptake) or excessive opening (increasing water loss), both of which reduce photosynthetic efficiency (Calatayud et al., 2011).

In a field study in Northern Italy, *Populus nigra* L. and *Quercus robur* L. individuals growing near highways showed up to 40% less shoot biomass compared to conspecifics in cleaner suburban areas, despite similar age and initial size. This discrepancy was linked to persistent stomatal injury, pigment degradation, and decreased nonstructural carbohydrate storage (Lorenzini et al., 2006).

Below-ground biomass reduction, though less visible, is equally significant. Roots may be affected both directly and indirectly. Acid deposition (driven by SO_2 and NO_x) lowers soil pH and alters microbial community structure, impairing symbiotic relationships such as mycorrhizae. This disturbance leads to reduced uptake of essential nutrients, particularly calcium, magnesium, and phosphorus (Sridhar et al., 2022). Moreover, under environmental stress, plants often reduce carbon allocation to roots, diverting photosynthates instead toward the shoot to repair damaged tissues or to maintain cell turgor by osmotic regulation (Dietze et al., 2014).

The root: shoot ratio is a dynamic index that reflects a plant's resource allocation strategy under varying environmental conditions. In pollution-stressed environments, this ratio may increase, decrease, or remain unchanged, depending on the severity, duration, and type of pollutant exposure. Increased root: shoot ratios are typically observed in crops when shoot tissues are more severely affected, for instance, during foliar ozone exposure, leading to a relative preservation of root development. This may be interpreted as a survival mechanism aimed at maximizing water and nutrient acquisition while shoot growth is curtailed. In contrast, when pollutants primarily impact root systems (e.g., due to soil acidification or heavy metal accumulation), the ratio decreases as above-ground tissues continue to develop at the expense of the roots (Qi et al., 2019). Fig. 3.2 illustrates the shift in biomass allocation under ozone exposure,

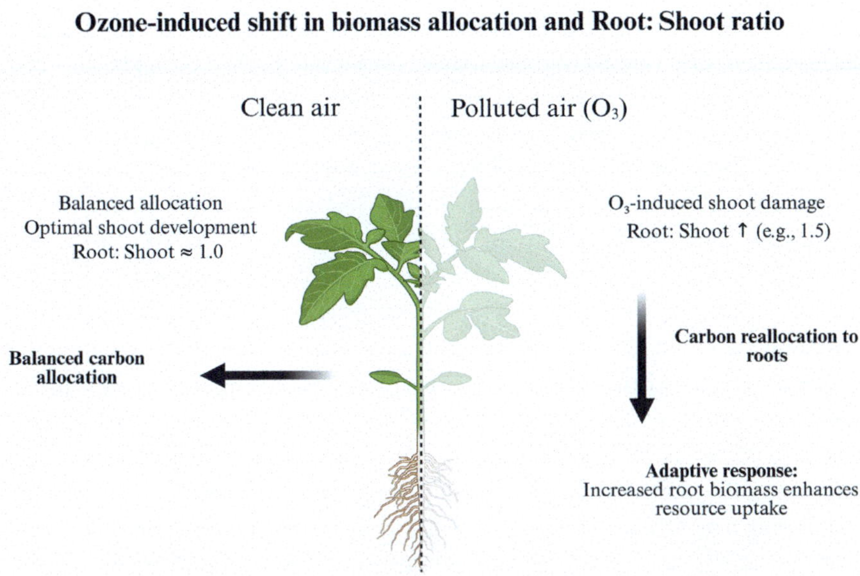

Figure 3.2 ***Schematic picture of the shift in biomass allocation under ozone exposure. Clean air supports a balanced shoot and root growth (Root: Shoot ≈ 1.0). Under polluted air (O_3), oxidative damage to foliage reduces shoot biomass and induces a compensatory increase in root allocation (Root: Shoot ↑, e.g., 1.5), enhancing nutrient and water uptake as an adaptive strategy. Note: Created in BioRender. Muscari Tomajoli, G. (2026) https://BioRender.com/pwjrbkg.*** Schematic picture of the shift in biomass allocation under ozone exposure. Clean air supports a balanced shoot and root growth (Root: Shoot ≈ 1.0). Under polluted air (O_3), oxidative damage to foliage reduces shoot biomass and induces a compensatory increase in root allocation (Root: Shoot ↑, e.g., 1.5), enhancing nutrient and water uptake as an adaptive strategy. Note: Created in BioRender. Muscari Tomajoli, G. (2026) https://BioRender.com/pwjrbkg. No permission required.

highlighting changes in root:shoot ratio as a result of shoot damage and adaptive carbon reallocation to roots.

Species also differ markedly in their plasticity. For example, *Populus tremula* exhibits significant shifts in biomass allocation under moderate pollution, adjusting root: shoot ratios within seasons. These adjustments, however, may not be sustainable in the long term, as resource imbalance often leads to progressive decline in either root support or canopy maintenance (Bussotti & Ferretti, 2009).

Woody species differ markedly in their responses to air pollution, due to inherent physiological, anatomical, and biochemical traits. Some taxa, such as *P. tremula* L. and *P. sylvestris* L., are recognized as particularly sensitive to ozone due to high stomatal conductance, thin cuticles, and low antioxidant capacity (Bussotti & Ferretti, 2009). In contrast, other species such as *Quercus ilex* L. or *Carpinus betulus* L. display higher tolerance, often linked to their ability to modulate stomatal aperture and accumulate detoxifying compounds such as flavonoids and ascorbate (Paoletti, 2009).

Species with high pollutant uptake and visible injury are often employed as bioindicators, while others capable of accumulating contaminants without showing external damage are classified as bioaccumulators (Arena et al., 2017). For instance, *Platanus* × *acerifolia* (Aiton) Willd develops characteristic necrotic lesions and leaf bronzing under urban ozone exposure, making it a useful indicator in air quality monitoring programs (Sæbø et al., 2012). Conversely, *Acer pseudoplatanus* L. can intake heavy metals with minimal visible symptoms, serving as a suitable bioaccumulator. In a research, a significant accumulation of lead (Pb), cadmium (Cd), and zinc (Zn) was demonstrated in seedling tissues of this species grown on polluted urban soils (Daghestani & Kolahi, 2024).

These interspecific differences are critical when selecting trees for urban greening projects. Planting tolerant species can help mitigate pollutant levels through phytoremediation, whereas sensitive species can provide early warnings of air quality degradation. Assessing growth inhibition in woody species under pollution requires an integration of field observations and controlled experiments. The OTCs and free-air ozone enrichment systems are widely used to simulate elevated pollutant exposure under semi-natural conditions. These setups allow for continuous monitoring of plant responses while controlling for confounding variables like drought or temperature fluctuations (Ainsworth & Long, 2005; Feng et al., 2022).

Standard metrics include shoot length, internode number, LAI, and dry weight biomass. More advanced techniques involve chlorophyll fluorescence emission, stable isotope analysis (δ13C, δ15N), and root scanning with rhizotrons or WinRHIZO software (Hodge et al., 2009). Biomass partitioning is typically quantified after destructive sampling, with dry weights of roots, stems, and leaves expressed as proportions of total biomass to calculate root: shoot ratios. Importantly, long-term monitoring is essential to understand cumulative effects and delayed responses, particularly in perennial species that may not show symptoms until multiple growing seasons have passed.

The implications of pollution-induced growth inhibition extend beyond the individual plant level, influencing forest structure, ecosystem services, and climate feedback. Declines in above-ground biomass reduce carbon sequestration potential, compromising the role of urban forests as carbon sinks (Nowak & Crane, 2002). Altered growth dynamics also shift interspecific competition, often favoring fast-growing, pollution-tolerant species at the expense of native or sensitive taxa (McKinney, 2008).

Furthermore, root damage reduces soil stabilization and weakens mutualistic relationships with fungi and rhizobacteria, contributing to biodiversity loss in urban soils (Van Der Heijden et al., 2008). Collectively, these changes threaten the ecological resilience of urban woodlands and challenge their capacity to buffer anthropogenic stressors such as heat islands, erosion, and air toxics.

Growth inhibition is among the most observable and ecologically meaningful responses of woody plants to air pollution (Bussotti & Ferretti, 2009; Feng et al., 2022; Wilkinson & Davies, 2010). It reflects a suite of physiological and biochemical disruptions that impair carbon assimilation, resource allocation, and developmental processes (Ainsworth & Long, 2005; Paoletti, 2009).

While the degree of inhibition varies among species and pollutants, the overall pattern reveals a consistent threat to tree health and forest functionality in polluted environments. A deeper understanding of species-specific responses, combined with improved monitoring techniques and mitigation strategies, is essential for protecting urban green spaces in the face of ongoing environmental change (Xu, 2021; Yan et al., 2024). Planting pollution-tolerant trees, reducing emissions, and restoring soil microbial health may help sustain woody species growth and preserve urban biodiversity (Arena et al., 2017; Daghestani & Kolahi, 2024).

3.2.4 Altered plant shape

Urban air pollution has a strong influence not only on the growth rate of woody species but also on their structural development and phenotype. These alterations, which included modification of crown symmetry, reductions in shoot elongation, leaf size, internode length, and branching complexity, are indicators of alterations in morphogenetic processes usually regulated by closely synchronized hormonal gradients (e.g., auxin, cytokinin, and gibberellin) and environmental, ecological variables such as light, water availability, and nutrient supply. Pollutants such as O_3, NO_x, SO_2, and PM affect photosynthesis by altering stomatal function and gas exchange, resulting in reduced carbon assimilation and ultimately limiting growth (Paoletti, 2009).

Pollutants may also induce oxidative stress, which triggers a cascade of metabolic imbalances responsible for early leaf senescence, altered root-to-shoot ratios, and damaged cambial activity. These outcomes are more pronounced in juvenile trees, which are more sensitive to environmental constraints. Morphological plasticity, a critical trait that confers on plants the capability to modify their structure in response to changing conditions, declined under chronic pollution stress. In such contexts, changes in plant phenotyping may not reflect an acclimation process but rather maladaptive responses responsible for overall fitness and survival decline (Escobedo et al., 2011; Sæbø et al., 2012).

Among different pollution agents, O_3, SO_2, and NO_x are known to interfere with hormonal signaling pathways, particularly those involving auxins and gibberellins, which coordinate apical dominance and shoot architecture. For instance, O_3 exposure has been shown to reduce indole-3-acetic acid concentrations in the shoot apex, weakening apical dominance and promoting irregular lateral branching (Booker & Fiscus, 2005). In *Betula pendula* Roth. and *Acer pseudoplatanus* L., this results in a denser, more rounded crown with reduced internode length and abnormal shoot angles (Percy & Ferretti, 2004; Samuelson & Edwards, 1993). Structural distortions are also reported in conifers. In *P. sylvestris*, chronic exposure to urban pollution has been associated with crown thinning, reduced needle retention, and altered shoot taper (Paoletti, 2009). These symptoms are not merely esthetic: trees with reduced taper and narrowed crowns are more prone to mechanical instability and wind damage due to unfavorable load distribution (Escobedo et al., 2011).

Changes in plant shape can also affect the canopy structure. For example, pollution-induced reductions in leaf insertion angle and petiole elongation alter the three-dimensional arrangement of foliage, impairing light interception and canopy gas exchange efficiency (Samuelson & Edwards, 1993). Long-term observational studies in *Q. robur* have shown a progressive shift toward more layered crowns in the presence of high pollutant concentrations, potentially linked to oxidative damage to shoot meristems and phyllotactic control systems (Ainsworth & Long, 2005).

Tree architectural changes are species-specific and often mediated by the plant's capacity to detoxify ROS. Species such as *Q. ilex* and *C. betulus* display limited structural changes under moderate pollution due to high antioxidant enzyme activity and waxy cuticles, whereas *P. tremula* and *Tilia cordata* Mill. show more pronounced deformities, likely ascribed to higher stomatal conductance and pollutant uptake rates (Escobedo et al., 2011; Gratani et al., 2019).

An altered plant architecture has critical ecological implications. Irregular crown morphology affects microclimatic buffering, competition for light, and reproductive site selection. In urban forestry, these changes may reduce a tree's ornamental value, longevity, and service provision, and should thus be considered when selecting species for green infrastructure planning (Escobedo et al., 2011).

At the urban ecosystem scale, plant-level changes may have deep repercussions on the community structure. The reduction in tree canopy density and loss of structural integrity can compromise key ecosystem services, including carbon sequestration, habitat provision, air filtration, and urban cooling (Escobedo et al., 2011; Locosselli et al., 2019). Trees with problems in structural development are also more susceptible to mechanical breakdown (Matyssek et al., 2012). Long-term exposure to urban air pollutants may also exert selective pressure on tree populations, potentially influencing community composition and favoring invasive species with higher tolerance thresholds, thus reducing biodiversity and resilience (McKinney, 2008).

3.2.5 Altered reproduction capacity

Air pollutants have strong effects on the reproductive biology of woody plants, influencing every stage from floral initiation to seed development. These disruptions are driven by both direct injury to reproductive structures and indirect physiological stress affecting hormone signaling, resource allocation, and phenological synchrony (Black et al., 2000).

Ozone (O_3) is particularly impactful, as it causes oxidative damage to floral tissues and reduces pollen viability, especially in sensitive broadleaved species. Even short-term ozone exposure during floral development significantly decreased pollen germination rates in species such as *B. pendula* and *Q. robur*. Air pollutants may also disrupt hormonal signaling networks involved in flowering. Abscisic acid and ethylene, both stress-related phytohormones, are upregulated under pollutant stress, contributing to premature flower senescence or the inhibition of anthesis. This hormonal imbalance may be particularly critical in gymnosperms, which have longer reproductive cycles and more vulnerable cones or strobili (Black et al., 2000; Grantz et al., 2003).

Nitrogen deposition and heavy metals, often associated with urban pollution, further impair reproductive performance by altering nutrient dynamics and DNA integrity in floral meristems. In species such as *Acer platanoides* L. and *P. nigra,* such pollutants have been linked to malformed flowers, reduced seed set, and low germination success (Ashmore, 2005). In many cases, resource reallocation under stress favors vegetative over reproductive growth (Arena et al., 2013; Karlsson et al., 2004). Trees exposed to chronic pollution may delay or suspend reproduction to prioritize leaf repair, osmotic balance, or lignification processes (Ashmore, 2005). While this response is adaptive in the short term, persistent reductions in flowering and seed production may lead to long-term consequences in population dynamics and gene flow, particularly in fragmented urban habitats (Duque et al., 2021).

Species-specific reproductive responses to air pollution differ markedly. For example, *Tilia cordata* and *Fraxinus excelsior* have demonstrated higher resilience, maintaining reproductive traits such as pollen viability and seed production under moderate pollution conditions (Arena et al., 2013; Karlsson et al., 2004). In contrast, *Prunus avium* and *Sorbus aucuparia* have shown reduced flowering and fruiting performance under similar environmental pressures (Bussotti & Ferretti, 2009; Duque et al., 2021). This divergence emphasizes the importance of considering species-specific reproductive sensitivity when planning urban greening and ecological restoration strategies.

3.2.6 Suspended flowering and fruiting

Air pollution can induce not only growth delay and reproductive impairment, but also the complete suspension of flowering and fruiting in woody species, with significant consequences for regeneration and biodiversity conservation in urban ecosystems. Several mechanisms contribute to this phenomenon: oxidative stress, hormonal imbalance (e.g., increased ethylene and abscisic acid), and competition between vegetative and reproductive organs for limited resources (Black et al., 2000). O_3 has been shown to alter phenological timing. For instance, moderate O_3 exposure can accelerate flowering in herbaceous species but desynchronize it from pollinator activity, resulting in ineffective pollination and fruiting (Duque et al., 2021).

In woody plants, these ozone-induced disruptions in hormonal regulation and carbon allocation may lead to reproductive failure, including the absence of fertile flowers or fruits. For example, in *Platanus acerifolia* and *B. pendula,* chronic exposure to O_3 and NO_x is associated with aborted flower buds and sporadic fruit production, with seed loss up to 60% in highly

Table 3.1 Summary of air pollution effects on plant physiology, anatomy, and growth. The arrows up and down indicate an increase or a decrease, respectively.

Pollutant	Entry route	Primary target(s)	Physiological effects	Morphological effects	Biomass allocation	Reproductive effects
O_3	Stomata (leaves)	Chloroplasts, stomata	Photosynthetic decline, ROS increase	Leaf necrosis, curling	Root: Shoot increase	Flower/fruit abortion
SO_2	Stomata	Mesophyll, chloroplasts	Acidification increase, Chlorophyll decline	Leaf bleaching	Variable	Reduced pollen viability
NO_x	Stomata, cuticle	Leaf apoplast, guard cells	Gas exchange reduction, ROS increase	Chlorosis, thinner cuticle	Shoot biomass decline	Altered flowering timing
PM	Cuticle, stomata	Surface tissues	Physical blockage, ROS increase	Leaf surface abrasion	Overall biomass decline	Delay in fruit set
HMs	Roots, cuticle	Vascular tissue	Inhibited nutrient uptake	Root browning, reduced length	Root growth reduction	Malformation of seeds

Note: O_3, SO_2, NO_x, PM, and HMs represent ground-level ozone, sulfur dioxide, nitrogen oxides, particulate matter, and heavy metals; *ROS*, reactive oxygen species.

No permission required.

urbanized areas (Grantz et al., 2003). Moreover, O_3 can degrade floral volatile organic compounds, making flowers less attractive to pollinators (Langford et al., 2023). The interruption of flowering results in incomplete seasonal reproductive cycles, compromising seed production and the genetic diversity of plant populations. In long-lived species like *Q. ilex*, the suspension of fruiting even for a few years can hinder natural regeneration and disrupt trophic interactions with frugivores and seed dispersers (Ashmore, 2005).

Cumulative effects of sterility or repeated nonfruiting flowering cycles in confined urban areas drastically reduce seed dispersal capacity, affecting colonization dynamics and resilience of green infrastructure. The main effects of different pollutants on plants are summarized in Table 3.1.

Given these cascading effects, understanding and mitigating the impact of urban air pollution on woody species is essential for sustaining green infrastructure in cities. Selecting tree species based on their physiological resilience to air pollution and their capacity to maintain ecosystem services, air quality improvement policies, and long-term ecological monitoring are critical components of adaptive urban forest management aimed at preserving both plant health and ecosystem functionality.

3.3 Organ-level responses

Air pollution determines severe injuries and changes in physiological processes of woody plant species, which might alter their organs at anatomical and morphological levels, leading to significant alterations in leaf, stem, root, and reproductive structure traits, and in functionality (Mehmood et al., 2024). In general, visible leaf alterations, depending on tree sensitivity, are the first indication of injury (Diana Grecia et al., 2025).

A reduced biomass growth is often attributed to both gaseous (mainly CO_2 and NO_2) and particulate pollutants (Mehmood et al., 2024) that, together with metals associated with exerting a stronger influence than climate on plant growth (Locosselli et al., 2019). An extensive description of the acute and chronic effects on plants exerted by single pollutants, both organic and inorganic, gaseous and particulate, with the indication of the more sensitive groups (angiosperms and conifers), plant ages, and organs. Anyway, in urban-industrial areas, woody plants are simultaneously exposed to multiple contaminations, affecting the overall morpho-anatomical characteristics of their organs. Albeit the extents of these alterations are species-specific and are related to plant age, climate conditions of the growing area, and season (Devkota et al., 2024; Gheorghe and Ion, 2011; Leghari and Zaidi, 2013; Shrestha et al., 2025), several recurrent responses can be highlighted in multicontaminated environments.

3.3.1 Leaves

The most susceptible organ of a plant to injury is the leaf due to the presence of abundant stomata, which permit the penetration of pollutants into leaf tissues and play a fundamental role in photosynthesis (Gupta, 2016). High pollutant loads, especially SO_2, O_3, H_2S, and NO_2, cause visible damage to leaves, including lesions, chlorosis, necrosis, and even leaf death (El-Khatib et al., 2012; Mehmood et al., 2024). Apart from these severe responses, a reduction in leaf number, surface, and thickness has been generally observed, with worldwide trees in polluted environments characterized by smaller and thinner leaves compared to those in natural areas (Fig. 3.3). In the following, a few examples of such alterations are reported, encompassing tree species characterized by different phylogenetic and taxonomic features and distributed in different geographical areas (Mehmood et al., 2024).

The reduction of leaf area often observed in more polluted sites might be due to the adverse effects of urban pollution on plant leaf morphology, representing one of the adaptation mechanisms of plant species to

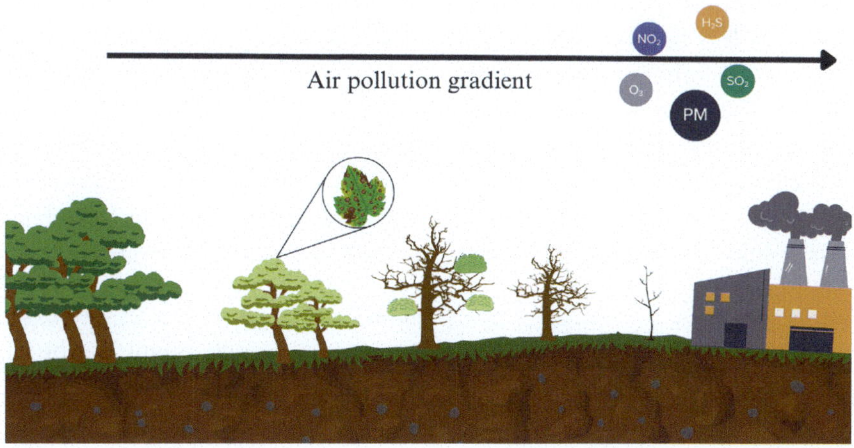

Figure 3.3 Influence of air pollution on leaves of woody plants along an urbanization gradient. Note: Image created with Canva by M. Napoletano. No permission required.

environmental stresses (Tiwari et al., 2006). One of the reasons for decreasing leaf area in polluted areas might be related to the effects of air pollution on gas exchange reductions, impairing photosynthesis and productivity. Overall, this reduction causes less contact between air pollutants and plant leaves and protects them from pollution (Assadi et al., 2011). Leaves of 13 common plant species in Quetta (Pakistan) exhibited changes in color and shape, as well as significant reductions in their leaf length, width, and area at polluted sites when compared with the same plant species from nonpolluted sites, with significant variations in the morphological responses according to both the season and plant age (Leghari & Zaidi, 2013). In particular, the heaviest effects were observed during Summer, followed by Autumn and Spring; in addition, the reduction in leaf features of plants from polluted sites also increased with the age of the plants (Leghari & Zaidi, 2013). Similarly, a significant decrease in leaf area has been described for *Cinnamomum camphora* (L.) J. Presl, *Lawsonia inermis* L., and *Bougainvillea spectabilis* Willd. growing in industrial sites in comparison with the same species growing in natural sites in Saudi Arabia (Al-Toukhy, 2015). The well-documented reduction in leaf surface in polluted areas reported for broadleaf species has been observed in conifers, too. For example, in *Pinus tabuliformis* Carrière, leaf width reduced significantly along a rural-urban gradient in Beijing (China), and the extent of the observed variations depended on the leaf age and land-use type (Su et al., 2021). It has been recently reported that the reduction of leaf area in more polluted sites helps plants to keep their water balance in leaf tissues, decreasing the water loss and increasing the plant resistance. In other words, the observed leaf alterations may be either a result of the adverse effects of pollution on the plant or an adaptive response of the plant to reduce the surface area exposed to pollutants. Differences in leaf size indicate the tolerance and sensitivity of different plant species (Devkota et al., 2024). Overall, the morpho-anatomical effects of air pollution on leaves of woody species reverberate in changes in their functions, resulting in a reduced photosynthetic capability as well as in tree ecological functions (Shrestha et al., 2025). *Bougainvillea glabra* Choisy., *Callistemon citrinus* (Curtis) Skeels, *Euphorbia pulcherrima* Willd, *Jasminum mesnyi* Hance, *Lagerstroemia indica* L., and *Nerium oleander* L.from different locations in Kathmandu Valley (Nepal) exhibited significant reductions in their leaf length, width, and area at polluted sites when compared with the same plant species from nonpolluted sites (Devkota et al., 2024). In the same geographical area (Koshi Province, Nepal), leaf morphological features were significantly reduced in 12 species, particularly in *Azadirachta indica* A. Juss., *Syzygium cumini* (L.) Skeels, *Saraca ascosa* (Roxb.) Willd. and *Tectona grandis* L.f., across a natural-urban-industrial gradient (Shrestha et al., 2025). A significant reduction in thickness and leaf mass per unit area has also been recently reported for *Acer pictum* Thunb., *Fraxinus chinensis* Roxb., *Koelreuteria paniculata* Laxm., *Salix babylonica* L., and *Sophora japonica* L. along a natural-rural-urban gradient in the Beijing metropolitan region of China (Yan et al., 2024).

SLA and leaf dry weight are morphological traits of leaves that usually represent the distribution of plant biomass. In the Mediterranean area, during the exposure of *Quercus ilex* L. to air pollution, SLA was increased with increasing the amount of pollution (Gratani et al., 2000). Leaf mass area (LMA) is used as a proxy for carbon assimilation from photosynthesis and for the representation of the relative growth rate of plants. Decrease of LMA in plants in air-polluted regions causes to higher leaf area for more carbon accumulation and photosynthesis activity. In addition, it is reported that it changes in relation to the physiological adaptation of plants to environmental pollution. Several studies reported that a decrease in LMA is a kind of strategy that urban plants use for adaptation to air pollution alongside capturing light and increasing the relative growth rate of plants (Yan et al., 2024). Leaf tissue density (LTD) is related to the defensive mechanisms of plants to environmental stresses. The greater the LTD, the plant water loss is lower (Shipley & Vu, 2002). It is reported that LTD values increase in air-polluted regions, and plant growth increases facilitating carbon storage for improving the defensive responses (Yan et al., 2024).

3.3.2 Stems

Less studied are the effects of air pollution on stems, even if it is known that air pollutants can significantly affect woody plants through both direct and indirect morphological and anatomical stem alteration. Indeed, apart from the direct effects on stem morphology and functionality, the reduced leaf surface and thickness in air-polluted environments can affect the stem growth, leading to changes in stem length and thickness (or in other morphological characteristics), as well as to a reduction in branch numbers (Mehmood et al., 2024) and changes in wood density. Overall, pollutants emitted from the industrial and vehicular exhausts exert a negative influence on stem anatomy, in the tropical plants *Croton sparsiflorus* Morong and *Cassia occidentalis* L., characterized by a reduction in stem circumference when growing in polluted sites (Sukumaran, 2014).

The studies on air pollution effects on stems have been generally carried out exposing woody plants to single or multiple pollutants, in order to evaluate also their additive, antagonistic, or synergistic effects (Landolt & Keller, 1985). It is reported that woody plant species exposed to PM can show a decrease in stem height and diameter, as well as changes in stem anatomy, and that the combined effects of PM and other pollutants depend on the species (Černiauskas et al., 2025). In this context, the combined exposure to PM, O_3, and CO_2 resulted in a reduced diameter growth for *Picea abies* (L.) H. Karst. and *B. pendula* seedlings compared to the unexposed controls (Černiauskas et al., 2025). Similar alterations in stem height and diameter have been described for *T. cordata*, *B. pendula*, *P. nigra*, and *Betula pubescens* Ehrh exposed to SO_2 and NO_2 (Whitmore & Freer-Smith, 1982). All these responses are mainly attributed to the effects of air pollution on annual growth rings, resulting in changes in ring width (Landolt & Keller, 1985). Similarly, the growth of *Tipuana tipu* (Benth.) O. Kuntze(one of the most common tree species in the metropolitan region of São Paulo, Brazil) is reduced under high concentrations of PM as well as of airborne aluminum (Al), barium (Ba), and zinc (Zn), which exert a dramatic influence on tree annual growth (Locosselli et al., 2019).

Not all woody plant species are equally sensitive to air pollution and their susceptibility to stem alterations depends on their natural tolerance to specific pollutants (Molnár et al., 2020). A meta-analysis on the growth of woody species to ground-level ozone highlighted that subtropical species are significantly less sensitive to O_3 than temperate ones, whereas deciduous broadleaf species are significantly more sensitive than evergreen broadleaf and needle-leaf species (Li et al., 2017).

3.3.3 Roots

Notwithstanding, it is not probable that gaseous air pollutants have a direct influence upon the roots; they may exert an indirect effect via the damage to the upper parts of the woody plants (Landolt & Keller, 1985). Indeed, reduced carbon allocation to roots caused by air pollution exposure results in reduced root growth and turnover. Overall, it seems that root growth is even more susceptible to air pollution than are leaf and stem growths (Landolt & Keller, 1985). This holds true not only for particulate pollutants but also for gaseous ones, such as in the case of deciduous broadleaf species, whose root biomass is much more affected by O_3 than shoot biomass (Li et al., 2017). Anyway, elevated O_3 concentrations decrease the root biomass and the number of root tips also in the case of evergreen broadleaf species, such as *Elaeocarpus sylvestris* (Lour.) Poir. and *Michelia chapensis* Dandy, two woody plants native to subtropical China (Pan et al., 2020).

Despite the heavier effects of pollutants (and other abiotic stresses) on root structure and development than those observed in the above-ground organs, below-ground organs have been significantly less studied due to the limited availability for root observations (Kul et al., 2020). This generates concerns, considering that apart from the case of gaseous pollutants, roots are directly exposed to all the inorganic and organic pollutants that may be deposited in soil and accumulated in roots (Oubohssaine & Dahmani, 2024), also as a defense mechanism applied by the plant to protect its leaves, stem, and fruits against pollutant toxicity (Kul et al., 2020).

Overall, chronic exposure of woody plants to pollutants may result in reduced allocation to roots and lower carbohydrate reserves, thereby increasing the dependence of root activity on current photosynthesis supply. Anyway, as a consequence of the reduced root growth, the uptake of nutrients and water is impaired too (Walden, 1989), and in susceptible woody plants, there is some evidence for root dieback as a probable result of reduced carbohydrate supply from the affected upper organs (Landolt & Keller, 1985). Plants exposed to high concentrations of HMs, for example, display stunted root elongation and growth, as well as browning and death (Kul et al., 2020).

3.3.4 Reproductive structures

Albeit few studies on the effects of air pollutants on the reproductive structures of woody plant species have been carried out, several pieces of evidence highlight that air pollution affects them, causing both morphological and anatomical changes that can impair fertility, other than the overall plant health. Indeed, pollutants can disrupt the development of flowers and fruits, potentially leading to reduced seed production and seedling survival (Dueck, 1991). Oxidizing pollutants like O_3, SO_2, and NO_2 can affect the size, shape, and number of flowers (Mehmood et al., 2024), decreasing floral yield and delaying fruiting (Mehmood et al., 2024). In addition, SO_2 causes tip necrosis (Gheorghe and Ion, 2011). Apart from the reduction in reproductive success and in the overall tree health, in several cases, air pollutants can cause premature senescence or death of reproductive organs (Savita, 2021), further reducing plant reproductive potential. Understanding these effects is crucial for developing strategies to properly protect trees and forests.

3.4 Responses at the tissue and cellular level

Plants growing in polluted regions showed thicker palisade tissue and thinner spongy tissue for their resistance to air pollution (Jahan & Iqbal, 1992). Whereas spongy tissues have a role in the entery of pollutants into the leaf, given the amount of intercellular spaces and permission of stomata to pollutants entry (Qin et al., 2014), palisade tissues have important roles in photosynthesis and in accelerating the transfer of photosynthesis production from leaves and avoiding the restriction of pollutants on photosynthesis (Rashidi et al., 2012). The increase of leaf thickness of *F. chinensis* in air-polluted regions may protect plants from air pollution effects (Singh et al., 2017).

Plants growing along the roadsides have modified leaf surface characters, including stomata and epidermal cells, due to the stress of automobile exhaust emissions with high traffic density in urban areas. Stomata and cuticle are the first receptors during the encountering pollutants. Stomata should be considered for the direct entery of gases into the leaf. Response of stomata to air pollutants is varying and species specific, as well as depends on pollutant type and concentration, as well as on the age of plants and environmental conditions (Abeyratne & Ileperuma, 2006b; Gupta, 2016).

In leaves exposed to air pollution, auxiliary alterations include smaller stomatal and epidermal cells, thickened cell walls, epicuticular wax deposition, and chlorosis. In determining each species' resistance/sensitivity to environmental pollutants, cuticle and epidermal features play a crucial role (Uka et al., 2017). Stomatal index can be considered as one of the good anatomical adaptations shown by the plant to air pollution (Ogunrotimi et al., 2017). Stomatal density increment showed the adaptation and tolerance of plants to air pollution (Yan et al., 2024). Low stomata number has been a sign of plant adaptation to air pollution, because of lower stomata number, lower of gaseous pollutants absorption from the air (Devkota et al., 2024). In polluted areas, the plant leaves had lower stomatal density and stomatal index (%) compared to less polluted areas, and this reduction may be related to the reduced leaf area (Saadabi, 2011). But the number of clogged stomata in leaves showed increments in polluted areas (Devkota et al., 2024).

3.5 Conclusions

Woody species respond to urban air pollution in different ways, according to the complexity of the considered level, spanning from whole plant responses to organ, tissue, and cell level responses. The whole response of woody species mainly includes growth inhibition, alteration in plant height and shoot growth, biomass reduction, and alteration in the root: shoot ratio, altered plant shape, altered reproduction capacity, as well as suspended flowering and fruiting. These alterations are attributable to those observed at the organ level, with a generally reported reduction in leaf surface, stem diameter, root length, and number and functionality of reproductive structures. Similarly, the observed variations observed in organ (mainly leaves) shapes and colors could be related to the alterations observed at the tissue and cell levels, mainly palisade and spongy tissues, as well as stomata-related properties (number and traits).

Overall, more investigations, above all at the cellular level (cell wall and membrane alterations, such as thickening cell wall and lipid peroxidation, plasmolysis and cell collapse) and intracellular level (disruption of vacuoles, chloroplasts, mitochondria, peroxisomes, abnormal nucleus shapes) are desirable, in order to know and prevent damages at the three levels and loss of ecosystem services associated with urban vegetation. It is also advisable more investigation about multiple influential parameters affecting several tree species to different levels of air pollution.

References

Abbasi, S., Hosseini, S. M., Khorasani, N., & Karbassi, A. (2018). Response of the morphological traits of Elm (*Ulmus minor* umbraculifera) leaves to air pollution in urban areas (a case study of Tehran metropolitian city, Iran). *Applied Ecology and Environmental Research, 16*(4), 4955–4968. https://doi.org/10.15666/aeer/1604_49554968.

Abeyratne, V., & Ileperuma, O. (2006). Impact of ambient air pollutants on the stomatal aperture of argyreia populifolia. *Ceylon. Journal of Science (Biological Science), 35*, 9–15.

Abeyratne, V., & Ileperuma, O. A. (2006b). Air pollution monitoring in the city of Kandy: Possible transboundary effects. *Journal of the National Science Foundation of Sri Lanka, 34*(3). https://doi.org/10.4038/jnsfsr.v34i3.3644.

Ainsworth, E. A., & Long, S. P. (2005). What have we learned from 15 years of free-air CO_2 enrichment (FACE)? A meta-analytic review of the responses of photosynthesis, canopy properties and plant production to rising CO_2. *New Phytologist, 165*(2), 351–372. https://doi.org/10.1111/j.1469-8137.2004.01224.x.

Ali, M. U., Liu, G., Yousaf, B., Ullah, H., Abbas, Q., Munir, M. A. M., & Irshad, S. (2019). Biomonitoring and health risks assessment of trace elements in various age- and gender-groups exposed to road dust in habitable urban-industrial areas of Hefei, China. *Environmental Pollution, 244*, 809–817. https://doi.org/10.1016/j.envpol.2018.10.084.

Al-Toukhy, A.-T. (2015). Effect of air pollution on leaf traits of three tree species growing in the industrial zone of Jeddah, Saudi Arabia. *Journal of King Abdulaziz University - Meteorology, Environment and Arid Land Agriculture Sciences, 26*(2), 33–38. https://doi.org/10.4197/met.26-2.4.

Antenozio, M. L., Caissutti, C., Caporusso, F. M., Marzi, D., & Brunetti, P. (2024). Urban air pollution and plant tolerance: Omics responses to ozone, nitrogen oxides, and particulate matter. *Plants, 13*(15). https://doi.org/10.3390/plants13152027, Article 15.

Arena, C., De Maio, A., De Nicola, F., Santorufo, L., Vitale, L., & Maisto, G. (2013). Assessment of eco-physiological performance of *Quercus ilex* L. leaves in urban area by an integrated approach. *Water, Air, & Soil Pollution, 225*(1), 1824. https://doi.org/10.1007/s11270-013-1824-6.

Arena, C., Santorufo, L., Cataletto, P. R., Memoli, V., Scudiero, R., & Maisto, G. (2017). Eco-physiological and antioxidant responses of holm oak (*Quercus ilex* L.) Leaves to Cd and Pb. *Water, Air, & Soil Pollution, 228*(12), 459. https://doi.org/10.1007/s11270-017-3638-4.

Ashmore, M. R. (2005). Assessing the future global impacts of ozone on vegetation. *Plant, Cell & Environment, 28*(8), 949–964. https://doi.org/10.1111/j.1365-3040.2005.01341.x.

Assadi, A., Ghasemi Pirbalouti, A., Malekpoor, F., Teimori, N., & Assadi, L. (2011). Impact of air pollution on physiological and morphological characteristics of *Eucalyptus camaldulensis* Den. *Journal of Food Agriculture and Environment, 9*, 676–679.

Black, V. J., Black, C. R., Roberts, J. A., & Stewart, C. A. (2000). Tansley review No. 115. *New Phytologist, 147*(3), 421–447. https://doi.org/10.1046/j.1469-8137.2000.00721.x.

Booker, F. L., & Fiscus, E. L. (2005). The role of ozone flux and antioxidants in the suppression of ozone injury by elevated CO_2 in soybean. *Journal of Experimental Botany, 56*(418), 2139–2151. https://doi.org/10.1093/jxb/eri214.

Bussotti, F., & Ferretti, M. (2009). Visible injury, crown condition, and growth responses of selected Italian forests in relation to ozone exposure. *Environmental Pollution, 157*(5), 1427–1437. https://doi.org/10.1016/j.envpol.2008.09.034.

Calatayud, P.-A., Gitau, C., Calatayud, S., Dupas, S., Le Ru, B., & Silvain, J.-F. (2011). Variability in the reproductive biology and in resistance against *Cotesia sesamiae* among two *Busseola fusca* populations. *Journal of Applied Entomology, 135*(6), 423–429. https://doi.org/10.1111/j.1439-0418.2010.01561.x.

Černiauskas, V., Varnagirytė-Kabašinskienė, I., Čėsnienė, I., Armoška, E., & Araminienė, V. (2025). Response of tree seedlings to a combined treatment of particulate matter, ground-level ozone, and carbon dioxide: Primary effects. *Plants, 14*(1). https://doi.org/10.3390/plants14010006, Article 1.

Daghestani, M., & Kolahi, M. (2024). Phytoremediation potential of seedlings: Comparing heavy metal accumulation in *Ailanthus, Acer,* and *Fraxinus* species. *Environmental Monitoring and Assessment, 196*(10), 920. https://doi.org/10.1007/s10661-024-13054-7.

Devkota, A., Shrestha, S., & Jha, P. (2024). Effect of air pollution on the leaf morphology and anatomy of common plant species of Kathmandu Valley. *European Journal of Ecology, 10*(1). https://doi.org/10.17161/eurojecol.v10i1.19487.

Diana Grecia, A.-M., Sergio Arturo, T.-S., & Marlenne, G.-R. (2025). Review: Implications of air pollution on trees located in urban areas. *Earth, 6*(2). https://doi.org/10.3390/earth6020038, Article 2.

Dietze, M. C., Sala, A., Carbone, M. S., Czimczik, C. I., Mantooth, J. A., Richardson, A. D., & Vargas, R. (2014). Nonstructural carbon in woody plants. *Annual Review of Plant Biology, 65*, 667–687. https://doi.org/10.1146/annurev-arplant-050213-040054.

Dueck, T. A. (1991). Air pollution and reproductive processes in natural plant species. In J. Rozema, & J. A. C. Verkleij (Eds.). *Ecological responses to environmental stresses. Tasks for vegetation science, vol 22* (pp. 200–209). Dordrecht: Springer. https://doi.org/10.1007/978-94-009-0599-3_18.

Duque, L., Poelman, E. H., & Steffan-Dewenter, I. (2021). Effects of ozone stress on flowering phenology, plant-pollinator interactions and plant reproductive success. *Environmental Pollution, 272*, 115953. https://doi.org/10.1016/j.envpol.2020.115953.

El-Khatib, A. A., Radwan, D. E., & Alramah-Said, A. A. (2012). Morpho-anatomical characteristics of olive (*Olea europaea* L.) Trees leaf as bio-indicator of cement dust air pollution in libya. *Journal of Environmental Studies, 9*(1), 65–72. https://doi.org/10.21608/jesj.2012.191533.

Escobedo, F. J., Kroeger, T., & Wagner, J. E. (2011). Urban forests and pollution mitigation: Analyzing ecosystem services and disservices. *Environmental Pollution, 159*(8), 2078–2087. https://doi.org/10.1016/j.envpol.2011.01.010.

Feng, H., Ding, Y., Zou, B., Cohen, J. B., Ye, S., Yang, Z., Qin, K., Liu, L., & Gu, X. (2022). Vegetation-related dry deposition of global PM2.5 from satellite observations. *Journal of Geographical Sciences, 32*(4), 589–604. https://doi.org/10.1007/s11442-022-1962-0.

Fiorentino, A., Rajput, F. Z., Di Serio, A., Baldi, V., Guarino, F., Baldantoni, D., Ronga, D., Mazzei, P., Motta, O., Falanga, M., Cicatelli, A., & Castiglione, S. (2025). Role of plants and urban soils in carbon stock: status, modulators, and sustainable management practices. *Plants, 14*(4), 546. 10.3390/plants14040546.

Frei, M. (2015). Breeding of ozone resistant rice: Relevance, approaches and challenges. *Environmental Pollution, 197*, 144–155. https://doi.org/10.1016/j.envpol.2014.12.011.

Ghahremaninejad, F., Hoseini, E., & Fereidounfar, S. (2021). Cities in drylands as artificial protected areas for plants. *Biodiversity and Conservation, 30*(1), 243–248. https://doi.org/10.1007/s10531-020-02079-2.

Gheorghe, I. F., & Ion, B. (2011). The effects of air pollutants on vegetation and the role of vegetation in reducing atmospheric pollution. In M. K. Khallaf (Ed.). *The impact of air pollution on health, economy, environment and agricultural sources,* IntechOpen, https://doi.org/10.5772/17660.

Gheshlaghpoor, S., Abedi, S. S., & Moghbel, M. (2023). The relationship between spatial patterns of urban land uses and air pollutants in the Tehran metropolis, Iran. *Landscape Ecology, 38*(2), 553–565. https://doi.org/10.1007/s10980-022-01549-y.

Grantz, D. A., Garner, J. H. B., & Johnson, D. W. (2003). Ecological effects of particulate matter. *Environment International, 29*(2), 213–239. https://doi.org/10.1016/S0160-4120(02)00181-2.

Gratani, L., Catoni, R., & Tarquini, F. (2019). Carbon dioxide sequestration capability of the botanical garden of rome: Environmental and economic benefits. *American Journal of Plant Sciences, 10*(8). https://doi.org/10.4236/ajps.2019.108090, Article 8.

Gratani, L., Crescente, M. F., & Petruzzi, M. (2000). Relationship between leaf life-span and photosynthetic activity of *Quercus ilex* in polluted urban areas (Rome). *Environmental Pollution, 110*(1), 19–28. https://doi.org/10.1016/S0269-7491(99)00285-7.

Gupta, A. (2016). Effect of air pollutants on plant gaseous exchange process: Effect on stomata and respiration. In U. Kulshrestha, & P. Saxena (Eds.). *Plant responses to air pollution* (pp. 85–92). Singapore: Springer. https://doi.org/10.1007/978-981-10-1201-3_8.

Hasnain, A., Dadkhah-Aghdash, H., Luqman, M., Muhammad, S., Sardar, A. A., Ali, S., Mehmood, F., Khan, U. A., Mehmood, Z., John, A., Binyameen, Khan, Z. I., Yang, H.-H., & Awan, M. U. F. (2024). Impact of automobile exhaust on biochemical and genomorphic characteristics of *Mimusops elengi* L. growing along roadsides of Lahore city, Pakistan. *Heliyon, 10*(6). https://doi.org/10.1016/j.heliyon.2024.e28157.

He, N., Liu, C., Tian, M., Li, M., Yang, H., Yu, G., Guo, D., Smith, M. D., Yu, Q., & Hou, J. (2018). Variation in leaf anatomical traits from tropical to cold-temperate forests and linkage to ecosystem functions. *Functional Ecology, 32*(1), 10–19. https://doi.org/10.1111/1365-2435.12934.

Hodge, A., Berta, G., Doussan, C., Merchan, F., & Crespi, M. (2009). Plant root growth, architecture and function. *Plant and Soil, 321*(1), 153–187. https://doi.org/10.1007/s11104-009-9929-9.

Horaginamani, S. M., & Ravichandran, M. (2010). Ambient air quality in an urban area and its effects on plants and human beings: A case study of Tiruchirappalli, India. *Kathmandu University Journal of Science, Engineering and Technology, 6*(2). https://doi.org/10.3126/kuset.v6i2.4007, Article 2.

Huttunen, S., & Manninen, S. (2013). A review of ozone responses in Scots pine (*Pinus sylvestris*). *Environmental and Experimental Botany, 90*, 17–31. 10.1016/j.envexpbot.2012.07.001.

Jahan, S., & Iqbal, M. Z. (1992). Morphological and anatomical studies of leaves of different plants affected by motor vehicles exhaust. *Medical Journal of Islamic World Academy of Sciences, 5*(1), 21–23.

Jochner, S., Markevych, I., Beck, I., Traidl-Hoffmann, C., Heinrich, J., & Menzel, A. (2015). The effects of short- and long-term air pollutants on plant phenology and leaf characteristics. *Environmental Pollution, 206*, 382–389. https://doi.org/10.1016/j.envpol.2015.07.040.

Kangasjärvi, J., Jaspers, P., & Kollist, H. (2005). Signalling and cell death in ozone-exposed plants. *Plant, Cell & Environment, 28*(8), 1021–1036. https://doi.org/10.1111/j.1365-3040.2005.01325.x.

Karlsson, P. E., Uddling, J., Braun, S., Broadmeadow, M., Elvira, S., Gimeno, B. S., Le Thiec, D., Oksanen, E., Vandermeiren, K., Wilkinson, M., & Emberson, L. (2004). New critical levels for ozone effects on young trees based on AOT40 and simulated cumulative leaf uptake of ozone. *Atmospheric Environment, 38*(15), 2283–2294. https://doi.org/10.1016/j.atmosenv.2004.01.027.

Khosropour, E., Attarod, P., Shirvany, A., Pypker, T. G., Bayramzadeh, V., Hakimi, L., & Moeinaddini, M. (2019). Response of *Platanus orientalis* leaves to urban pollution by heavy metals. *Journal of Forestry Research, 30*(4), 1437–1445. https://doi.org/10.1007/s11676-018-0692-8.

Kul, R., Ekinci, M., Turan, M., Ors, S., Yildirim, E., Kul, R., Ekinci, M., Turan, M., Ors, S., & Yildirim, E. (2020). How abiotic stress conditions affects plant roots. In E. Yildirim, M. Turan, & M. Ekinci (Eds.). *Plant Roots,* IntechOpen, https://doi.org/10.5772/intechopen.95286.

Kuster, V. C., Castro, S. A. B.de, & Vale, F. H. A. (2016). Photosynthetic and anatomical responses of three plant species at two altitudinal levels in the Neotropical savannah. *Australian Journal of Botany, 64*(8), 696–703. https://doi.org/10.1071/BT15280.

Landolt, K., & Keller, T. (1985). Uptake and effects of air pollutants on woody plants. *Experientia, 41*, 301–310.

Langford, B., Ryalls, J. M. W., Mullinger, N. J., Hayden, P., Nemitz, E., Pfrang, C., Robins, A., Touhami, D., Bromfield, L. M., & Girling, R. D. (2023). Mapping the effects of ozone pollution and mixing on floral odour plumes and their impact on plant-pollinator interactions. *Environmental Pollution, 336*, 122336. https://doi.org/10.1016/j.envpol.2023.122336.

Leghari, S. K., & Zaidi, M. A. (2013). Effect of air pollution on the leaf morphology of common plant species of quetta city. *Pakistan Journal of Botany, 45*, 447–454.

Li, P., Feng, Z., Catalayud, V., Yuan, X., Xu, Y., & Paoletti, E. (2017). A meta-analysis on growth, physiological, and biochemical responses of woody species to ground-level ozone highlights the role of plant functional types. *Plant, Cell & Environment, 40*(10), 2369–2380. https://doi.org/10.1111/pce.13043.

Li, X., Zhang, T., Sun, F., Song, X., Zhang, Y., Huang, F., Yuan, C., Yu, H., Zhang, G., Qi, F., & Shao, F. (2021). The relationship between particulate matter retention capacity and leaf surface micromorphology of ten tree species in Hangzhou, China. *Science of the Total Environment, 771*, 144812. https://doi.org/10.1016/j.scitotenv.2020.144812.

Liu, Y., & Ding, H. (2008). Variation in air pollution tolerance index of plants near a steel factory: Implications for landscape-plant species selection for industrial areas. *Wseas Transactions on Environment and Development, 4*(1), 24–32.

Locosselli, G. M., Camargo, E. P. D., Moreira, T. C. L., Todesco, E., Andrade, M. D. F., André, C. D. S. D., André, P. A. D., Singer, J. M., Ferreira, L. S., Saldiva, P. H. N., & Buckeridge, M. S. (2019). The role of air pollution and climate on the growth of urban trees. *Science of the Total Environment, 666*, 652–661. https://doi.org/10.1016/j.scitotenv.2019.02.291.

Lorenzini, G., Grassi, C., Nali, C., Petiti, A., Loppi, S., & Tognotti, L. (2006). Leaves of *Pittosporum tobira* as indicators of airborne trace element and PM10 distribution in central Italy. *Atmospheric Environment, 40*(22), 4025–4036. https://doi.org/10.1016/j.atmosenv.2006.03.032.

Matyssek, R., Wieser, G., Calfapietra, C., de Vries, W., Dizengremel, P., Ernst, D., Jolivet, Y., Mikkelsen, T. N., Mohren, G. M. J., Le Thiec, D., Tuovinen, J.-P., Weatherall, A., & Paoletti, E. (2012). Forests under climate change and air pollution: Gaps in understanding and future directions for research. *Environmental Pollution, 160*, 57–65. https://doi.org/10.1016/j.envpol.2011.07.007.

McKinney, M. L. (2008). Effects of urbanization on species richness: A review of plants and animals. *Urban Ecosystems, 11*(2), 161–176. https://doi.org/10.1007/s11252-007-0045-4.

Mehmood, Z., Yang, H.-H., Awan, M., Ahmed, U., Hasnain, A., Luqman, M., Muhammad, S., Sardar, A., Chan, T.-Y., & Sharjeel, A. (2024). Effects of air pollution on morphological, biochemical, DNA, and tolerance ability of roadside plant species. *Sustainability, 16*(8), 3427. https://doi.org/10.3390/su16083427.

Molnár, V. E., Simon, E., Tóthmérész, B., Ninsawat, S., & Szabó, S. (2020). Air pollution induced vegetation stress -the air pollution tolerance index as a quick tool for city health evaluation. *Ecological Indicators, 113*, 106234. 10.1016/j.ecolind.2020.106234.

Mousavi Javardi, N. S., Pakravan, M., Panahi, P., & Zarei, R. (2023). Assessment of the relationships between leaf characteristics with air pollutants: A case study on oriental plane (*Platanus orientalis* L.) and Caucasian hackberry (*Celtis caucasica* Willd.). *Arboriculture & Urban Forestry*. https://doi.org/10.48044/jauf.2023.023, jauf.2023.023.

Muscari Tomajoli, G. (2026). https://BioRender.com/tx6s1d8.

Neinhuis, C., & Barthlott, W. (1998). Seasonal changes of leaf surface contamination in beech, oak, and ginkgo in relation to leaf micromorphology and wettability. *New Phytologist, 138*(1), 91–98. https://doi.org/10.1046/j.1469-8137.1998.00882.x.

Nowak, D. J., & Crane, D. E. (2002). Carbon storage and sequestration by urban trees in the USA. *Environmental Pollution, 116*(3), 381–389. https://doi.org/10.1016/S0269-7491(01)00214-7.

Ogunrotimi, D. G., Adereti, F. K., Eludoyin, A. O., & Awotoye, O. O. (2017). Urban air pollution control: Selection of trees for ecological monitoring using anticipated performance indices in a medium-size urban area in Southwest Nigeria. *Interdisciplinary Environmental Review, 18*(1), 40–54. https://doi.org/10.1504/IER.2017.084193.

Oubohssaine, M., & Dahmani, I. (2024). Phytoremediation: Harnessing plant power and innovative technologies for effective soil remediation. *Plant Stress, 14*, 100578. https://doi.org/10.1016/j.stress.2024.100578.

Pan, L., Lin, W., Yu, M., Lie, G., Xue, L., & Chen, H. (2020). Effects of elevated ozone concentrations on root characteristics and soil properties of *Elaeocarpus sylvestris* and *Michelia chapensis*. *Bulletin of Environmental Contamination and Toxicology, 104*(5), 682–688. https://doi.org/10.1007/s00128-020-02832-x.

Paoletti, E. (2009). Ozone and urban forests in Italy. *Environmental Pollution, 157*(5), 1506–1512. https://doi.org/10.1016/j.envpol.2008.09.019.

Percy, K. E., & Ferretti, M. (2004). Air pollution and forest health: Toward new monitoring concepts. *Environmental Pollution, 130*(1), 113–126. https://doi.org/10.1016/j.envpol.2003.10.034.

Poorter, H., Niinemets, Ü., Poorter, L., Wright, I. J., & Villar, R. (2009). Causes and consequences of variation in leaf mass per area (LMA): A meta-analysis. *New Phytologist, 182*(3), 565–588. https://doi.org/10.1111/j.1469-8137.2009.02830.x.

Qi, Y., Wei, W., Chen, C., & Chen, L. (2019). Plant root-shoot biomass allocation over diverse biomes: A global synthesis. *Global Ecology and Conservation, 18*, e00606. https://doi.org/10.1016/j.gecco.2019.e00606.

Qin, X., Sun, N., Ma, L., Chang, Y., & Mu, L. (2014). Anatomical and physiological responses of Colorado blue spruce to vehicle exhausts. *Environmental Science and Pollution Research, 21*(18), 11094–11098. https://doi.org/10.1007/s11356-014-3015-4.

Rashidi, F., Jalili, A., Babaie Kafaki, S., Sagheb-Talebi, K., & Hodgson, J. (2012). Anatomical responses of leaves of black locust (*Robinia pseudoacacia* L.) to urban pollutant gases and climatic factors. *Trees, 26*(2), 363–375. https://doi.org/10.1007/s00468-011-0598-y.

Roy, A., Mandal, M., Das, S., Popek, R., Rakwal, R., Agrawal, G. K., Awasthi, A., & Sarkar, A. (2024). The cellular consequences of particulate matter pollutants in plants: Safeguarding the harmonious integration of structure and function. *Science of the Total Environment, 914*, 169763. https://doi.org/10.1016/j.scitotenv.2023.169763.

Saadabi, A. M. A. (2011). Effects of auto-exhaust pollution on the micro-morphology and leaf epidermal features of ornamental plants in Khartoum, Sudan. *Journal of Applied Sciences Research,* 7(3), 270–273.

Sæbø, A., Popek, R., Nawrot, B., Hanslin, H. M., Gawronska, H., & Gawronski, S. W. (2012). Plant species differences in particulate matter accumulation on leaf surfaces. *Science of the Total Environment, 427–428*, 347–354. https://doi.org/10.1016/j.scitotenv.2012.03.084.

Salimi, A., & Dadkhah Aghdash, H. (2019). Air pollution tolerance index (APTI) of three tree species of *Morus alba* L., Ailanthus altissima (Mill.) *Swingle* and *Salix babylonica* L. in different areas of Tehran city. *Journal of Environmental Studies, 45*(3), 513–524. https://doi.org/10.22059/jes.2019.259202.1007685.

Samuelson, L. J., & Edwards, G. S. (1993). A comparison of sensitivity to ozone in seedlings and trees of *Quercus rubra* L. *New Phytologist, 125*(2), 373–379. https://doi.org/10.1111/j.1469-8137.1993.tb03889.x.

Savita (2021). Impacts of air pollutants on forest ecosystem and role in ecological imbalance. In In. S. Tiwari, & P. Saxena (Eds.). *Air pollution and its complications: From the regional to the global scale* (pp. 129–142). Springer International Publishing. https://doi.org/10.1007/978-3-030-70509-1_8.

Shipley, B., & Vu, T.-T. (2002). Dry matter content as a measure of dry matter concentration in plants and their parts. *New Phytologist, 153*(2), 359–364. https://doi.org/10.1046/j.0028-646X.2001.00320.x.

Shrestha, U., Rijal, S., Chettri, M. K., Acharya, B. D., Paudel, M. R., Shrestha, A., Adhikari, P., & Devkota, A. (2025). Effects of dust pollution on leaf morphology and chlorophyll content: A comparative study across two seasons in Biratnagar, Nepal. *Banko Janakari, 35*(1). https://doi.org/10.3126/banko.v35i1.73301, Article 1.

Singh, H., Verma, A., Kumar, M., Sharma, R., Gupta, R., Kaur, M., Negi, M., & Sharma, S. K. (2017). Phytoremediation: A green technology to clean up the sites with low and moderate level of heavy metals. *Austin Biochemistry, 2*(2), 1012. https://austinpublishinggroup.com/biochemistry/fulltext/biochemistry-v2-id1012.php.

Sridhar, B., Lawrence, G. B., Debenport, S. J., Fahey, T. J., Buckley, D. H., Wilhelm, R. C., & Goodale, C. L. (2022). Watershed-scale liming reveals the short- and long-term effects of pH on the forest soil microbiome and carbon cycling. *Environmental Microbiology, 24*(12), 6184–6199. https://doi.org/10.1111/1462-2920.16119.

Su, Y., Cui, B., Luo, Y., Wang, J., Wang, X., Ouyang, Z., & Wang, X. (2021). Leaf functional traits vary in urban environments: Influences of leaf age, land-use type, and urban–rural Gradient. *Frontiers in Ecology and Evolution, 9*. https://doi.org/10.3389/fevo.2021.681959.

Sukumaran, D. (2014). Effect of air pollution on the anatomy some tropical plants. *Applied Ecology and Environmental Sciences, 2*(1). https://doi.org/10.12691/aees-2-1-5, Article 1.

Taylor, G. E. (1978). Plant and leaf resistance to gaseous air pollution stress. *New Phytologist, 80*(3), 523–534. https://doi.org/10.1111/j.1469-8137.1978.tb01584.x.

Tiwari, S., Agrawal, M., & Marshall, F. M. (2006). Evaluation Of ambient air pollution impact on carrot plants at a sub urban site using open top chambers. *Environmental Monitoring and Assessment, 119*(1), 15–30. https://doi.org/10.1007/s10661-005-9001-z.

Uka, U. N., Hogarh, J., & Belford, E. J. D. (2017). Morpho-anatomical and biochemical responses of plants to air pollution. *International Journal of Modern Botany, 7*(1), 1–11.

Van Der Heijden, M. G. A., Bardgett, R. D., & Van Straalen, N. M. (2008). The unseen majority: Soil microbes as drivers of plant diversity and productivity in terrestrial ecosystems. *Ecology Letters, 11*(3), 296–310. https://doi.org/10.1111/j.1461-0248.2007.01139.x.

Walden, R. (1989). *Genetic Transformation in Plants*. N.J: Englewood Cliffs.

Whitmore, M. E., & Freer-Smith, P. H. (1982). Growth effects of SO_2 and/or NO_2 on woody plants and grasses during spring and summer. *Nature, 300*(5887), 55–57. https://doi.org/10.1038/300055a0.

Wilkinson, S., & Davies, W. J. (2010). Drought, ozone, ABA and ethylene: New insights from cell to plant to community. *Plant, Cell & Environment, 33*(4), 510–525. https://doi.org/10.1111/j.1365-3040.2009.02052.x.

Xu, X. (2021). Recent advances in studies of ozone pollution and impacts in China: A short review. *Current Opinion in Environmental Science & Health, 19*, 100225. https://doi.org/10.1016/j.coesh.2020.100225.

Yan, X., Li, P., Wu, X., Wang, J., Wang, Z., Xu, J., Hou, X., Fan, D., Yan, Z., & Du, E. (2024). Variations in the leaf economics spectrum, anatomical, ultrastructural, and stomatal traits of five tree species in the urban-rural air pollution environment. *Journal of Environmental Sciences, 155*, 177–192. https://doi.org/10.1016/j.jes.2024.08.009.

CHAPTER FOUR

Biochemical and molecular responses of plants to urban air pollution

Richa Rai
Plant Ecology and Climate Change Science Division, CSIR-National Botanical Research Institute, Lucknow, Uttar Pradesh, India

4.1 Introduction

Urban environments face growing challenges, including the unplanned expansion o f urban areas, population migration from rural areas to cities, and increased transportation, which affect air quality. These factors contribute to the emissions of atmospheric pollutants, including nitrogen dioxide (NO_2), sulfur dioxide (SO_2), particulate matter (PM), volatile organic compounds, and the secondary air pollutant tropospheric ozone (O_3). Air pollution is increasingly recognized as a major abiotic stress affecting plant growth, productivity, and survival, particularly in urban and peri-urban environments (Oksanen & Kontunen-Soppela, 2021). These pollutants, along with their degradation of air quality, also pose a threat to vegetation by altering its physiological, biochemical, and molecular processes (Table 4.1).

Green infrastructure enhances air quality by capturing gaseous air pollutants (SO_2, NO_2, O_3), greenhouse gases, and particulate pollution (PM_{10} and $PM_{2.5}$) (Semeraro et al., 2021). Initiatives such as green roofs and walls effectively reduce air pollution and promote urban cooling. Nature-based solutions, such as urban forestry, green infrastructure, and engineered bio-systems, offer a sustainable and multifunctional alternative to traditional gray infrastructure for improving air quality management (Kumar et al., 2024). Plants, being sessile organisms, are continuously exposed to ambient air pollutants throughout their life cycle (Bhardwaj et al., 2026). The impact of air pollution depends mainly on pollutant type, concentration, exposure duration, and prevailing environmental conditions, as well as on intrinsic plant traits such as leaf morphology, stomatal

Urban Air Pollution and Plant Responses
DOI: https://doi.org/10.1016/B978-0-443-40320-0.00003-6

Table 4.1 Major urban air pollutants, their sources, and their impact on plants.

Air pollutant	Sources	Mode of entry in plants	Impact on plants
Tropospheric ozone (O_3)	Secondary air pollutant	Stomata	Causes oxidative damage in plants
Oxides of nitrogen	Vehicular	Stomatal/ Cuticular	Lowering of pH, ROS formation
SO_2	Industrial	Stomatal	Lowering of pH, formation of sulfite and bi-sulfite ions leading to formation of ROS
Particulate pollution (PM_{10} and $PM_{2.5}$)	Dust/Combustion	Deposition	Stomatal blockage, affecting gas exchange parameters, induces oxidative stress

ROS, reactive oxygen species.

behavior, antioxidant capacity, and genetic makeup (Oksanen & Kontunen-Soppela, 2021). Acute exposure to the pollutants may result in visible injury and tissue damage. In contrast, exposure to sublethal concentrations for a longer duration causes chronic injury, leading to subtle biochemical and molecular alterations that cumulatively affect plant growth and yield (Oksanen & Kontunen-Soppela, 2021).

Understanding the biochemical and molecular responses of plants to air pollution is essential, as these responses determine the degree of tolerance or sensitivity of plant species. Second, biochemical and molecular markers often respond earlier than visible symptoms, making them valuable tools for biomonitoring. Third, such knowledge provides a scientific basis for selecting pollution-tolerant species for urban greening, green infrastructure, and climate-resilient landscape planning. This chapter provides a comprehensive synthesis of current knowledge on how plants perceive air pollutants, how they disrupt cellular homeostasis, and how plants activate biochemical and molecular defense mechanisms to cope with pollution-induced stress. It also highlights oxidative stress caused by air pollutants, antioxidant defense systems, hormonal regulation, and changes in gene expression, which collectively shape plant tolerance strategies.

4.2 Pollutant uptake and reactive oxygen species formation within plants

Air pollutants primarily enter plant tissues through stomatal openings, which serve as the principal pathway for gas exchange (Rai & Agrawal, 2012). Pollutants such as O_3, NO_2, and SO_2 diffuse into the leaf interior along concentration gradients, with stomatal conductance playing a pivotal role in regulating pollutant flux (Oksanen & Kontunen-Soppela, 2021). Environmental factors—including light intensity, vapor pressure deficit, temperature, and soil moisture—significantly influence stomatal behavior and, consequently, the uptake of pollutants (Fig. 4.1).

The phytotoxic effects of air pollutants are influenced by the pattern of ambient exposure, the extent to which pollutants diffuse into leaves and enter the cellular liquid phase, and their reactivity with cellular components (Hasan et al., 2021). Leaves are the main route of uptake for gaseous pollutants and the primary site of particulate matter deposition (Rao et al., 2025). Gaseous uptake of pollutants is controlled by stomatal aperture and gas-diffusion conductance. Particulate matter deposition blocks stomata,

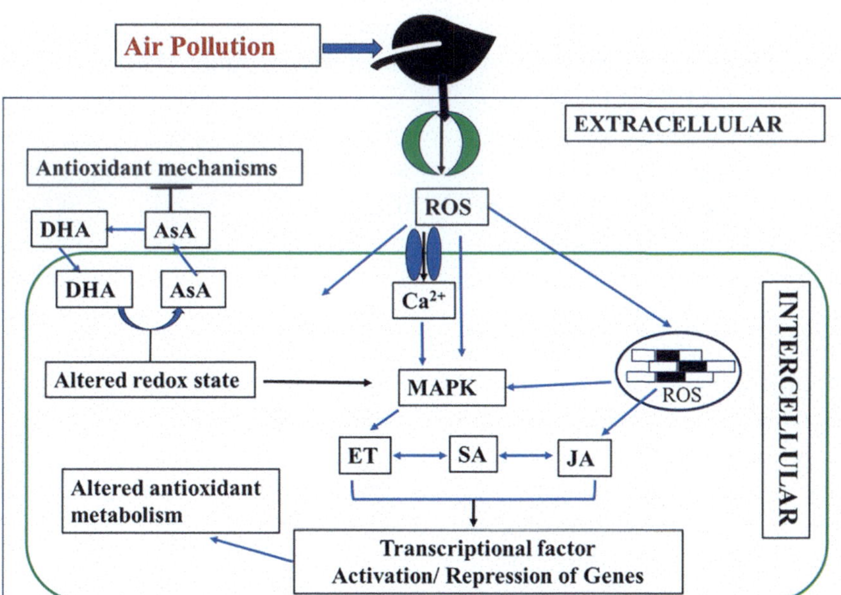

Figure 4.1 A diagrammatic representation of air pollutants' uptake, ROS formation, their signaling, and antioxidant mechanisms. *ROS*, reactive oxygen species.

affecting stomatal conductance (Bhardwaj et al., 2026). Leaf conductance is determined by boundary-layer conductance and stomatal conductance. Various modeling results indicate that a major factor affecting plant response to air pollutants is stomatal conductance (Pleijel et al., 2004).

Factors that influence stomatal conductance also affect plant responses to air pollutants, including leaf temperature, the water vapor pressure difference between the leaf and the surrounding air (VPD), photosynthetic photon flux density (PPFD), soil water availability, and atmospheric CO_2 concentration (Emberson et al., 2009). O_3 flux varies within the plant canopy due to variations in environmental factors such as PPFD, leaf temperature, and VPD within the leaf canopy (Fuhrer, 2009). The absorption of air pollutants is driven by a chemical potential gradient between the atmosphere and the site of deposition, whether on the foliar surface or in the leaf interior (Taylor et al., 1982). In addition, the boundary layer resistance and stomatal resistance are also important factors that influence air pollutant uptake (Monteith & Unsworth, 1990). Boundary layer resistance is primarily influenced by leaf morphological traits, including size, shape, epidermal features, and trichome density, as well as leaf orientation and ambient wind speed. It arises from the thin, relatively stagnant layer of air that surrounds the leaf, which acts as a barrier to the diffusion of pollutants such as ozone (O_3), sulfur dioxide (SO_2), nitrogen oxides (NO_x), and particulate matter.

It has been noted that at low concentrations, SO_2 stimulates stomatal opening and at higher concentrations it may close them (Darrall, 1989). Black and Unsworth (1980) reported that low concentrations of SO_2 in *Vicia faba* L. stimulated stomatal conductance within 15 min of exposure, which persisted for several days, which may be attributed to the destruction of epidermal cells adjacent to stomata and the accumulation of sulfur within guard cells. Larger stomatal apertures led to greater pollutant uptake but also enhanced water loss through unrestricted transpiration. SO_2 is extremely soluble in aqueous medium, which suggests that SO_2 dissolves mainly on the lower inner surfaces of the guard cells, where most of the H_2O is present (Heath, 1988). Guard cells are more vulnerable to SO_2 injury as these are deprived of the ability to convert SO_2 to H_2S using energy from photosynthetic electron transport.

Nitrogen oxides (NO and NO_2) increase internal resistance to CO_2 uptake, and the stomata close at high NO_2 concentration. The entry of NO_2 into leaves is similar to that of SO_2; however, entry through the cuticle is higher due to lower cuticular resistance against NO_2 as compared to SO_2 or O_3 (Kerstiens et al., 1992). At higher O_3 concentrations, stomatal closure was reported in the majority of species (Darrall, 1989). The direct effect of O_3 on stomatal function is correlated with its action on cell permeability (Heath, 1980). Ozone increased membrane permeability to K+ ions (Dominy and Heath, 1985) mediated through a membrane-bound K^+ ATPase activity. Low O_3 concentrations affected the membrane permeability of subsidiary cells, resulting in water flow into the guard cells, which became more turgid and led to wider opening of stomata (Robinson et al., 1998). It can be concluded that control of stomatal aperture will regulate air pollutant flux into leaves, and it can be considered as a sort of "first line of defence." The stomatal pore size is regulated by the activity of guard cell ion channels and transporters that control the transport of osmotically active solutes (Hedrich, 2012). Activation of different types of guard ion channels, which are required for stomatal closure, is also essential for the restriction of the influx of O_3 to leaves. Reversible protein phosphorylation, mediated by protein kinases and phosphatases, plays a crucial role in controlling the channels' activity. An increase in cytosolic calcium is one of the earliest responses to O3, observed within seconds of exposure onset (Evans et al., 2005). Calcium is an important second messenger in plants and a key component of divergent signaling cascades regulating plant stress and developmental responses (Batistic & Kudla, 2012) and also stomatal aperture (Huang et al., 2019). The second important family of ion channels responsible for the regulation of the stomatal aperture is the guard cell anion channels, rapidly deactivating (R-type) and slowly deactivating (S-type), which are activated in response to factors causing stomatal closure (Kollist et al., 2019). The S-type SLOW ANION CHANNEL 1 (SLAC1) identified in Arabidopsis mutant screens for O_3 sensitivity (Vahisalu et al., 2008) is essential for stomatal closure induced by several different stimuli. SLAC1 is essential for the process of stomatal closure in response to not only ozone (O_3) but also a range of factors, including abscisic acid (ABA) and calcium ions (Rao et al., 2025).

SLAC1 is activated by the protein kinase OPEN STOMATA 1, which phosphorylates SLAC1 at multiple residues in the N-terminal region (Vahisalu et al., 2010). A receptor-like protein kinase known as GUARD CELL HYDROGEN PEROXIDE-RESISTANT 1 (GHR1), along with several calcium-dependent protein kinases (CPKs), including CPK3, CPK6, CPK21, and CPK23, has also been reported to phosphorylate the N-terminal region of SLAC-1. This highlights the diverse signaling pathways involved in the activation of SLAC1 (Scherzer et al., 2012).

The OST1 activity, as well as the activity of CPKs and GHR1in SLAC1 phosphorylation, is controlled by the PP2C phosphatases ABA INSENSITIVE (ABI)1 and (ABI)2, the key elements in ABA signaling (Hua et al., 2012). In the presence of ABA, the cytosolic ABA receptors PYRABACTIN RESISTANCE 1 (PYR1)/PYR1-like (PYL)/ REGULATORY COMPONENTS OF ABA RECEPTORS (RCAR) form a ternary complex between ABA, PYR/PYL/RCAR and PP2C, thus inhibiting PP2C activity (Park et al., 2009). This deactivation enables OST1 and other kinases to phosphorylate SLAC1, leading to the activation of anion efflux and stomatal closure. The activity of ABI1 and ABI2 was also inhibited by reactive oxygen species (ROS) in vitro (Meinhard et al., 2002), suggesting an ABA-independent pathway for OST1 activation. Studies conducted by Vahisalu et al. (2010) and Merilo et al. (2013) using Arabidopsis mutants confirmed that stomatal closure was caused mainly by signaling through PYR/PYL/RCAR under O3 exposure. OST1 and SLAC1 are pivotal components involved in mediating stomatal closure in response to ozone (O3). OST1 regulates the activity of multiple ion channels within guard cells and occupies a central position in ABA-dependent stomatal responses. ROS are significantly engaged in the regulation of stomatal closure, influencing various signaling pathways either directly or indirectly.

4.3 Reactive oxygen species formation

Air pollutants such as tropospheric ozone, SO_2, and NO_2 swiftly penetrate the leaf through stomata and decompose within the apoplast, producing ROS, including superoxide radicals, hydrogen peroxide, and hydroxyl radicals. The study also showed distinguishable differences in the regulation of ROS-related genes under NO_2 and O_3 stress. While O_3 exposure increased respiratory burst oxidase homolog (RBOH) expression, a key gene involved in ROS production, NO_2 treatment suppressed it. Despite this difference, substantial evidence confirms that NO_2 causes oxidative stress.

Reactions of formation of ROS are given in the equation below.

(i) ROS formation in plant cells due to O3

$O_3 + OH^{\boxdot -} O_2^{\boxdot}$ (superoxide radical)

${}^{\boxdot}O_2^{-} + H^{+}\ HO_2$ (peroxy radical)

$\boxdot HO_2 + {}^{\square}HO_2\ H_2O_2$ (hydrogen peroxide) $+ O_2$

${}^{\square}O_2^{-} + {}^{\square}HO_2 + H_2O\ H_2O_2 + OH^{-} + O_2$

(ii) Production of ROS due to SO2 absorption within the cellular space (Bartosz, 1997)

$HSO_3^{-} + {}^{\square}O_2^{-} + 2H^{+}\ HSO_3^{\square}$ (bisulfate radical)$+ 2^{\square}OH$

$HSO_3^{-} + {}^{\square}OH + H^{+}\ HSO_3^{-} + H_2O$

$HSO_3^{\square} + {}^{\square}O_2\ SO_3^{2-} + O_2^{\square -} + H^{+}$

$HSO_3^{\square} + {}^{\square}OH\ SO_3^{2-} + H_2O$

$2HSO_3^{\square}\ SO_3^{2-} + HSO_3^{-} + H^{+}$

$SO_3 + H_2O\ SO_4^{2-} + 2H^{+}$

${}^{\square}O_2^{-} + {}^{\square}O_2^{-} + 2H^{+}\ O_2 + H_2O_2$

${}^{\square}OH + {}^{\square}OH\ H_2O_2$

(iii) Production of ROS due to NO_2 absorption within the cellular space

$NO_2 + H_2O\ HNO_3$

$NO + H_2O\ HNO_2$

$HNO_3 + HNO_2\ NO_3^{2-} + NO_2^{2-} + H^{+}$

NO_2 is also capable of initiating peroxidation processes within lipid membranes, acting like a free radical (Ramge et al., 1993).

4.3.1 Reactive oxygen species sensing by plants

ROS serve as early signaling molecules, triggering downstream biochemical and molecular responses. Plants induces ROS during various metabolic process like light reactions in chlorophyll or during electron transport in mitochondria, however, under stress their excess production may lead either oxidative damage to cellular components or act as a signaling molecule, which initiates a cascade of defense response that includes mitogen-activated protein kinase (MAPK) pathways, calcium signaling and hormonal network such as ABA, jasmonic acid, (JA) and ethylene (ET) (Niekerk et al., 2024; Angon et al., 2024).

(i) MAPK signaling pathway: In plants, MAPK signaling cascades are activated in response to air pollution (Colcombet & Hirt, 2008; Pitzschke et al., 2009; Rodriguez et al., 2010). Apoplastic ROS, such as hydrogen peroxide, can be transported into the mesophyll cells through aquaporins (Dynowski et al., 2008), and the apoplastic ROS signal is rapidly relayed to the chloroplast (Vainonen & Kangasjärvi, 2015). Transport of ROS and oxidized ascorbate from the apoplast to the intercellular space alters the redox balance of the cell and activates mitogen-activated protein kinase signaling cascades (Ahlfors et al., 2004). This, in turn, enhances the activity of hormone biosynthesis genes, leading to changes in hormone levels such as ethylene, jasmonic acid, salicylic acid, and ABA after acute O3 exposure. (Ainsworth, 2017) (Fig. 4.1).

A MAPK signaling module comprises three protein kinases that sequentially phosphorylate and activate one another, with a MAPK kinase kinase (MEKK) activating a MAPK kinase (MKK), ultimately resulting in the phosphorylation and activation of a MAPK (MPK). MPKs phosphorylate a diverse array of substrates, including metabolic enzymes and transcription factors; therefore, the activity of MPKs must be meticulously regulated to ensure a precise response (Khan et al., 2024). Specific MAPK signaling modules MEKK-MKK-MPK (or MKK-MPK) have been assigned to specific stresses (Colcombet and Hirt 2008; Pitzschke et al., 2009; Rodriguez et al., 2010); however, understanding of the regulatory mechanisms and interactions involved is far from being complete. Ahlfors et al. (2004) showed activation of MAPKs, MPK3, and MPK6 was rapid and transient (0.5–2 h) after exposure to ozone in Arabidopsis genotypes. The study also showed that activation of MPK3/6 led to the translocation of the proteins to the nucleus, where they further interacted with transcription factors and regulated gene expression. Miles et al. (2005) highlighted an inverse regulatory mechanism between MPK3 and MPK6, suggesting that suppression of MPK6 results in increased and prolonged activation of MPK3, while a decrease in MPK3 causes greater activation of MPK6 in air pollution-sensitive Arabidopsis genotypes (Overmyer et al., 2005). The above-mentioned studies clearly highlighted a fine balance between MPK3/MPK6 role for appropriate ozone response because any changes in its strength or activation time led to enhanced ozone sensitivity. Overmyer et al. (2005) exposed wild and mutant Arabidopsis (*rcd1*) to 300 ppb O3 and found that activation of MPK3/MPK6 was prolonged in ozone-sensitive Arabidopsis, inducing cell death and ethylene biosynthesis in comparison to the wild-type plants.

4.4 Phytohormones (jasmonic acid, salicylic acid, and ethylene)

Plant adaptation to air pollution involves a complex regulatory network of phytohormone signaling pathways. The main hormones are ABA, JA, salicylic acid (SA), ET, cytokinin, brassinosteroids (BRs), and auxin, all of which together influence how plants respond to air pollution (Antenozio et al., 2024; Rao et al., 2025). Phytohormone signaling is profoundly influenced by air pollution stress, which significantly alters plant responses to stress and defense mechanisms via complex biochemical and molecular pathways.

4.4.1 Jasmonic acid signaling modulation by air pollutants

Under air pollution stress, JA biosynthesis and signaling pathways are differentially modulated by the type of pollutant, its concentration, and duration of exposure. Studies have shown that acute ozone exposure can stimulate JA accumulation and activate JA-responsive genes, enhancing the plant's defensive capacity (Ruan et al., 2019). Conversely, prolonged exposure to pollutants may interfere with JA biosynthesis enzymes or signaling components, leading to suppressed JA responses. This suppression may reduce the plant's ability to initiate an effective defense response against herbivores and pathogens. Furthermore, exposure to air pollution may affect the crosstalk between JA and other hormones, affecting the overall defense strategy of plants.

4.4.2 Salicylic acid signaling under air pollution stress

Air pollutants like ozone (O_3), nitrogen oxides (NO_x), sulfur dioxide (SO_2), and particulate matter cause oxidative stress within plant cells by producing ROS, which may further triggers an increase in biosynthesis of SA. SA plays an important role in regulating redox balance, activating defense-related genes, and modulating oxidative stress signaling during ozone exposure.

Elevated SA levels activate systemic acquired resistance, a critical defense mechanism that enhances the expression of pathogenesis-related (PR) genes and protects the plant against biotic stressors, such as pathogens. SA signaling also modulates antioxidant enzyme activities, helping to mitigate oxidative damage caused by air pollution (Li et al., 2006; Morales et al., 2021). However, exposure to higher concentrations of air pollutants may disrupt SA homeostasis, leading to impaired signaling and reduced SAR effectiveness. This imbalance may result in increased susceptibility to pathogens or compromised stress tolerance (Modesti et al., 2023).

SA biosynthesis in plants occurs mainly via two metabolic routes: the phenylpropanoid pathway and the isochorismate synthase (ICS) pathway, although their relative contributions differ among plant species. In the phenylpropanoid pathway, phenylalanine ammonia-lyase (PAL) is the key enzyme that plays a main role in SA formation (Huang & van der Hoorn, 2025; Yalpani et al., 1993). Earlier studies have suggested that SA accumulation is also influenced by ethylene signaling under air pollution exposure, and it modulates the expression of PAL and chorismate mutase genes in tobacco (Modesti et al., 2023).

Evidence from isotope-labeling experiments using ^{14}C-labeled benzoic acid as a precursor has highlighted clear species-specific differences in the preferred SA biosynthetic pathway (Huang & van der Hoorn, 2025; Yalpani et al., 1993). In tobacco, SA production is largely associated with the phenylalanine-derived benzoic acid pathway, and limited activation of the ICS pathway under air pollution exposure (Huang & van der Hoorn, 2025; Yalpani et al., 1993). In contrast, *Arabidopsis* mainly relies on the ICS pathway for SA synthesis, and it was confirmed using salicylic acid induction-deficient 2 (*sid2*) mutant of *Arabidopsis*, which lacks functional *ICS1* exposed to air pollution. Result showed lower SA accumulation following air pollution exposure (Morales et al., 2021; Ogawa et al., 2005). Taken together, these findings indicate that tobacco mainly synthesizes SA through the phenylalanine/benzoic acid route, whereas the ICS pathway is the primary source of SA in *Arabidopsis*.

Several studies have demonstrated that SA accumulation is closely associated with the extent of ozone-induced leaf damage, showing a dose-dependent relationship with lesion formation (Modesti et al., 2023). For instance, the tobacco cultivar "Xanthi" expressing the bacterial *NahG* gene, which degrades SA, accumulates substantially lower SA levels and consequently develops fewer lesions under ozone stress. Interestingly, both SA-deficient plants and those that accumulate excessively high levels of SA show increased sensitivity to ozone, suggesting that maintaining SA within an optimal range is important for effective defense responses (Rao and Davis,1999). A similar trend has been reported in *Arabidopsis*, in the ozone-tolerant Col-0 ecotype transformed with *NahG*, which becomes more susceptible to ozone than the wild type, indicating the importance of endogenous SA in stress tolerance. Conversely, the ozone-sensitive Cvi-0 ecotype accumulates approximately three times more SA than Col-0, suggesting that excessive SA may also contribute to increased sensitivity (Morales et al., 2021). These observations imply that ozone tolerance is associated with balanced SA homeostasis rather than simply elevated SA levels, possibly through its influence on PR gene expression and cellular redox regulation. Supporting this idea, PR1 (AT2G14610) expression was strongly induced in Cvi-0 plants following ozone exposure, with transcript levels exceeding those observed in Col-0 even under control conditions (Modesti et al., 2023).

While SA plays an important role in regulating plant defense responses under ozone stress, increasing evidence suggests that ethylene also acts as a key signaling molecule in mediating ozone-induced physiological and molecular responses. The interaction between SA and ethylene signaling pathways is particularly important in determining the extent of oxidative damage and the activation of defense mechanisms during ozone exposure.

4.4.3 Ethylene signaling

In the ethylene biosynthetic pathway, two major enzymes, 1-aminocyclopropane-1-carboxylic acid synthase (ACS) and ACC oxidase (ACO), regulate ethylene formation (Khan et al., 2024). Ethylene synthesis begins when S-adenosyl-L-methionine is converted into 1-aminocyclopropane-1-carboxylic acid (ACC) by ACS. Then, ACO catalyzes the oxidation of ACC to produce ethylene, marking the final step of the pathway (Houben and Poel2019). The genes encoding *ACS* and *ACO* belong to highly conserved gene families present across diverse plant species, where they perform similar biochemical roles in ethylene production (Khan et al., 2024).

In *Arabidopsis*, nine *ACS* genes have been identified; however, only one appears to function as the primary regulator of ethylene accumulation during ozone exposure. A consistent association has also been observed between ozone-induced ethylene production and the severity of leaf injury in different plant species (Li et al., 2006). For example, under ozone stress, the ozone-tolerant tobacco cultivar "Bel-B" produced considerably lower ethylene levels than the ozone-sensitive cultivar "Bel-W3." Additionally, using ethylene biosynthesis inhibitors during ozone treatment greatly reduced leaf damage, emphasizing ethylene's role in ozone-related injury (Bandurska et al., 2009). The increased leaf damage associated with higher ethylene levels may be caused by the formation of reactive free radicals and toxic aldehydes resulting from chemical reactions between ozone and ethylene (Gupta et al., 2018; Altaf et al., 2024). Interestingly, certain plant-associated microorganisms possess ACC deaminase activity, which can mitigate damage by degrading the ethylene precursor ACC, thereby decreasing stress-induced ethylene accumulation and improving plant resilience to environmental pollutants (Kantharaj et al., 2024). At the molecular level, ethylene functions as a signaling molecule by binding to specific receptors, and existing evidence indicates that ozone exposure activates ethylene-dependent pathways that ultimately lead to leaf injury symptoms.

The accumulation of ROS acts as an initial indicator that triggers various hormonal pathways, especially SA and ethylene signaling networks (Rao et al., 2025). These signaling pathways do not operate alone; instead, they interact to regulate plant defense responses to air pollutants. SA is recognized for inducing the expression of PR genes and enhancing antioxidant defense mechanisms, thus helping to regulate redox homeostasis. In contrast, ethylene often acts in concert with ROS to regulate stress signaling and programmed cell death associated with ozone-induced lesions. Several studies indicate that ROS-triggered signaling can boost both SA levels and ethylene production, and these hormones then influence ROS generation through feedback loops involving antioxidant enzymes and defense gene activation (Margarete et al., 2005; Vainonen & Kangasjärvi, 2015). This SA–ethylene–ROS signaling network ultimately decides whether plants trigger protective mechanisms or suffer oxidative damage and cell death. Hence, the balance and interaction among these signaling elements are crucial in determining plant tolerance or sensitivity to air pollution stress. Leppälä et al. (2022) conducted a study involving approximately 372 *Arabidopsis thaliana* accessions exposed to NO_2 and ozone (O_3), demonstrating significant genotypic variation in tolerance, as revealed by genome-wide association analysis. Transcriptomic and microarray data revealed that both pollutants trigger comparable molecular responses, particularly through the activation of hormone-mediated signaling pathways. Notably, genes associated with JA, ET, and SA signaling—such as *CEJ1*, *GRX480*, and *FMO1*—were markedly upregulated under pollutant exposure, highlighting the importance of transcriptional regulation in plant stress adaptation.

Further insights from transcriptomic profiling of *Ambrosia artemisiifolia* pollen exposed to prolonged NO_2 and O_3 fumigation indicated enrichment of gene ontology categories associated with both abiotic and biotic stress responses, including JA biosynthesis and phosphate regulation. Additionally, pathways related to ethylene perception, ABA, and auxin signaling were notably activated, emphasizing the central role of phytohormones—especially JA—in mediating responses to atmospheric conditions and pollutants (Zhao et al., 2017).

4.4.4 Abscisic acid signaling responses to air pollution

Air pollution stress often triggers increased ABA production as a defense response to check water loss and restrict pollutant uptake through stomatal control (Rao et al., 2025). Elevated levels of ABA enhance the antioxidative defense mechanism, mitigating oxidative damage caused by pollutants. However, higher accumulation of ABA may adversely affect plant growth and development, resulting in inhibited cell division and elongation, delayed flowering, and impaired photosynthesis. Furthermore, alterations in ABA signaling induced by pollution could disrupt hormonal equilibrium, thereby impacting overall plant response to air pollution (Hsu et al., 2021).

4.5 Reactive oxygen species detoxification by nonenzymatic antioxidants and antioxidative enzymes

A critical outcome of the signaling transduction cascade is the activation of the plant's antioxidant defense system, which includes both enzymatic components (superoxide dismutase, catalase, and peroxidases) and nonenzymatic components (ascorbate, glutathione, and tocopherols) (Rao et al., 2025) (Fig. 4.1). Plants upregulate their antioxidative machinery by inducing gene expression and posttranslational modifications of key antioxidative enzymes, which help them in developing adaptive responses to reduce oxidative damage at cellular levels (such as proteins, lipids, membranes, and nucleic acids) caused by air pollution (Rathod et al., 2024). Plants have an antioxidant defense system (nonenzymatic and enzymatic antioxidants) in cellular organelles that scavenges various ROS generated by air pollution stress (Table 4.2). Enzymatic antioxidative defense system consists of ascorbate peroxidase (APX), monodehydroascorbate reductase (MDHAR), dehydroascorbate reductase (DHAR), glutathione reductase (GR), superoxide dismutase (SOD), catalase (CAT), glutathione peroxidase (GPX), glutathione S-transferase, and peroxiredoxin, while nonenzymatic antioxidant comprises of ascorbate, glutathione (GSH), carotenoids, tocopherols, flavonoids, etc. (Molina & Segura, 2021). Among all of these, ascorbic acid (AsA), GSH, APX, MDHAR, DHAR, and GR comprise the AsA-GSH cycle or Halliwell-Asada pathway. Ascorbate is one of the most powerful substrates for scavenging H_2O_2. Ascorbate maintains the reduced state of α-tocopherol. These defenses are not restricted to the intracellular compartment, but are also found in the apoplast to a limited extent.

The level of ROS produced after exposure to air pollution needs to be carefully regulated to facilitate a positive cellular response through the signaling cascade without causing uncontrolled cell death (Pastori & Foyer, 2002; Dietz, 2007). ROS are physiologically generated by various

Table 4.2 Function and localization of different enzymatic and nonenzymatic antioxidants.

Antioxidants	Type	Function	Location
Superoxide dismutase (SOD)	Enzymatic	$^{\boxdot}O_2^{-}$ to H_2O_2	Chloroplast
Catalase (CAT)	Enzymatic	H_2O_2=H_2O + ½ O_2	Peroxisome
Ascorbate peroxidase (APX)	Enzymatic	H_2O_2 + AA=H_2O + DHA	Chloroplast
Monohydroascorbate reductase (MDHAR)	Enzymatic	MDHA + NAD(P) H=AA + NAD(P)+	Chloroplast
Dehydroascorbate reductase (DHAR)	Enzymatic	DHA + 2GSH=AA + GSSH	Chloroplast
Glutathione reductase (GR)	Enzymatic	GSSH + NAD(P) H=2GSH + NAD(P)+	Chloroplast
Ascorbate	Nonenzymatic	Reactive oxygen species scavenging	Cytosol
Glutathione	Nonenzymatic	Redox balance	Mitochondria

processes, such as cellular metabolism, and plants have developed highly efficient enzymatic and nonenzymatic antioxidant systems to detoxify them. The antioxidant defense systems are important in determining cell fate, not only by maintaining ROS levels within a controlled range but also by serving as a central component of cellular redox balance and signal modulation.

4.5.1 Antioxidative enzymes

Enzymatic antioxidant defense mainly detoxifies the H_2O_2 in a plant cell. Superoxide dismutase catalyzes the disproportionation of superoxide ($^{\boxdot}O_2^{-}$) into molecular oxygen and hydrogen peroxide (H_2O_2). Additionally, SOD mitigates the risk of hydroxyl radical formation by scavenging $^{\boxdot}O_2^{-}$ through the metal-catalyzed Haber-Weiss-type reaction (Jomova et al., 2024). There are three distinct isoforms of SOD—namely, Mn-SOD, Cu/Zn-SOD, and Fe-SOD—each localized within different cellular organelles. Mn-SOD is predominantly present in mitochondria and peroxisomes, while Cu/Zn-SOD is located in the cytosol and chloroplasts, and Fe-SOD is primarily found within chloroplasts.

Apart from AsA and GSH, its enzymes include APX, MDHAR, DHAR, and GR (Hasanuzzaman et al., 2019), having significant roles in detoxification of H_2O_2. Both AsA and GSH are present in the cytosol, nucleus, chloroplast, mitochondria, and peroxisomes, performing their roles with the help of the above-mentioned four enzymes. Each enzyme has multiple isoforms, which are localized in various cellular compartments (Foyer & Kunert, 2024). Both AsA and GSH are present in cell organelles at millimolar levels, with their concentrations varying between organelles. In *Arabidopsis thaliana*, for example, AsA reaches its peak concentration of 22.8 mM in the peroxisome, while GSH is more abundant in the mitochondria, with a concentration of 14.9 mM (Hasanuzzaman et al., 2017; Zechmann, 2014).

APX scavenges H_2O_2 by utilizing reduced ascorbate, and for the regeneration of reduced ascorbate, two other enzymes are involved, namely mono-dehydroascorbate reductase and dehydroascorbate reductase. Mono-dehydroascorbate is reduced to ascorbate by electrons from the light reaction in photosystem II and photosystem I, which clearly suggests regeneration of ascorbate within the chloroplast is regulated by electron transport. Glutathione plays a role in the regeneration of ascorbate from dehydroascorbate, with the enzyme involved being DHAR. Over time, GSH is oxidized to glutathione disulfide, and GSH is subsequently regenerated by glutathione reductase in a reaction that depends on NADPH. Kumar et al. (2022) evaluated the response of roadside plants toward urban air pollutants. Results showed that higher catalase and APX activities were found in tree species *Grevillea robusta*, *Cassia fistula*, and *Quercus leucotrichophora*. Karmakar and Padhy (2019) also reported similar findings in *Shorea robusta* Gaertn. f., deciduous trees due to PM-associated heavy metal exposure. Lyu et al. (2023) reported that particulate pollution significantly increased SOD and peroxidase (POD) expression in three evergreen climbers, named *Hedera nepalensis* K. Koch., *Ficus pumila* L., and *Euonymus fortunei* (Turcz.) Hand. Mazz. The study found increased SOD expression in winter, attributed to higher particulate pollution.

Air pollution exposure increases transcript levels and activity of several crucial antioxidative enzymes in the Halliwell and Asada pathway. Li & Yi (2012) analyzed the antioxidant response of *Arabidopsis* sps to SO_2 stress and showed that the transcript levels of POD, GPX, and SOD genes were enhanced after exposure to 30 mg m^{-3} SO_2 for 72 h of SO_2 exposure using Genechip and real time PCR (RT-PCR) analysis. In-gel enzymatic assays showed that SOD (Fe-SOD and Cu/Zn SOD) and POD isoforms increased after 72 h of SO_2 exposure, while two CAT isoforms (CAT2 and CAT3) decreased. Elevated O_3 levels boost the expression of GR genes, helping to replenish the GSH pool needed for regenerating ascorbate (Castagna and Ranieri, 2009).

Studies have shown significant variation in tolerance to air pollution mediated by antioxidative enzymes. Similarly, *Bougainvillea spectabilis* seedlings exposed to acute NO_2 exposure showed visible leaf damage and increased activities of antioxidant enzymes, including POD, SOD, and CAT, indicating the activation of defense mechanisms (Sheng et al., 2022). In addition to gaseous pollutants, particulate matter (PM) also significantly affects the gene expression of plants. Plants develop various adaptive mechanisms at both transcriptional and posttranscriptional stages to withstand particulate pollution stress. A study conducted on *Photinia* × *fraseri* revealed that numerous redox-related genes involved in encoding cytochrome b5, protein disulfide isomerase, ascorbate peroxidases, and superoxide dismutases were downregulated under pollution conditions (Vergata et al., 2023). It further showed that genes associated with primary metabolism, such as mitochondrial electron transport and ATP synthesis, may initially be upregulated under low pollution levels, but they tend to decline under severe exposure.

4.5.2 Nonenzymatic antioxidants and secondary metabolites to scavenge reactive oxygen species

4.5.2.1 Nonenzymatic antioxidants

4.5.2.1.1 Ascorbate and glutathione

The detoxification of ROS generated by air pollutants within the apoplast can therefore be regarded as the first line of defense against them (Castagna & Ranieri, 2009). Ascorbate is the main redox buffer and ROS scavenger within the apoplast (Foyer & Noctor, 2009). Accumulation of extracellular ROS stimulates the supply of ascorbate to the apoplast. The active transportation of ascorbate across the plasma membrane indicates the presence of specific ascorbate transporters. The antioxidative capacity of apoplastic ascorbate (ASC) depends on the rate of transport across the plasma membrane of ASC into the cell and of the oxidized form (dehydroascorbate, DHA) into the symplast for re-reducing it back. The antioxidant role of ascorbate is mainly dependent on the ability of the cell to maintain it in a reduced state, and it occurs at the cost of reduced GSH by monodehydroascorbate or DHAR. Cytosolic DHAR is an essential enzyme for recycling ascorbate in the apoplast. *Arabidopsis dhar* mutants, which are sensitive to ozone (O_3), have reduced ascorbate levels in the apoplast (Yoshida et al., 2006). Glutathione's antioxidative properties depend on the reduced form, which is regenerated by GR at the expense of nicotinamide adenine dinucleotide phosphate (NADPH) oxidation in the Halliwell-Asada pathway, which operates in chloroplasts and cytosol (Noctor and Foyer, 1998).

4.5.3 Phenylpropanoid pathways

Air pollution significantly affects plant metabolism, often leading to the accumulation of secondary metabolites that enhance stress tolerance. Phenolics, flavonoids, phenylpropanoids, and terpenes help in protection by scavenging ROS during air pollution exposure. Air pollution has been generally found to stimulate the phenylpropanoid pathway (Cabané et al., 2004) (Fig. 4.2). Many phenylpropanoid compounds play a role in plant defense against different environmental stress conditions thanks to their barrier and antioxidant effects (Booker & Miller, 1998; Cabané et al., 2004; Dixon & Paiva,1995). Phenylpropanoids are highly diverse secondary metabolites derived from the shikimate pathway originating from the glycolysis and the pentose phosphate pathway routes (Kumar et al., 2024). The phenylpropanoid pathway further branches into two pathways of producing lignin and flavonoid-related metabolites. The first enzyme of this pathway is 3-deoxy-d arabino-heptulosonate 7-phosphate synthase (DAHPS), which catalyzes the condensation of phosphoenol-pyruvate (PEP) and erythrose-4-phosphate, yielding DAHP and inorganic

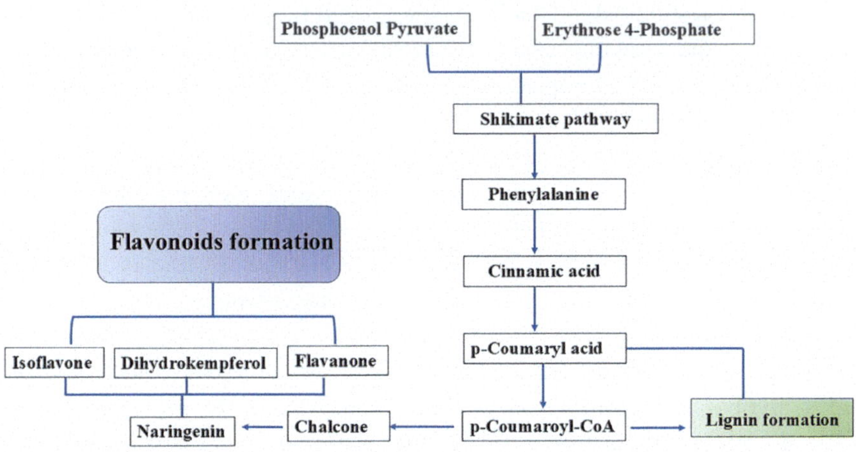

Figure 4.2 A diagrammatic representation of the phenylpropanoid pathway for the formation of flavonoids and lignin.

phosphate (Pi). PEP is localized at the intersection of four metabolic routes: mitochondrial respiration, isoprene biosynthesis (MEP pathway), shikimate pathway (e.g., aromatic amino acids, flavonoids), and anaplerotic pathway (Dizengremel et al., 2012). The hydroxycinnamic acids and lignin glycosides act as building blocks and precursors for lignin biosynthesis, affecting the structure of the cell wall as well as the growth of the whole plant (Terashima et al., 2016). The changes in the lignin biosynthesis are a result of balancing growth and development, as well as the defense responses to air pollutants (König et al., 2014). Flavonoids are one of the major secondary metabolite groups in plants acting as a defense against air pollution (Falcone Ferreyra et al., 2012) (Fig. 4.2).

Exposure to ozone (O_3) has been shown to activate enzymes involved in phenolic metabolism. Study conducted on *Melissa officinalis* revealed increased activity of cinnamyl alcohol dehydrogenase (CAD), which helps plants to survive oxidative stress (Fatima et al., 2018; Tonelli et al., 2015) showed genetic differences in wheat varieties exposed to elevated O_3 and tolerant cultivars such as Kharchiya 65 showed higher activity of enzymes of phenylpropanoid pathways like phenylalanine lyase (PAL), 4-Coumarate: coenzyme A ligase (4CL), CAD, and Chalcone isomerase, and accumulate more α-tocopherol, while sensitive cultivars rely more on flavonoid production to reduce ROS damage. Likewise, increased PAL activity and accumulation of phenolic compounds, such as rutin and caffeic acid, have been reported in *Liriodendron tulipifera* during extended O_3 exposure. Similarly, an increase in the activity of shikimate dehydrogenase (SKDH), phenylalanine ammonia lyase (PAL), and cinnamyl alcohol dehydrogenase (CAD) was observed in both Eridano and I-214 poplar clones following 15 days of O_3 exposure (60 ppb, 5 h d 1), a higher PAL activity being associated with a greater tolerance to O_3 (Di Baccio et al., 2008). Increased O_3 levels often lead to higher phenolic content and the accumulation of specific metabolites, such as isofraxidin and trans-chalcone, as seen in *Fraxinus excelsior*. In addition, combinations of environmental stresses, such as high light and O_3 exposure, can promote the accumulation of polyphenols in leaf tissues, thereby enhancing stress tolerance.

Air pollution exposure also triggers lignin biosynthesis, especially in aboveground plant tissues, forming a structural barrier against oxidative stress, and it involves the activation of key enzymes, PAL of the phenylpropanoid pathway, indicating stress-induced defense responses. Increased lignin levels and changes in lignin composition have been widely observed in plants exposed to air pollutants. However, high pollutant levels may inhibit certain metabolic pathways, as seen in *Laurus nobilis*, where phenylpropanoid levels decreased. Conversely, in *Psidium guajava*, flavonoid content did not vary significantly under air pollution exposure, highlighting species-specific responses and the need for further research (Vergata et al., 2023)

Terpenoid metabolism is another critical component of plant defense under air pollution stress (Lahiri & Krishna, 2024). In *Ginkgo biloba*, enhanced terpene synthesis has been associated with increased antioxidant capacity (Matsuura et al., 2026). Compounds such as carotenoids, zeaxanthin, and tocopherols play significant roles in photoprotection and ROS scavenging, while isoprene helps maintain membrane stability, particularly in chloroplasts (Chatterjee et al., 2024).

4.6 Conclusion

Air pollution significantly impacts plant systems by disrupting cellular balance and initiating complex biochemical and molecular responses. A key part of this response is the production of reactive oxygen species, which serve both as signaling molecules and as sources of oxidative damage. Plants defend against this stress through a coordinated system that includes antioxidant defenses, hormonal regulation, and metabolic reprogramming. Overall, plant responses to air pollutants involve coordinated regulation of antioxidant systems, including increased expression of genes encoding enzymes such as APX, MDHAR, DHAR, and GR, along with enhanced synthesis of nonenzymatic antioxidants like ascorbate and glutathione. These integrated defense mechanisms play a crucial role in mitigating oxidative damage and maintaining cellular homeostasis under pollution stress. The study highlights that plant tolerance or sensitivity depends on the interaction among ROS signaling pathways, MAPK cascades, phytohormone networks, the defense mechanism to combat ROS, and the induction of secondary metabolites along with activation of stress-responsive genes. The chapter also highlights significant variation in tolerance among species and genotypes, which needs to be identified for plantations across different urban greening areas. In-depth studies are necessary to translate mechanistic insights into species selection strategies for urban green infrastructure. Although there has been significant progress, the current understanding remains incomplete and is primarily confined to controlled experimental environments. Bridging the gap between laboratory findings and real-world applications requires comprehensive, multiscale research strategies to identify air pollution-tolerant species.

References

Ahlfors, R., Macioszek, V., Rudd, J., Brosché, M., Schlichting, R., Scheel, D., & Kangasjärvi, J. (2004). Stress hormone-independent activation and nuclear translocation of mitogen-activated protein kinases in *Arabidopsis thaliana* during ozone exposure. *The Plant Journal, 40*, 512–522.

Ainsworth, E. A. (2017). Understanding and improving global crop response to ozone pollution. *Plant Journal, 90*(5), 886–897. https://doi.org/10.1111/tpj.13298.

Altaf, F., Parveen, S., Lone, M. L., Haq, A. U., Farooq, S., Tahir, I., Mansoor, S., & Alyemeni, M. N. (2024). Modulation of Leaf Senescence in Bergenia ciliata (Haw.) Sternb. through the Supplementation of Kinetin and Methyl Jasmonate. *Pak. J. Bot. 56*, 667–677 2024.

Angon, P. B., Islam, M. S., Das, A., Anjum, N., Poudel, A., & Suchi, S. A. (2024). Effects and present perspectives of heavy metals contamination: soil, plants and huma food chain. *Heliyon, 10*, e28357.

Antenozio, M. L., Caissutti, C., Caporusso, F. M., Marzi, D., & Brunetti, P. (2024). Urban air pollution and plant tolerance: Omics responses to ozone, nitrogen oxides, and particulate matter. *Plants, 13*(15), 2027. https://doi.org/10.3390/plants13152027.

Bandurska, H., Borowiak, K., & Miara, M. J. A. B. C. S. B. (2009). Effect of two different ambient ozone concentrations on antioxidative enzymes in leaves of two tobacco cultivars with contrasting ozone sensitivity. *Acta Biol. Cracov. Ser. Bot. 51*, 37–44.

Bartosz, G. (1997). Oxidative stress in plants. *Acta Physiologiae Plantarum, 19*, 47–64.

Batistič, O., & Kudla, J. (2012). Analysis of calcium signaling pathways in plants. *Biochimica et Biophysica Acta, 1820*, 1283–1293.

Bhardwaj, M. K., Anto, S., Singh, K. J., Behera, S. K., & Rai, R. (2026). Physiological and biochemical traits mediate species-specific air pollution tolerance in ornamental plants. *International journal of phytoremediation.* https://doi.org/10.1080/15226514.2026.2635531.

Black, V. J., & Unsworth, M. H. (1980). Stomatal responses to sulphur dioxide and vapor pressure deficit. *Journal of Experimental Botany, 31*, 667–677.

Booker, F. L., & Miller, J. E. (1998). Phenylpropanoid metabolism and phenolic composition of soybean [Glycine max (L.) Merr.] leaves following exposure to ozone. *Journal of Experimental Botany, 49*, 1191–1202.

Cabané, M., Pireaux, J. C., Léger, E., Weber, E., Dizengremel, P., Pollet, B., & Lapierre, C. (2004). Condensed lignins are synthesized in poplar leaves exposed to ozone. *Plant Physiology, 134*(2), 586–594.

Castagna, A., & Ranieri, A. (2009). Detoxification and repair process of ozone injury: from O_3 uptake to gene expression adjustment. *Environmental Pollution, 157*, 1461–1469.

Chatterjee, S., Mandal, M., Kakkar, M., Basak, G., Khan, N. B., Chakraborty, R., Popek, R., Sarkar, A., & Barman, C. (2024). An insight to strategical responses of particulate pollution in plants: From phenome to genome. *Plant Stress, 14*, 100671. https://doi.org/10.1016/j.stress.2024.100671.

Colcombet, J., & Hirt, H. (2008). *Arabidopsis* MAPKs: a complex signalling network involved in multiple biological processes. *The Biochemical Journal, 413*, 217–226.

Darrall, N. M. (1989). The effect of air pollutants on physiological processes in plants. *Plant Cell and Environment, 12*, 1–30.

Di Baccio, D., Castagna, A., Paoletti, E., Sebastiani, L., & Ranieri, A. (2008). Could the differences in O3 sensitivity between two poplar clones be related to a difference in antioxidant defense and secondary metabolic response to O3 influx? *Tree Physiology, 28*, 1761–1772.

Dietz, K.-J. (2007). Plant thiol enzymes and thiol homestasis in relation to thiol-dependent redox regulation and oxidative stress. In N. Smirnoff (Ed.). *Antioxidants and reactive oxygen species in plants,* (pp. 25–52). Wiley. https://doi.org/10.1002/9780470988565.ch2.

Dixon, R. A., & Paiva, N. L. (1995). Stress-induced phenylpropanoid metabolism. *The Plant Cell, 7*, 1085–1097.

Dizengremel, P., Vaultier, M. N., Le Thiec, D., Cabané, M., Bagard, M., Gérant, D., Gérard, J., Dghim, A. A., Richet, N., Afif, D., & Pireaux, J. C. (2012). Phosphoenolpyruvate is at the crossroads of leaf metabolic responses to ozone stress. *New Phytologist, 195*(3), 512–517.

Dominy, P. J., & Heath, R. L. (1985). Inhibition of the K-stimulated ATPase of the plasmalemma of pinto bean leaves by ozone. *Plant Physiology, 77*, 43–45.

Dynowski, M., Schaaf, G., Loque, D., Moran, O., & Ludewig, U. (2008). Plant plasma membrane water channels conduct the signalling molecule H_2O_2. *The Biochemical Journal, 414*, 53–61.

Emberson, L. D., Büker, P., Ashmore, M. R., Mills, G., Jackson, L. S., Agrawal, M., Atikuzzaman, M. D., Cinderby, S., Engardt, M., Jamir, C., Kobayashi, K., Oanh, N. T. K., Quadir, Q. F., & Wahid, A. (2009). A comparison of North American and Asian exposure–response data for ozone effects on crop yields. *Atmospheric Environment, 43*(12), 1945–1953. https://doi.org/10.1016/j.atmosenv.2009.01.005.

Evans, N. H., McAinsh, M. R., Hetherington, A. M., & Knight, M. R. (2005). ROS perception in *Arabidopsis thaliana*: the ozone-induced calcium response. *The Plant Journal, 41*(4), 615–626.

Falcone Ferreyra, M. L., Rius, S., & Casati, P. (2012). Flavonoids: biosynthesis, biological functions, and biotechnological applications. *Frontiers in Plant Science, 3*, 222.

Fatima, A., Singh, A. A., Mukherjee, A., Agrawal, M., & Agrawal, S. B. (2018). Variability in defence mechanism operating in three wheat cultivars having different levels of sensitivity against elevated ozone. *Environmental and Experimental Botany, 155*, 66–78. https://doi.org/10.1016/j.envexpbot.2018.06.015.

Foyer, C. H., & Noctor, G. (2009). Redox regulation in photosynthetic organisms: signaling, acclimation, and practical implications. *Antioxidants & Redox Signaling, 11*(4), 861–905.

Foyer, C. H., & Kunert, K. (2024). The ascorbate-glutathione cycle coming of age. *Journal of Experimental Botany, 75*(9), 2682–2699. https://doi.org/10.1093/jxb/erae023.

Fuhrer, J. (2009). Ozone risk for crops and pastures in present and future climates. *Die Naturwissenschaften, 96*(2), 173–194. https://doi.org/10.1007/s00114-008-0468-7.

Gupta, S. K., Sharma, M., Majumder, B., Maurya, V. K., Lohani, M., Deeba, F., & Pandey, V. (2018). Impact of Ethylene Diurea (EDU) on Growth, Yield and Proteome of Two Winter Wheat Varieties under High Ambient Ozone Phytotoxicity. *Chemosphere, 196*, 161–173.

Hasan, M. M., Rahman, M. A., Skalicky, M., Alabdallah, N. M., Waseem, M., Jahan, M. S., Ahammed, G. J., El-Mogy, M. M., El-Yazied, A. A., Ibrahim, M. F. M., & Fang, X. W. (2021). Ozone induced stomatal regulations, MAPK and phytohormone signaling in plants. *International Journal of Molecular Sciences, 22*, 6304. https://doi.org/10.3390/ijms22126304.

Hasanuzzaman, M., Bhuyan, M. H. M. B., Anee, T. I., Parvin, K., Nahar, K., Mahmud, J. A., & Fujita, M. (2019). Regulation of ascorbate-glutathione pathway in mitigating oxidative damage in plants under abiotic stress. *Antioxidants, 8*(9), 384. https://doi.org/10.3390/antiox8090384.

Hasanuzzaman, M., Nahar, K., Anee, T. I., & Fujita, M. (2017). Glutathione in plants: Biosynthesis and physiological role in environmental stress tolerance. *PMBP, 23*, 249–268.

Heath, R. L. (1980). Initial events in injury to plants by air pollutants. *Annual Review of Plant Physiology, 31*, 395–431.

Heath, R. L. (1988). Biochemical mechanisms of pollutant stress. In W. W. Heck, J. Cliff, O. C. Taylor, & D. T. Tingey (Eds.). *Assessment of crop loss from air pollutants. Proceedings of an International Conference* (pp. 259–286). New York: Elsevier Applied Sciences 1988.

Hedrich, R. (2012). Ion channels in plants. *Physiological Reviews, 92*(4), 1777–1811.

Houben, M., & Van de Poel, B. (2019). 1-Aminocyclopropane-1-Carboxylic Acid Oxidase (ACO): The Enzyme That Makes the Plant Hormone Ethylene. *Front. Plant Sci. 10*, 695.

Hua, D., Wang, C., He, J., Liao, H., Duan, Y., Zhu, Z., Guo, Y., Chen, Z., & Gong, Z. (2012). A plasma membrane receptor kinase, GHR1, mediates abscisic acid-and hydrogen peroxide-regulated stomatal movement in Arabidopsis. *The Plant Cell, 24*(6), 2546–2561.

Huang, S., Waadt, R., Nuhkat, M., Kollist, H., Hedrich, R., & Roelfsema, M. R. G. (2019). Calcium signals in guard cells enhance the efficiency by which abscisic acid triggers stomatal closure. *The New Phytologist, 224*, 177–187.

Hsu, P. K., Dubeaux, G., Takahashi, Y., & Schroeder, J. I. (2021). Signaling mechanisms in abscisic acid-mediated stomatal closure. *Plant Journal, 105*(2), 307–321. https://doi.org/10.1111/tpj.15067.

Huang, J., & van der Hoorn, R. A. L. (2025). The ancestral salicylic acid biosynthesis pathway in plants. *Trends in Plant Science, 31*(1), 17–19. https://doi.org/10.1016/j.tplants.2025.08.004 Aug 25.

Jomova, K., Alomar, S. Y., Alwasel, S. H., Nepovimova, E., Kuca, K., & Valko, M. (2024). Several lines of antioxidant defense against oxidative stress: Antioxidant enzymes, nanomaterials with multiple enzyme-mimicking activities, and low-molecular-weight antioxidants. *Archives of Toxicology, 98*(5), 1323–1367. https://doi.org/10.1007/s00204-024-03696-4.

Kantharaj, V., Ramasamy, N. K., Yoon, Y.-E., Lee, K.-A., Kumar, V., Choe, H., Chohra, H., Kim, Y.-N., & Lee, Y. B. (2024). Regulatory Response of Rice Seedlings to Exogenously Applied Kinetin During Oxidative Stress. *J. Plant Growth Regulator, 43*, 4680–4690.

Karmakar, D., & Padhy, P. K. (2019). Air pollution tolerance, anticipated performance, and metal accumulation indices of plant species for greenbelt development in urban industrial area. *Chemosphere, 237*, 124522.

Kersteins, G., Federholzner, R., & Lendzian, K. J. (1992). Dry deposition and cuticular uptake of pollutant gases. *Agriculture, Ecosystem and Environment, 42*, 239–253.

Khan, A., Shah, S. T., Basit, A., Mohamed, H. I., & Li, Y. (2024). Mitogen-activated protein kinase: A potent signaling protein that combats biotic and abiotic stress in plants. *Journal of Plant Growth Regulation, 43*(6), 1762–1786. https://doi.org/10.1007/s00344-024-11239-5.

Kollist, H., Zandalinas, S. I., Sengupta, S., Nuhkat, M., Kangasjärvi, J., & Mittler, R. (2019). Rapid Responses to Abiotic Stress: Priming the Landscape for the Signal Transduction Network. *Trends in Plant Science, 24*, 25–37.

König, S., Feussner, K., Kaever, A., Landesfeind, M., Thurow, C., Karlovsky, P., Gatz, C., Polle, A., & Feussner, I. (2014). Soluble phenylpropanoids are involved in the defense response of Arabidopsis against Verticillium longisporum. *New Phytologist, 202*(3), 823–837.

Kumar, A., Singh, H., Kumari, G., Bisht, S., Malik, A., Kumar, N., Singh, M., Raturi, A., Barthwal, S., Thakur, A., & Kaushal, R. (2022). Adaptive resilience of roadside trees to vehicular emissions via leaf enzymatic, physiological, and anatomical trait modulations. *Environmental Pollution, 313*, 120191.

Kumar, P., Corada, K., Debele, S. E., Emygdio, A. P. M., Abhijith, K. V., Hassan, H., Broomandi, P., Baldauf, R., Calvillo, N., Cao, S. J., Desrivières, S., Feng, Z., Gallagher, J., Kjeldsen, T. R., Khan, A. A., Khare, M., Kota, S. H., Li, B., Malham, S. K., ... Jones, L. (2024). Air pollution abatement from green-blue-grey infrastructure. *Innovation Geoscience, 2*(4). https://doi.org/10.59717/j.xinn-geo.2024.100100.

Lahiri, M., & Krishna, K. (2024). Effect of air pollution on plant secondary metabolites in selected trees of Delhi. *Environmental Quality Management, 33*(3), 399–409. https://doi.org/10.1002/tqem.22130.

Leppälä, J., Gaupels, F., Xu, E., Morales, L. O., Durner, J., & Brosché, M. (2022). Ozone and nitrogen dioxide regulate similar gene expression responses in *Arabidopsis* but natural variation in the extent of cell death is likely controlled by different genetic loci. *Frontiers in Plant Science, 13*. https://doi.org/10.3389/fpls.2022.994779.

Li, P., Mane, S. P., Sioson, A. A., Robinet, C. V., Heath, L. S., Bohnert, H. J., & Grene, R. (2006). Effects of chronic ozone exposure on gene expression in *Arabidopsis thaliana* ecotypes and in Thellungiella halophila. *Plant, Cell and Environment, 29*(5), 854–868. https://doi.org/10.1111/j.1365-3040.2005.01465.x.

Li, L., & Yi, H. (2012). Effect of sulfur dioxide on ROS production, gene expression and antioxidant enzyme activity in Arabidopsis plants. *Plant Physiology and Biochemistry, 58*, 46–53.

Lyu, X., Chang, L., Lu, Z., & Li, J. (2023). The ability of three climbing plant species to capture particulate matter and their physiological responses at different environmental sampling sites. *Frontiers in Environmental Science, 10*, 1084902.

Margarete, B., Andrea, K., Dortje, G., & Dietz, K.-J. (2005). Oxidative stress and ozone: Perception, signalling and response. *Plant, Cell & Environment. 28*(8), 1012–1020. https://doi.org/10.1111/j.1365-3040.2005.01326.x.

Matsuura, T., Okugawa, S., Yamakita, E., Kiyomizu, T., Tsutsui, Y., Kume, A., & Hanba, Y. T. (2026). Evaluation of the photosynthetic response of *Ginkgo biloba* as an urban tree to air pollution, soil salinity, and excess humidity. *Frontiers in Plant Science, 17*. https://doi.org/10.3389/fpls.2026.1746328.

Meinhard, M., Rodriguez, P. L., & Grill, E. (2002). The sensitivity of ABI2 to hydrogen peroxide links the abscisic acid-response regulator to redox signalling. *Planta, 214*, 775–782.

Merilo, E., Laanemets, K., Hu, H., Xue, S., Jakobson, L., Tulva, I., & Kollist, H. (2013). PYR/RCAR receptors contribute to ozone-, reduced air humidity-, darkness-, and CO_2-induced stomatal regulation. *Plant Physiology, 162*, 1652–1668.

Miles, G. P., Samuel, M. A., Zhang, Y., & Ellis, B. E. (2005). RNA interference-based (RNAi) suppression of AtMPK6, an *Arabidopsis* mitogen-activated protein kinase, results in hypersensitivity to ozone and misregulation of AtMPK3. *Environmental Pollution, 138*, 230–237.

Modesti, M., Marchica, A., Pisuttu, C., Risoli, S., Pellegrini, E., Bellincontro, A., Mencarelli, F., Tonutti, P., & Nali, C. (2023). Ozone-induced biochemical and molecular changes in *Vitis vinifera* leaves and responses to *Botrytis cinerea* infections. *Antioxidants, 12*(2), 343. https://doi.org/10.3390/antiox12020343.

Monteith, J. L., & Unsworth, M. H. (1990). *Principles of environmental physics*. London: Arnold, 229.

Molina, L., & Segura, A. (2021). Biochemical and metabolic plant responses toward polycyclic aromatic hydrocarbons and heavy metals present in atmospheric pollution. *Plants, 10*(11), 2305. https://doi.org/10.3390/plants10112305.

Morales, L. O., Shapiguzov, A., Safronov, O., Leppälä, J., Vaahtera, L., Yarmolinsky, D., Kollist, H., & Brosché, M. (2021). Ozone responses in *Arabidopsis*: Beyond stomatal conductance. *Plant Physiology, 186*(1), 180–192. https://doi.org/10.1093/plphys/kiab097.

Niekerk, L., Gokul, A., Basson, G., Badiwe, M., Nkomo, M., Klein, A., & Keyster, M. (2024). Heavy metal stress and mitogen activated kinase transcription factors in plants: exploring heavy metal-ROS influences on plant signalling pathways. *Plant, Cell & Environment, 47*, 2793–2810.

Noctor, G., & Foyer, C. H. (1998). Ascorbate and Glutathione: Keeping active oxygen under control. *Annual Review Plant Physiology Plant Molecular Biology, 49*, 249–279.

Ogawa, D., Nakajima, N., Sano, T., Tamaoki, M., Aono, M., Kubo, A., Kanna, M., Ioki, M., Kamada, H., & Saji, H. (2005). Salicylic acid accumulation under O_3 exposure is regulated by ethylene in tobacco plants. *Plant and Cell Physiology, 46*(7), 1062–1072. https://doi.org/10.1093/pcp/pci118.

Oksanen, E., & Kontunen-Soppela, S. (2021). Plants have different strategies to defend against air pollutants. *Current Opinion in Environmental Science & Health, 19*, 100222. https://doi.org/10.1016/j.coesh.2020.10.010.

Overmyer, K., Brosché, M., Pellinen, R., Kuittinen, T., Tuominen, H., Ahlfors, R., Keinänen, M., Saarma, M., Scheel, D., & Kangasjärvi, J. (2005). Ozone-induced programmed cell death in the Arabidopsis radical-induced cell death1 mutant. *Plant Physiology, 137*(3), 1092–1104.

Park, S. Y., Fung, P., Nishimura, N., Jensen, D. R., Fujii, H., Zhao, Y., & Cutler, S. R. (2009). Abscisic acid inhibits type 2C protein phosphatases via the PYR/PYL family of START proteins. *Science (New York, N.Y.), 324*, 1068–1071.

Pastori, G. M., & Foyer, C. H. (2002). Common components, networks, and pathways of cross-tolerance to stress. The central role of "redox" and abscisic acid-mediated controls. *Plant Physiology, 129*(2), 460–468. https://doi.org/10.1104/pp.011021.

Pitzschke, A., Schikora, A., & Hirt, H. (2009). MAPK cascade signalling networks in plant defence. *Current Opinion in Plant Biology, 12*, 421–426.

Pleijel, H., Danielsson, H., Ojanperä, K., Temmerman, L., Högy, P., Badiani, M., & Karlsson, P. E. (2004). Relationships between ozone exposure and yield loss in European wheat and potato—a comparison of concentration- and flux-based exposure indices. *Atmospheric Environment, 38*(15), 2259–2269. https://doi.org/10.1016/j.atmosenv.2003.09.076.

Rai, R., & Agrawal, M. (2012). Impact of tropospheric ozone on crop plants. *Section B: Biological Sciences. 82*(2), 241–257. https://doi.org/10.1007/s40011-012-0032-2.

Ramge, P., Badeck, F.-W., Plochl, M., & Kohlmaier, G. H. (1993). Apoplastic antioxidants as decisive elimination factors within the uptake process of nitrogen dioxide into leaf tissues. *New Phytologist, 125*, 771–785.

Rao, M. J., Duan, M., Ikram, M., & Zheng, B. (2025). ROS regulation and antioxidant responses in plants under air pollution: molecular signaling, metabolic adaptation, and biotechnological solutions. *Antioxidants, 14*(8), 907.

Rao, M. V., & Davis, K. R. (1999). Ozone-induced cell death occurs via two distinct mechanisms in Arabidopsis: the role of salicylic acid. *The Plant Journal, 17*, 603–614.

Rao, M. J., Duan, M., Zhou, C., Jiao, J., Cheng, P., Yang, L., Wei, W., Shen, Q., Ji, P., Yang, Y., Conteh, O., Yan, D., Yuan, H., Rauf, A., Ai, J., & Zheng, B. (2025). Antioxidant defense system in plants: Reactive oxygen species production, signaling, and scavenging during abiotic stress-induced oxidative damage. *Horticulturae, 11*(5), 477. https://doi.org/10.3390/horticulturae11050477.

Rathod, S. V., Saras, P., & Gondaliya, S. M. (2024). *Environmental pollution:Threats and challenges for management. In Eco-restoration of polluted environment.* FL, USA: CRCPress:BocaRaton, 1–34.

Robinson, M. F., Heath, J., & Mansfield, T. A. (1998). Disturbances in stomatal behaviour caused by air pollutants. *Journal of Experimental Botany, 49*, 461–469.

Rodriguez, M. C., Petersen, M., & Mundy, J. (2010). Mitogen-activated protein kinase signaling in plants. *Annual Review of Plant Biology, 61*, 621–649.

Ruan, J., Zhou, Y., Zhou, M., Yan, J., Khurshid, M., Weng, W., Cheng, J., & Zhang, K. (2019). Jasmonic acid signaling pathway in plants. *International Journal of Molecular Sciences, 20*(10), 2479. https://doi.org/10.3390/ijms20102479.

Scherzer, S., Maierhofer, T., Al-Rasheid, K. A., Geiger, D., & Hedrich, R. (2012). Multiple calcium-dependent kinases modulate ABA-activated guard cell anion channels. *Molecular Plant, 5*, 1409–1412.

Semeraro, T., Scarano, A., Buccolieri, R., Santino, A., & Aarrevaara, E. (2021). Planning of urban green spaces: An ecological perspective on human benefits. *Land, 10*, 105.

Sheng, Q., Zhou, C., Liang, Y., Zhang, H., Song, M., & Zhu, Z. (2022). Elevated NO_2 induces leaf defensive mechanisms in Bougainvillea spectabilis seedlings. *Ecotoxicology and Environmental Safety, 248*, 114292. https://doi.org/10.1016/j.ecoenv.2022.114292.

Taylor, G. E., Jr., Tingey, D. T., & Ratsch, H. C. (1982). Ozone flux in *Glycine max*.L Merr.: Sites of regulation and relationship to leaf injury. *Oceologia, 53*, 179–186.

Terashima, N., Ko, C., Matshusita, Y., & Westermark, U. (2016). Monolignol glucosides as intermediate compounds in lignin biosynthesis. Revisiting the cell wall lignification and new 13C-tracer experiments with Ginkgo biloba and Magnolia liliiflora. *Holzforschung, 70*(9), 801–810.

Tonelli, M., Pellegrini, E., D'Angiolillo, F., Petersen, M., Nali, C., Pistelli, L., & Lorenzini, G. (2015). Ozone-elicited secondary metabolites in shoot cultures of *Melissa officinalis* L. *Plant Cell, Tissue and Organ Culture (PCTOC, 120*(2), 617–629. https://doi.org/10.1007/s11240-014-0628-8.

Vahisalu, T., Kollist, H., Wang, Y.-F., Nishimura, N., Chan, W.-Y., Valerio, G., Lamminmäki, A., Brosché, M., Moldau, H., & Desikan, R. (2008). SLAC1 is required for plant guard cell S-type anion channel function in stomatal signalling. *Nature, 452*, 487–491.

Vahisalu, T., Puzorjova, I., Brosche, M., Valk, E., Lepiku, M., Moldau, H., Pechter, P., Wang, Y. S., Lindgren, O., & Salojarvi, J. (2010). Ozone-triggered rapid stomatal response involves the production of reactive oxygen species, and is controlled by SLAC1 and OST1. *Plant Journal, 62*, 442–453.

Vainonen, J. P., & Kangasjärvi, J. (2015). Plant signalling in acute ozone exposure. *Plant, Cell & Environment, 38*(2), 240–252.

Vergata, C., Contaldi, F., Baccelli, I., Basso, M. F., Santini, A., Pecori, F., Buti, M., Mengoni, A., Vaccaro, F., Basso Moura, B., Ferrini, F., & Martinelli, F. (2023). How does particulate matter affect plant transcriptome and microbiome? *Environmental and Experimental Botany, 209*, 105313. https://doi.org/10.1016/j.envexpbot.2023.105313.

Vergata, C., Contaldi, F., Baccelli, I., Buti, M., Vangelisti, A., Giordani, T., Moura, B., Ferrini, F., & Martinelli, F. (2023). The transcriptional mechanism responding to air particulate matter in Laurus nobilis (L.). *Environmental and Experimental Botany, 210*, 105304.

Yalpani, N., Leon, J., Lawton, M. A., & Raskin, I. (1993). Pathway of salicylic acid biosynthesis in healthy and virus-inoculated tobacco. *Plant Physiology, 103*(2), 315–321. https://doi.org/10.1104/pp.103.2.315.

Yoshida, S., Tamaoki, M., Shikano, T., Nakajima, N., Ogawa, D., Ioki, M., Aono, M., Kubo, A., Kamada, H., Inoue, Y., & Saji, H. (2006). Cytosolic dehydroascorbate reductase is important for ozone tolerance in Arabidopsis thaliana. *Plant and cell physiology, 47*(2), 304–308.

Zechmann, B. (2014). Compartment-specific importance of glutathione during abiotic and biotic stress. *Front. Plant Sci. 5*, 566.

Zhao, F., Durner, J., Winkler, J. B., Traidl-Hoffmann, C., Strom, T. M., Ernst, D., & Frank, U. (2017). Pollen of common ragweed (*Ambrosia artemisiifolia* L.): Illumina-based de novo sequencing and differential transcript expression upon elevated NO2/O3. *Environmental Pollution, 224*, 503–514. https://doi.org/10.1016/j.envpol.2017.02.032.

CHAPTER FIVE

Impact of urban air pollution on agriculture

Avanish Prajapati, Saumya Mishra, Umesh Kumar, Prince Kumar Singh, and Rajesh Kumar Sharma
Department of Botany, Banaras Hindu University, Varanasi, Uttar Pradesh, India

5.1 Introduction

Urban air pollution has arisen as a serious environmental concern in the era of growing urbanization and industrialization (Huebner, 2020). The intricate interactions between human activities in urban regions greatly worsen air quality, resulting in a negative effect on ecosystem health and human health. Urban regions are becoming the hubs of cultural and economic activity, drawing people who are looking for more options (Marson et al., 2023). Air pollution in cities is mostly caused by industrial processes, which discharge a variety of pollutants into the atmosphere (Kousehlar & Widom, 2020). Industries that engage in manufacturing processes release a variety of air pollutants, including sulfur dioxide (SO_2), volatile organic compounds (VOCs), heavy metals, nitrogen oxides (NOx), particulate matter (PM), carbon monoxide (CO), ozone (O_3), and polycyclic aromatic hydrocarbons (PAHs) (Akomolafe et al., 2024). One of the main sources of the pollutants is the combustion of fossil fuels such as coal and oil. It has been found that excessive crop yield losses and excessive damage to vegetation are caused by air pollution (Mills et al., 2018). $PM_{2.5}$ and O_3 have the highest monetary value when it comes to removing air pollutants and improving plant and tree health (Nowak et al., 2014). In urban regions with the largest population densities, ambient air pollution concentrations are typically found to be highest. Human pollution exposure as a result is increased per capita (Kumar et al., 2022; Ottosen & Kumar, 2020).

The three major gaseous air pollutants that have the biggest effects on plants are SO_2, NOx, and O_3 (Baciak et al., 2015). Abiotic stress from SO_2 results in a decrease in biomass growth (Noe et al., 2011). Conversely,

Urban Air Pollution and Plant Responses
DOI: https://doi.org/10.1016/B978-0-443-40320-0.00005-X

NO_2 is transformed into nitrous and nitric acids through the atmospheric chemical reactions that harm cell membranes and break down chlorophyll, when exposed (Jiechen et al., 2021). Furthermore, prior research has found that leaf chlorophyll content is impacted by NO_2 exposure (Sheng & Zhu, 2019). Further, it has been shown that NO_2 damages plants oxidatively, and antioxidants are essential for protecting plants against this harm (Sheng & Zhu, 2019). Exposure to pollutants was shown to cause plants to produce reactive oxygen species (ROS) (Oksanen & Kontunen-Soppela, 2021). ROS are the central component of a complex, multitiered network of antioxidant systems (AOS) in plants that function to counteract toxic ROS. Plant cellular homeostasis is maintained by the predominance of AOS over ROS (Dumanović et al., 2021). In contrast to other pollutants, O_3 has a very distinct effect on plants due to its oxidative nature. Exposure to O_3 typically results in the emergence of indications of damage, including white patches that later develop into necrotic brown areas. According to Lee et al. (2022), wilting, increased senescence, and the emergence of unusual leaf colors and patterns (e.g., bright green or pale coloration, bronzing patterns in interveinal areas, bleaching, reddening, etc.) are among the apparent foliar symptoms caused by O_3. Crop species from Africa that are exposed to high O_3 levels showed apparent leaf symptoms, which were shown graphically and quantified by Sharps et al. (2021). O_3 has a wide range of consequences, such as apparent damage, chlorotic mottles on leaves, stipple-like necrotic lesions, deterioration of chlorophyll, altered crop quality, improved gas exchange and stomatal closure, and altered carbon allocation. Additional consequences include decreased yield, stomatal conductance, and net photosynthesis being uncoupled, photosynthesis being inhibited, accelerated senescence, genetic changes, metabolic diseases, impaired resistance to disease and pests, oxidative stress, and so on (Arab et al., 2024). The losses in biomass production and reduced yield caused by plant-level damage ultimately lead to economic losses. When particulate debris of different sizes accumulates on leaf foliage, it may obstruct stomatal apertures (Anake et al., 2022). Photosynthetically active radiation's (PAR) optical absorption and reflection by leaves may be altered by deposited PM or even by higher ambient PM concentrations. This can exacerbate secondary stressors in plants and interfere with physiological processes such as photosynthetic rate and normal metabolism (Rai, 2016).

Plants exposed to air pollutants experience disruptions in their regular functions, such as effects on regular operations, disturbances in metabolism, and reduced capacity to improve yield (Mori et al., 2018). Utilizing plants as "green infrastructure" (GI) is one of the legal and technological solutions proposed to lessen air pollution and human exposure to it. Recently, this concept has become more significant throughout urban areas worldwide (Tomson et al., 2021). Using "nature-based solutions (NBS)" is one of the sustainable and increasingly endorsed/supported air pollution mitigation alternatives. According to Prigioniero et al. (2021), "nature-based solutions" in the context of air pollution include: (1) Absorbing, adsorbing, and catabolizing nearly all air pollutants; (2) Effectively depositing pollutants into vegetative surfaces (instead of artificial surfaces). Furthermore, it has been demonstrated that trees with roadside GI reduce pollutants (Khan et al., 2022). Gallagher et al. (2015) concluded that the use of different types of GI in porous media barriers can aid in the improvement of air quality after taking into account a number of passive techniques. According to Barwise and Kumar (2020), some biophysical characteristics can improve air pollution abatement. Consequently, it becomes critical to look into how air pollution affects plants and vice versa.

5.2 Sources of urban air pollution

The health and well-being of urban populations are greatly impacted by urban air pollutants, a complex environmental problem that results from the intricate interaction of several human and natural causes. Combustion activities, especially the burning of fossil fuels in automobiles, factories, and power plants, are the main offenders as they emit a variety of dangerous chemicals into the atmosphere (Lahieb Faisal et al., 2020). The major sources of air pollution in cities include a range of direct and indirect activities that emit various pollutants into the environment. During industrial production processes, heavy metals, SO_2, NO_x, PM, and VOCs are all emitted. The combustion of fossil fuels, such as coal and oil, is a major contributor to these emissions. The dispersion of these pollutants is sometimes made worse by the concentration of industry in metropolitan areas, which worsens the overall impact on air quality. There are two categories of sources of air pollution: natural and anthropogenic (Mallik, 2019).

5.2.1 Anthropogenic sources

Anthropogenic sources like vehicular emissions, industrial wastes, construction activity, energy production, waste management, residential heating and cooking, etc., are major causes of urban air pollution, which are Vehicular emissions, Industrial pollutants, Construction activities, Waste burning, and Household emissions, etc. (Verma et al., 2021). Motorcycles, trucks, buses, and cars all contribute significantly to urban air pollution. They release pollutants such as PM, NO_x, CO, and VOCs (Sadiq, 2022). The majority of potentially hazardous gases and their gaseous waste are added daily by emissions from fossil fuel-powered automobiles. CO, methane (CH_4), SO_x, and NO_x, and other gases are released as fossil fuels burn (Choudhary & Garg, 2013). Numerous factors, frequently associated with infrastructure and human activity, contribute to urban air pollution. Smog and acid rain can be caused by a variety of pollutants released by factories and power plants, such as SO_2, NO_x, and PM (Edo et al., 2024). Dust and other kinds of particles can be released into the atmosphere from construction sites, particularly in metropolitan areas that are growing quickly (Xu et al., 2020). This leads to poor air quality, particularly when the weather is dry. Air pollutants such as CO_2, SO_2, and NO_x are released when fossil fuels (such as coal, oil, and natural gas) are used for heating and power generation in metropolitan areas.

Air pollution in certain urban centers is caused by adjacent agricultural activity. This includes the discharge of dust from plouwing, ammonia (NH_3) and insecticides, particularly in periurban areas. In landfills, methane CH_4 and other gases are released when organic garbage breaks down. Dioxins and furans are among the air contaminants that might be released during the combustion of waste products (Losada et al., 2018). In many metropolitan areas, particularly in low-income districts, people utilize coal, wood, or other fuels for cooking and heating, and also use heaters, stoves, or open flames indoors (Liao et al., 2017). Poor indoor air quality results from homes being exposed to dangerous pollutants, including PM, CO, and VOCs. This can lead to chronic respiratory conditions, especially in young individuals and the elderly. In India, home air pollution has been a chronic health issue, continuing to rank as the second-highest risk factor for lost disability life years and causing 1.0 million premature deaths every year (Institute for Health Metrics and Evaluation, 2013). Ports and airports have the potential to contribute to air pollution due to the emissions from ships and airplanes. Air quality can be significantly impacted by the sulfur and nitrogen compounds released by ships and the nitrogen oxides and particles released by high-altitude aircraft (Suner, 2024).

5.2.2 Natural sources

Urban air pollution is mostly caused by human activities, but it can also come from natural causes like wildfires, pollen, and dust storms, particularly in places close to woods or deserts (Odubo & Kosoe, 2024). Volcanic activity, which produces sulfur, chlorine, and ash particulates, also adds pollutants to the air. The combination of the above sources in dense urban areas can lead to serious air quality problems, affecting human health and the environment. Efforts to reduce urban air pollution often focus on improving transportation, industrial processes, and energy systems, while encouraging sustainable urban planning.

5.3 Types of air pollutants

Urban air pollutants can be categorized into several types based on their chemical composition and their sources. Here are some major types.

5.3.1 Particulate matters

PM can be divided into $PM_{2.5}$ and PM_{10} categories based on size. PM_{10} is the category for PM with a diameter of 10 micrometers or less. It can cause health issues, including bronchitis, asthma, and cardiovascular disorders, which can be brought on by inhaling them into the respiratory system. $PM_{2.5}$ are finer particles with a diameter of 2.5 micrometers or smaller. Their small size allows them to pass through circulatory systems and go deeper into the lungs, which can result in serious health problems such as respiratory infections, lung cancer, and heart disease. Vehicle exhaust, industrial emissions, construction dust, burning of fossil fuels, and wild and forest fires contribute to the formation of PM (Li et al., 2015).

5.3.2 Nitrogen oxides

The two primary nitrogen oxides present in urban air pollutants are NO and NO_2. They are dangerous pollutants that contribute to acid rain, fine $PM_{2.5}$, and ground-level ozone. The burning of fossil fuels releases both NO and NO_2, which are important air pollutants that contribute to respiratory illnesses, smog, and environmental deterioration. To lessen their detrimental impact on the environment and human health, emissions from power plants, industry, and automobiles must be reduced. Transport emissions, industrial processes, power plants, and other combustion processes are major sources (Antenozio et al., 2024).

5.3.3 Sulfur dioxide

The SO_2 is a colorless, strong-smelling gas. The main contributors of SO_2 emissions are industrial processes and the burning of fossil fuels (Shi, 2019). The SO_2 standard is 0.03 ppm per annum. It has an impact on plant, animal, and human life. Those at increased risk of harm include youngsters, the elderly, and those with lung conditions. Since SO_2 is a sensory irritant that enters the lungs deeply, transforms into bisulfite, and interacts with sensory receptors to cause bronchoconstriction, the main health issues linked to SO_2 emissions in industrialized areas are respiratory irritation, bronchitis, mucus production, and bronchospasm.

5.3.4 Carbon monoxide

This gas is very toxic, colorless, odorless, and nonirritating. It is a byproduct of fuels like coal, wood, or natural gas burning incompletely. Major sources of carbon monoxide are vehicle exhaust, industrial processes, and the burning of fossil fuels in residential heating or cooking. Headache, dizziness, weakness, nausea, vomiting, and ultimately unconsciousness are signs of CO poisoning. Compared to oxygen, CO has a greater affinity for hamoglobin. Accordingly, prolonged exposure to elevated CO levels might result in severe poisoning (Balogun et al., 2014). Cardiovascular disease, ischemia, and hypoxia are seen as a result of the oxygen loss due to CO's competitive binding.

5.3.5 Volatile organic compounds

VOCs are a group of organic chemicals that can easily evaporate into the air, and they are the most prominent air contaminants outside. They are often divided into the distinct categories of nonmethane (NMVOCs) and CH_4 in this domain (Deng et al., 2022). CH_4 is a very effective greenhouse gas that contributes to global warming. Although the impact varies based on local air quality, other hydrocarbon VOCs are also important greenhouse gases because of their contributions to the production of ozone and the extension of methane's half-life in the atmosphere. Benzene, toluene, and xylene are aromatic chemicals found in NMVOCs that are thought to be carcinogenic and can cause leukemia if exposed for an extended period of time. 1, 3-butadiene is another hazardous substance that is frequently involved in industrial applications. Their sources include paint, solvents, cleaning products, fuel vapors, etc. (Li et al., 2022).

5.3.6 Ground-level ozone

NO_x and VOCs combine to generate ground-level O_3. One important component of the troposphere is ozone. Additionally, it is a crucial component of the Ozone layer, which is a term used to describe certain stratospheric areas. Many of the chemical activities that take place in the atmosphere both during the day and at night are powered by photochemical and chemical interactions involving it. It is a pollutant and a component of smog at excessively high concentrations caused by human activity, that is, mostly the burning of fossil fuels (Liu et al., 2018).

5.3.7 Ammonia

NH_3 is frequently released during agricultural activities. It is typically a colorless gas with a strong, distinctive, and pungent smell. NH_3 also serves as a building element for the manufacture of several medications, either directly or indirectly (Reche et al., 2012). Despite its widespread usage, NH_3 is dangerous and corrosive. Ammonia is able to contribute to the formation of small particulate matter when it interacts with other pollutants like NO_x and SO_2.

5.3.8 Lead

Lead chloride ($PbCl_2$) serves as the way the lead (Pb) is seen in the exhaust. The Pb compound discharged has a range of particle sizes, from less than 0.01 micron to several millimeters. Over a prolonged period, Pb becomes a dangerous element; it may accumulate in the body and cause major health problems, especially in children (neurological damage). Pb still arrives from other sources even while leaded gasoline has been phased out (Ye et al., 2022).

5.3.9 Polycyclic aromatic hydrocarbons and hydrogen sulfide

A class of chemical molecules known as PAHs is created when organic matter burns incompletely. They are regarded as harmful to human health and carcinogenic. Major sources include wood burning, cigarette smoke, industrial pollution, and vehicle exhaust (Valavanidis et al., 2006). The gas hydrogen sulfide (H_2S) is poisonous and smells strongly like rotten eggs. Higher doses may result in respiratory discomfort and it may irritate the eyes, nose, and throat (Mohammed & Devi, 2021). Landfills, sewage treatment facilities, and industrial operations (oil and gas refineries) are the main sources. These pollutants often interact and create secondary pollutants, such as smog and acid rain, which can further degrade air quality and harm both human health and the environment. Efforts to monitor, reduce emissions, and improve urban planning are essential for reducing the impact of these pollutants on cities.

5.4 Effects of urban air pollution on agriculture

Air pollution has become an important environmental stressor that can pose a risk to agricultural productivity and food security globally. O_3, SO_2, NO_x, CO, PM, and VOCs are different atmospheric pollutants derived from anthropogenic sources such as fossil fuel combustion, vehicular exhausts, industrial effluents, biomass burning, and agriculture (Onivefu & Imarhiagbe, 2024). These air pollutants can directly and indirectly affect crops, influencing physiological, morphological, biochemical, and yield and productivity, as shown in Fig. 5.1. The two major mechanisms through which air pollutants interact with agricultural crops are through foliar (leaf-mediated) and root-mediated uptake pathways. Foliar uptake is the major pathway for gaseous pollutants. These gases, including O_3, SO_2, and NO_2, diffuse into the leaf tissues through stomata, minute pores involved in gas exchange, and get dissolved in the aqueous phase of the substomatal cavity (Husen, 2021).

Inside, they get involved in different biochemical reactions, which result in the production of ROS, such as superoxide anion (O_2^-), hydrogen peroxide (H_2O_2), and hydroxyl radicals (•OH). These ROS destabilize cell membranes, unwind proteins, and harm nucleic acids, thus inducing

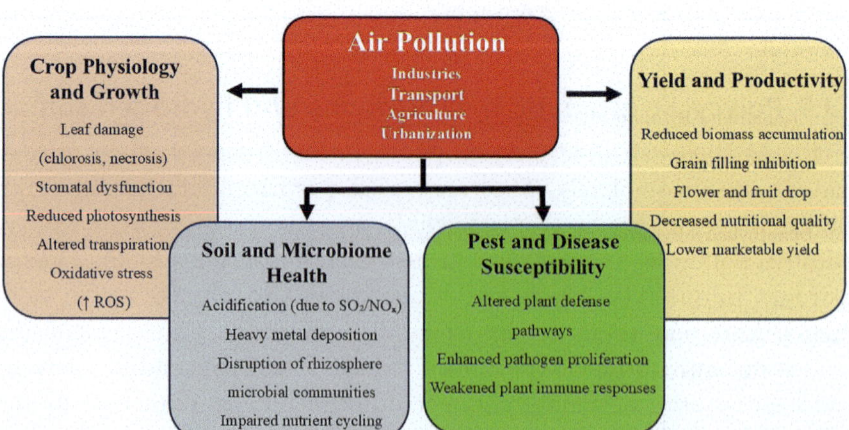

Figure 5.1 Effect of air pollution on agriculture. No permission required.

oxidative stress that inhibits photosynthesis, respiration, and other metabolic processes (Ozougwu, 2016). O_3 is especially phytotoxic because of its powerful oxidative character, causing visible foliar damage such as stippling, flecking, and early senescence. It disrupts the photosynthetic process by breaking down chlorophyll, causing thylakoid membrane damage, and suppression of Rubisco enzyme activity (Ghosh et al., 2018). SO_2, upon dissolution in the apoplastic water of the leaf, produces bisulfite and sulfite ions that disrupt metabolic processes, suppress photosynthetic enzymes, and cause chlorosis and necrosis (Jha & Yadav, 2023). Likewise, NO_x gases can inhibit nitrogen assimilation and decrease chlorophyll biosynthesis. VOCs, particularly aromatic hydrocarbons such as benzene, toluene, and xylene, can cause lipid peroxidation and interfere with membrane integrity. In contrast, PM, such as $PM_{2.5}$ and PM_{10}, disrupts plant functions by physically depositing on leaf surfaces, decreasing light interception, clogging stomata, and interfering with gas exchange (Uzoho, 2025). Different air pollutants have a multitude of mechanisms of action, which have been discussed in Table 5.1.

Heavy metals such as Pb, cadmium (Cd), and arsenic (As) adsorbed on PM also find their way into the plant system through leaf surfaces or after washing into the soil, being finally absorbed by roots (Shahid et al., 2017). Root uptake becomes relevant as pollutants are deposited on the soil via dry or wet deposition. For example, acid rain generated by SO_2 and NO_x causes acidification of soil, which changes nutrient bioavailability, contaminates microbial communities, and increases the mobility of toxic metals. Such modifications negatively impact root physiology by lowering root biomass and nutrient uptake efficiency. Heavy metals such as Cd and Pb are taken up by plants through their roots, where they compete with other essential nutrients such as calcium (Ca), magnesium (Mg), and zinc (Zn) for absorption sites (Angon et al., 2024). Once within the plant, these metals are transported to shoots and leaves through the xylem and become toxic by changing membrane permeability, inhibiting enzyme activities, and disrupting hormonal regulation. These pollutants, whether taken up via leaves or roots, induce severe physiological and biochemical disruptions that result in diminished growth and yield. One of the most important effects is inhibition of photosynthesis, which directly decreases the energy available for biomass production and reproductive growth.

Pollutants such as O_3 have been reported to decrease net photosynthetic rate, stomatal conductance, and water-use efficiency. Long-term exposure results in declining leaf area index, premature leaf fall, and lower

Table 5.1 Mechanisms of action and key impacts of air pollutants on different crops.

Air pollutants	Mechanism of action	Impacts	Crops	References
Ozone	Enters through stomata; causes oxidative stress; destroys chloroplasts and membranes	Reduced photosynthesis, early senescence, and decreased crop yield	Wheat, soybean, rice	Tiwari et al. (2018)
Sulfur dioxide	Condenses to sulfurous acid on wet leaf surfaces, inhibiting enzyme activity	Leaf injury, reduced growth, impaired photosynthesis	Spinach, alfalfa, beans	Robert et al. (2018)
Nitrogen oxides	Adds to acid rain and the secondary formation of O_3	Soil acidification, altered nutrient uptake, and indirect ozone stress	Maize, barley, rice	Rafie-Rad et al. (2024)
Particulate matter	Deposits on the leaf surface, inhibiting light absorption and plugging stomata	Inhibited photosynthesis, increased leaf temperature, and lower productivity	Tea, tomato, lettuce	Singh and Verma (2023)
Carbon monoxide	Binds to plant cytochrome oxidase, inhibiting respiration	Delayed flowering, lower energy production, suppressed growth	Corn, peas	Feng et al. (2023)
Ammonia	Volatilized from fertilizers/ livestock; injures leaf tissues; alters soil pH	Foliar damage, nutrient imbalance, reduced quality	Cabbage, wheat	Jalota et al. (2018)
Cadmium	Accumulates in soil and plant tissues; inhibits metabolic processes	Reduced germination, chlorosis, and contamination of edible parts	Leafy vegetables, cereals	Huang et al. (2022)
Lead	Inhibits photosynthesis and enzyme functions	Growth inhibition, reduced yield, contamination of produce	Spinach, carrot, radish	Rani et al. (2024)
Mercury	Inhibits cellular function and photosynthesis	Chlorosis, necrosis, reduced biomass	Rice, wheat	Guo et al. (2021)

Table 5.1 (Continued)

Air pollutants	Mechanism of action	Impacts	Crops	References
Fluorides	Accumulate in leaf tissues, leading to necrosis	Leaf tip burn, reduced photosynthesis, and yield loss	Grapes, apricots	Panda (2015)
Volatile organic compounds	Precursor to the formation of O_3; some are directly toxic to plants	O_3 indirectly damages, causes direct toxicity, altered growth	Soybean, cotton	Pinto et al. (2010)
Black carbon	Depositions on leaf surfaces, lowering albedo, and enhancing leaf temperature	Accelerated senescence, reduced photosynthesis	Wheat, maize	Anda et al. (2013)
Hydrogen fluoride	Penetrates plant tissues, leading to cell damage	Leaf necrosis, reduced growth, and yield loss	Citrus, corn	Bukhari et al. (2023)
Peroxyacetyl nitrate (PAN)	Secondary pollutant created from VOCs and NO_x, induces oxidative stress	Leaf bronzing, reduced photosynthesis, and growth inhibition	Bean, lettuce	Khan et al. (2022)
Chlorine gas (Cl_2)	Causes oxidative damage to plant tissues	Leaf bleaching, necrosis, and reduced growth	Beans, peas	Endress et al. (1978)
Nitric acid (HNO_3)	Forms from NO_x, leading to acid deposition	Soil acidification, nutrient leaching, and root damage	Wheat, rice	Hakeem et al. (2017)
Ammonium sulfate ($(NH_4)_2SO_4$)	Result of atmospheric reaction; deposits on leaves	Leaf injury, altered nutrient balance	Alfalfa	Yang et al. (2024)

No permission required.

chlorophyll content (Masutomi et al., 2019). Excessive ROS production triggers antioxidant defense mechanisms, such as enzymes such as superoxide dismutase, catalase, and peroxidases, which shift energy and resources away from growth and yield formation. These physiological stresses result in decreased filling of grains, abortion of flowers, deformity of fruits, and a lower harvest index. Yield loss depends on crop species, developmental stage at exposure, concentration of the pollutant, and duration of exposure. C3 plants like wheat, rice, and soybean are especially sensitive to O_3 exposure. For example, modeling techniques and field observations estimated the loss in global wheat due to ambient O_3 pollution from 3.9% to 15% while soybean and maize endured losses of 3.1%–16.6% and 2.2%–5.5%, respectively (Schauberger et al., 2019). In developing regions such as South Asia, where background concentrations of O_3 are rising rapidly as a consequence of industrial expansion and increased vehicle, the problem is more pervasive.

Fischer (2019) indicated that O_3 pollution may lead to wheat yield losses of as much as 52% in some regions of India, which amounts to 12–35 million metric tons of grain loss every year. The economic losses from such yield reductions are significant, threatening food security, rural livelihoods, and national economies. The effects of air pollution extend beyond cereal crops. Horticultural produce, fruits, and vegetables are also affected by quality deterioration, including decreased vitamin levels, poor coloration, and changed flavor due to secondary metabolism alterations. Pollutants can influence pollination efficiency by decreasing flower viability and changing nectar content, thereby influencing bee and insect activity (Ryalls et al., 2022). Crop plants such as tomato, lettuce, spinach, and grapes have been found to have decreased fruit and leaf quality under high-pollution conditions (Moretti et al., 2010). Moreover, the potential for residues of pollutants, particularly heavy metals, accumulated in edible portions is a severe threat to food safety and human health. Aside from direct crop impacts, air pollution also interferes with soil-plant-microbe interactions. Microbial diversity and functioning in soils under acidified and metal-polluted soils decrease, affecting nutrient cycling and organic matter decomposition. Mycorrhizal associations, important for nutrient acquisition and resistance to drought, are adversely affected by elevated metal content and soil acidity. Consequently, plant nutrient content decreases, and abiotic and biotic stress susceptibility rise (Naz et al., 2022).

5.5 Mitigation strategies

Numerous tactics are being explored and employed to mitigate these adverse consequences. Agronomic interventions such as employment of pollution-resistant crop varieties, changed dates of sowing to miss peak pollutant periods, and use of foliar sprays that include antioxidants such as ascorbic acid and salicylic acid may mitigate injury. Technological interventions, such as greenhouses, antipollution screens, and precision agriculture equipment, are under exploration to protect crops from exposure to pollution. Tree and shrub buffer zones can be used as biofilters for PM and gaseous pollutants. At the policy level, emission reductions through tighter environmental controls on industry, power stations, and transportation networks are vital. Global coordination under climate conventions, air monitoring, and inventories of emissions are steps to long-term mitigation. Advances in genomics and molecular biology hold the potential to create pollution-tolerant crops through genetic manipulation and marker-assisted selection. In addition, remote sensing and crop simulation models are increasingly employed to track crop responses to air pollution at the regional and global levels, informing policymakers and farmers in decision-making.

5.5.1 Use of pollution-resistant crop varieties

The use of pollution-resistant crop varieties is a critical approach in protecting agriculture from environmental pollutants. The genetically modified crops are engineered to be resistant to pollutants such as heavy metals, O_3, and PM, which are common in industrial and urban environments (Awaad, 2021). Through the incorporation of genes that provide resistance to these pollutants, such as improved antioxidant mechanisms or altered uptake pathways, these crops are able to sustain productivity and quality even under stressful conditions. This method not only maintains food security but also decreases the use of chemical interventions, thus encouraging sustainable agricultural practices (Rai et al., 2020).

5.5.2 Role of urban planning in reducing agricultural exposure to pollutants

Urban planning is also important in minimizing agricultural exposure to pollutants. Proper zoning can isolate industrial processes from agricultural fields, reducing the drift of air pollutants (Lin et al., 2024). Buffer zones, in the form of green belts or vegetative buffers, can also be used to reduce the transfer of contaminants. Additionally, urban planners can integrate policies that restrict vehicle emissions within proximity to agricultural areas and encourage the adoption of cleaner technologies. These steps together result in a healthier crop culture environment and less buildup of toxic materials in the food chain (Glazener & Khreis, 2019).

5.5.3 Use of barriers and green infrastructure

The combination of barriers and green infrastructure is a powerful way to protect agricultural land from pollution (Dhakal & Chevalier, 2017). Wetlands, for example, are natural filters that clean water sources of pollutants before they reach farms. Green walls and vegetative buffers also have the ability to capture PM in the air and absorb gaseous pollutants, cleaning the air surrounding agricultural land. Such green infrastructures, apart from providing protection to the crops, promote biodiversity and are also responsible for the overall robustness of the agricultural ecosystem (Vitaliano et al., 2024).

5.5.4 Technological solutions: vertical farming and indoor agriculture

Technological advancements have given rise to novel methods of, agriculture such as vertical farming, indoor agriculture, and roof plantation providing, controlled environments (Benke & Tomkins, 2017). Vertical farming is a process of growing crops in layers inside controlled environments, usually with the help of hydroponic or aeroponic systems. This technique minimizes exposure of crops to external contaminants to a great extent and provides year-round cultivation with efficiency in resource utilization (Chole et al., 2021). Indoor agriculture, in the same vein, offers a protected space where light, temperature, and humidity are all carefully controlled, providing ideal growing conditions. Such technologies not only reduce the effect of pollution but also help resolve issues of land shortages and climate change, positioning them as part of the fundamental building blocks of sustainable agriculture (Hati & Singh, 2021).

5.6 Socioeconomic impacts of urban air pollution

5.6.1 Economic losses due to reduced crop productivity

One of the most direct and quantifiable effects of urban air pollution is the decrease in agricultural productivity, resulting in high economic losses.

Pollutants like ground-level O_3, SO_2, and PM negatively impact plant physiology, suppressing photosynthesis, decreasing chlorophyll content, and affecting stomatal function (Dhir, 2016). Of these, O_3 is especially detrimental to major crops such as wheat, rice, and soybean. Research has shown that even moderate exposure to O_3 can decrease crop yields by 10%–40%, varying with the type of crop and duration of exposure (Ainsworth, 2017). PM and heavy metals on the surface of crops deteriorate product quality, causing market rejection and lower prices. Vegetables and fruits are particularly prone to esthetic damage, with values corresponding to consumer preference and export potential. These losses in yields and product quality aggregate into lower revenues for farmers and national economic losses, especially for countries where agriculture has high contributions to GDP and employment (Das et al., 2021). The indirect economic effects are similarly significant. Low yields require additional inputs (such as fertilizers, irrigation, and pest control), raising production costs. In addition, risk and insurance costs rise in high-pollution regions, with implications for farm investment and credit status. Consequently, this inhibits long-term agricultural growth and threatens food system resilience.

5.6.2 Impact on food security and supply chains

Urban air pollution significantly erodes food security that rests on four pillars: availability, access, utilization, and stability (Mrabet, 2023). Air pollution lowers the availability of food through the lowered productivity of crops and their quality. Since staple crops and horticulture crops come in contact with growing levels of O_3, SO_2, and $PM_{2.5}$, their yields drop, creating gaps and driving prices higher, particularly in urban areas where demands are increasing (Tai et al., 2014). Air pollution interferes with food supply chains. Logistics in polluted urban centers are frequently delayed because of traffic jams, vehicle bans during pollution alerts, and absenteeism of workers from pollution-related illnesses. These conditions contribute to higher postharvest losses, particularly for perishable items such as vegetables, milk, and fruits. Supply chain disruptions also drive up the prices of food and cause shortages in low-income communities, further worsening food insecurity (Crippa et al., 2022). Crops that are cultivated close to traffic routes or industrial zones uptake airborne pollutants, including PAHs and heavy metals, deposited on leaves and fruits (Mahida et al., 2023). Observations from urban locations such as Kathmandu, Delhi, and Accra indicate that leafy greens cultivated in the city tend to exceed WHO safety standards for heavy metals, posing food safety risks and undermining consumer confidence in locally grown produce (Nabulo et al., 2010).

5.6.3 Public health concerns linked to contaminated crops

Contaminants such as Cd, Pb, As, and Hg are frequently present in urban air and may be deposited on or absorbed by plants. These heavy metals and organic contaminants, such as PAHs, are bioaccumulative and carcinogenic, presenting chronic health risks when consumed through food (Guo et al., 2019). For example, exposure to Pb via vegetables can damage the neurodevelopment in children and exacerbate cardiovascular disease in adults (Gundacker et al., 2021). Cd exposure has been linked to kidney damage, bone deformation, and reproductive toxicity. These food pollutants aggravate the load of noncommunicable diseases like cancer, diabetes, and hypertension. The vulnerable populations, such as children, pregnant women, and immunocompromised patients, are at increased risk due to exposure. Since diet is a repeated day-to-day route of exposure, repeated consumption over many years with moderately contaminated food can cause cumulative toxicity (Yang et al., 2025). Food contamination erodes public confidence in farm markets and urban agriculture programs. In addition, inadequate food safety surveillance in developing nations permits contaminated food to enter the retail chain without restriction. This is particularly concerning in informal food markets, where price determines demand but safety regulations are weakly enforced.

5.6.4 Effects on livelihoods of farmers

Farmer's livelihoods, particularly those within periurban and industrial areas, are greatly impacted by urban air pollution. The combined effects of decreased crop yields and contamination risks translate into huge revenue losses (Agrawal et al., 2003). Excessive levels of O_3, PM, and HMs cause reduced yields and crop rejections, lowering revenue and raising production risks for smallholders (Dewan & Lakhani, 2024). Farmers cultivating crops in contaminated areas are exposed to reputational threats, with consumers and traders refusing to buy visibly damaged or potentially contaminated produce. This impacts those who cultivate high-value perishable crops like leafy greens, fruits, and vegetables that are susceptible to surface deposition of contaminants. Illnesses further aggravate the situation. Farmworkers in contaminated areas are exposed to respiratory irritants, pesticide residues, and PM regularly, and their cases of asthma, bronchitis, and fatigue are on the rise. All these health conditions decrease work potential and raise health costs, thus subjecting more financial burden to farming households. Air pollution also speeds up the degradation of soil, further threatening long-term land productivity. Soil subjected to acid rain or contaminated irrigation water becomes increasingly acidic and nutrient-poor, needing more inputs to sustain yields, increasing costs and diminishing profitability. As farming becomes increasingly uneconomical, numerous rural households go into debt or distress migration, commonly leaving farming for uncertain livelihoods in urban slums (Sivakumar, 2011).

5.7 Policy responses

Periurban and rural agriculture systems are increasingly at risk due to the fast growth of metropolitan regions and industrial development. This study critically evaluates these international and Indian policy measures, comparing global best practices with national strategies to assess their effectiveness in reducing the impact of urban air pollution on agriculture and promoting long-term food and environmental security.

5.7.1 International policies

Urban air pollution poses a growing threat to the agricultural ecosystem by reducing crop productivity and degrading soil quality. This study examines international policies aimed at mitigating these impacts, highlighting strategies for protecting agricultural productivity amid rapid urbanization and industrial growth. China has adopted a comprehensive strategy to combat air pollution, targeting industrial, vehicular, and dust emissions (Yang et al., 2018); Vehicular emissions have been addressed by enforcing stricter standards, eliminating "yellow-marked" vehicles, and improving fuel quality (Qiu et al., 2016); Key policies like the Air Pollution Control Action Plan and Blue-Sky Defense War promote green industry, clean energy, sustainable transport, and environmental taxation to ensure long-term air quality improvements (Jiang et al., 2021). In the United States, the Clean Air Act of 1990 has been implemented through multiple phases to enhance air quality management; Phase 1 (January 2004–August 2005) focused on expanding local enforcement capabilities; Phase 2 (September 2005–July 2006) advanced efforts by moving beyond traditional enforcement authority; Finally, Phase 3 (August 2006–2008) introduced alternate compliance and enforcement strategies to improve regulatory effectiveness (Bruhl et al., 2013). South Africa's air pollution regulations include the Atmospheric Pollution Prevention Act, 1965, and in 1998, the National Environmental Management Act, alongside the 2004/2005 Air Quality Act, which together strengthen environmental governance and establish comprehensive air quality standards (Naiker et al., 2012). In Taiwan, the Air Pollution Control Act (APCA) 1975, amended in 2006, established standards for MSW incinerator emissions (1997), mandated upgrades to pollution control equipment, and supported local governments with subsidies to implement clean air zones and promote environmentally friendly transportation (Tsai, 2016). Bhutan reduces vehicular pollution by banning leaded gasoline, restricting second-hand car imports, and enforcing emission standards, setting a model for sustainable transport in developing countries (Phuntsho & Kanitpong, 2010).

5.7.2 Indian policies

India has implemented several national policies and regulations to mitigate urban air pollution and its adverse effects on agriculture. These measures focus on controlling emissions, promoting cleaner technologies, and enhancing monitoring systems to protect crop health and soil quality, thereby ensuring sustainable agricultural productivity amidst rapid urbanization and industrial growth.

5.7.2.1 National ambient air quality standards

Major contaminants that contribute to air pollution in urban areas have limitations established by the NAAQS. First announced in 1994, these guidelines have since undergone revisions, the latest being in 2009 and 2014. The major contaminants and the permissible limits for each are as follows: $PM_{2.5}$, 24-hour average 60 μg/m^3 and Annual average 40 μg/m^3; For PM_{10}, 24-hour average 100 μg/m^3 and Annual average 60 μg/m^3; NO_2, 24-hour average 80 μg/m^3 and Annual average 40 μg/m^3; SO_2, 24-hour average 80 μg/m^3 and Annual average 50 μg/m^3; O_3, 8-hour average 180 μg/m^3 and 1-hour average 100 μg/m^3; NH_3, 24-hour average 400 μg/m^3; Pb, 24-hour average 0.5 μg/m^3; CO, 8-hour average 2 mg/m^3 and 1-hour average 4 mg/m^3; PAHs, Annual average 6 ng/m^3; Benzene, 24-hour average 5 μg/m^3; Benzo(a) pyrene (BaP), Annual average 1 ng/m^3 (Hossain & Horowitz, 2025).

5.7.2.2 Air quality index

The public is informed about the quality of the air through the air quality index (AQI). With an emphasis on PM2.5 and PM10 levels, it is based on the concentration of certain pollutants. People may learn about the health concerns linked with the present air quality and how filthy the air is by using the AQI. There are six categories on the AQI scale. (1) Good (0–50): There is little to no risk from air pollution, and the quality of the air is deemed good. (2) Satisfactory (51–100): The air quality is fine, but certain pollutants can harm sensitive people's health. (3) Moderately Polluted (101–200): People with respiratory conditions may feel uncomfortable, and those who are sensitive should be on health alert. (4) Poor (201–300): Everyone should be aware of the health risks, and sensitive groups may be particularly affected. (5) Very Poor (301–400): Serious health consequences that should be taken seriously by the entire community. (6) Severe (401–500): Serious respiratory problems might occur, and the entire population could experience a health crisis (Hossain et al., 2021).

5.7.2.3 Regulations for vehicular emissions

Bharat Stage (BS) Emission Standards: The BS standards control the emissions of air pollutants from automobiles, just as the Euro emission standards do. By April 2020, India has gradually advanced from BS-IV (Euro-IV equivalent) to BS-VI (Euro-VI equivalent), the highest degree of vehicle pollution control technology. In order to lower emissions from high-polluting cars, particularly in metropolitan areas, the Indian government implemented the Vehicle Scrappage Policy, which calls for the phase-out of outdated automobiles (Gajbhiye et al., 2023).

5.7.2.4 Central pollution control board and state pollution control boards

Industries must implement pollution control strategies and follow the Central pollution control board (CPCB's) emission guidelines. This entails setting strict restrictions on the quantity of PM, SOx, NOx, and other hazardous pollutants released by power plants, industrial facilities, and other significant sources of pollution. The CPCB and State pollution control boards are in charge of establishing monitoring stations, keeping an eye on the quality of the air, and regularly verifying that the requirements are being followed. Penalties and remedial measures may result from violations (Mahajan et al., 2023).

5.7.2.5 National clean air programme

The National Clean Air Programme was established in 2019 with the goal of lowering PM concentrations by 20%–30% by 2024 in order to improve air quality in Indian cities. In addition to raising awareness and creating action plans tailored to each city, the program's main objectives are to enhance air quality monitoring, data analysis, and enforcement (Guttikunda et al., 2023).

5.8 Summary and future prospects

Many different sources of anthropogenic activities are the foundation of the complex problem of urban air pollution. The deterioration of air quality in metropolitan areas is caused by a variety of factors, including agricultural practices, construction activities, vehicular exhaust, domestic burnings, and industrial pollutants, etc. In order to address these sources, a comprehensive and cooperative strategy combining technical advancements, community involvement, and governmental legislation is urgently required. Implementing stricter emission limits, promoting cleaner technology, creating sustainable urban planning projects, and implementing eco-friendly practices across a range of industries should all be part of the effort to reduce urban air pollution. The implementation of modern monitoring systems together with public awareness campaigns is crucial in cultivating a shared commitment toward the enhancement of air quality. The demand for practical solutions to reduce urban air pollution is growing as urbanization keeps speeding up. Cities can strive to strike a balance between environmental sustainability and development, protecting the health and welfare of their residents for future generations by recognizing the particulars of each source of pollution and putting comprehensive strategies into place.

References

Agrawal, M., Singh, B., Rajput, M., Marshall, F., & Bell, J. N. B. (2003). Effect of air pollution on peri-urban agriculture: A case study. *Environmental Pollution, 126*(3), 323–329.

Ainsworth, E. A. (2017). Understanding and improving global crop response to ozone pollution. *The Plant Journal, 90*(5), 886–897.

Akomolafe, O. O., Olorunsogo, T., Anyanwu, E. C., Osasona, F., Ogugua, J. O., & Daraojimba, O. H. (2024). Air quality and public health: A review of urban pollution sources and mitigation measures. *Engineering Science & Technology Journal, 5*(2), 259–271.

Anake, W. U., Bayode, F. O., Jonathan, H. O., Omonhinmin, C. A., Odetunmibi, O. A., & Anake, T. A. (2022). Screening of plant species response and performance for green belt development: Implications for semi-urban ecosystem restoration. *Sustainability, 14*(7), 3968.

Anda, A., Jakusch, P., & Kocsis, T. (2013). Effect of black caarbon on the growth, development and evapotranspriration of maize. *Applied Ecology and Environmental Research, 11*(4), 541–555.

Angon, P. B., Islam, M. S., Das, A., Anjum, N., Poudel, A., & Suchi, S. A. (2024). Sources, effects and present perspectives of heavy metals contamination: Soil, plants and human food chain. *Heliyon, 10*(7).

Antenozio, M. L., Caissutti, C., Caporusso, F. M., Marzi, D., & Brunetti, P. (2024). Urban air pollution and plant tolerance: Omics responses to ozone, nitrogen oxides, and particulate matter. *Plants, 13*(15), 2027.

Arab, L., Hoshika, Y., Müller, H., Cotrozzi, L., Nali, C., Tonelli, M., Ache, P., Paoletti, E., Alfarraj, S., Albasher, G., & Hedrich, R. (2024). Chronic ozone exposure preferentially modifies root rather than foliar metabolism of date palm (*Phoenix dactylifera*) saplings. *ESS Open Archive eprints, 572*, 57252039.

Awaad, H. A. (2021). Environmental pollution tolerance in crop plants. *Mitigating Environmental Stresses for Agricultural Sustainability in Egypt,* 141–173.

Baciak, M., Warmiński, K., & Bęś, A. (2015). *The Effect of Selected Gaseous Air Pollutants on Woody Plants.*

Balogun, I. A., Balogun, A. A., & Adegoke, J. (2014). Carbon monoxide concentration monitoring in Akure—A comparison between urban and rural environment. *Journal of Environmental Protection, 2014.*

Barwise, Y., & Kumar, P. (2020). Designing vegetation barriers for urban air pollution abatement: A practical review for appropriate plant species selection. *Npj Climate and Atmospheric Science, 3*(1), 12.

Benke, K., & Tomkins, B. (2017). Future food-production systems: Vertical farming and controlled-environment agriculture. *Sustainability: Science, Practice and Policy, 13*(1), 13–26.

Bruhl, R. J., Linder, S. H., & Sexton, K. (2013). Case study of municipal air pollution policies: Houston's air toxic control strategy under the white administration, 2004–2009. *Environmental Science & Technology, 47*(9), 4022–4028.

Bukhari, S., Ahmed, S., Azhar, A., Sardar, R., Hassan, S., & Ahmad, M. N. (2023). Environmental Fluoride Exposure and its Toxic Effects on Plants: A Review. *Fluoride, 56.*

Chole, A. S., Jadhav, A. R., & Shinde, V. (2021). Vertical farming: Controlled environment agriculture. *Just Agriculture, 1*, 249–256.

Choudhary, M. P., & Garg, V. (2013). Causes, consequences and control of air pollution. *All India Seminar on Methodologies for Air Pollution Control, Held at MNIT.*

Crippa, M., Solazzo, E., Guizzardi, D., Van Dingenen, R., & Leip, A. (2022). Air pollutant emissions from global food systems are responsible for environmental impacts, crop losses and mortality. *Nature Food, 3*(11), 942–956.

Das, S., Pal, D., & Sarkar, A. (2021). Particulate matter pollution and global agricultural productivity. *Sustainable Agriculture Reviews 50: Emerging Contaminants in Agriculture,* 79–107.

Deng, J., Zhang, Z. Y., Yang, Q., & Wu, X. D. (2022). Overview of non-methane volatile organic compounds for world economy: From emission source to consumption sink. *Energy Nexus, 6*, 100064.

Dewan, S., & Lakhani, A. (2024). Impact of ozone pollution on crop yield, human health, and associated economic costs in the Indo-Gangetic plains. *Science of the Total Environment, 945*, 173820.

Dhakal, K. P., & Chevalier, L. R. (2017). Managing urban stormwater for urban sustainability: Barriers and policy solutions for green infrastructure application. *Journal of Environmental Management, 203*, 171–181.

Dhir, B. (2016). Air pollutants and photosynthetic efficiency of plants. In *Plant responses to air pollution.* Singapore: Springer Singapore, 71–84.

Dumanović, J., Nepovimova, E., Natić, M., Kuča, K., & Jaćević, V. (2021). The significance of reactive oxygen species and antioxidant defense system in plants: A concise overview. *Frontiers in Plant Science, 11*, 552969.

Edo, G. I., Itoje-akpokiniovo, L. O., Obasohan, P., Ikpekoro, V. O., Samuel, P. O., Jikah, A. N., Nosu, L. C., Ekokotu, H. A., Ugbune, U., Oghroro, E. E. A., & Emakpor, O. L. (2024). Impact of environmental pollution from human activities on water, air quality and climate change. *Ecological Frontiers*.

Endress, A. G., Swiecki, T. J., & Taylor, O. C. (1978). Foliar and microscopic observations of bean leaves exposed to hydrogen chloride gas. *Environmental and Experimental Botany, 18*(2), 139–149.

Feng, L., Wei, L., Liu, Y., Ren, J., & Liao, W. (2023). Carbon monoxide/heme oxygenase system in plant: Roles in abiotic stress response and crosstalk with other signals molecules. *Nitric Oxide, 138*, 51–63.

Fischer, T. (2019). Wheat yield losses in India due to ozone and aerosol pollution and their alleviation: A critical review. *Outlook on Agriculture, 48*(3), 181–189.

Gajbhiye, M. D., Lakshmanan, S., Aggarwal, R., Kumar, N., & Bhattacharya, S. (2023). Evolution and mitigation of vehicular emissions due to India's Bharat Stage Emission Standards–A case study from Delhi. *Environmental Development, 45*, 100803.

Gallagher, J., Baldauf, R., Fuller, C. H., Kumar, P., Gill, L. W., & McNabola, A. (2015). Passive methods for improving air quality in the built environment: A review of porous and solid barriers. *Atmospheric Environment, 120*, 61–70.

Ghosh, A., Singh, A. A., Agrawal, M., & Agrawal, S. B. (2018). Ozone toxicity and remediation in crop plants. *Sustainable Agriculture Reviews, 27*, 129–169.

Glazener, A., & Khreis, H. (2019). Transforming our cities: Best practices towards clean air and active transportation. *Current environmental health reports, 6*, 22–37.

Gundacker, C., Forsthuber, M., Szigeti, T., Kakucs, R., Mustieles, V., Fernandez, M. F., Bengtsen, E., Vogel, U., Hougaard, K. S., & Saber, A. T. (2021). Lead (Pb) and neurodevelopment: A review on exposure and biomarkers of effect (BDNF, HDL) and susceptibility. *International Journal of Hygiene and Environmental Health, 238*, 113855.

Guo, P., Du, H., Wang, D., & Ma, M. (2021). Effects of mercury stress on methylmercury production in rice rhizosphere, methylmercury uptake in rice and physiological changes of leaves. *Science of the Total Environment, 765*, 142682.

Guo, W., Pan, B., Sakkiah, S., Yavas, G., Ge, W., Zou, W., Tong, W., & Hong, H. (2019). Persistent organic pollutants in food: Contamination sources, health effects and detection methods. *International journal of environmental research and public health, 16*(22), 4361.

Guttikunda, S., Ka, N., Ganguly, T., & Jawahar, P. (2023). Plugging the ambient air monitoring gaps in India's national clean air programme (NCAP) airsheds. *Atmospheric Environment, 301*, 119712.

Hakeem, K. R., Sabir, M., Ozturk, M., Akhtar, M. S., & Ibrahim, F. H. (2017). Nitrate and nitrogen oxides: Sources, health effects and their remediation. *Reviews of Environmental Contamination and Toxicology, 242*, 183–217.

Hati, A. J., & Singh, R. R. (2021). Smart indoor farms: Leveraging technological advancements to power a sustainable agricultural revolution. *AgriEngineering, 3*(4), 728–767.

Hossain, I., Rahman, M. S., Sattar, S., Haque, M., Mullick, A. R., Siraj, S., Sultana, N., Ahsan-Uz-Zaman, M., Samima, I. S., Haidar, A., & Khan, M. H. (2021). Environmental overview of air quality index (AQI) in Bangladesh: Characteristics and challenges in present era. *International Journal of Research in Engineering and Technology, 4*, 110–115.

Hossain, M., & Horowitz, H. M. (2025). Assessing compliance with updated NAAQS PM2. 5 annual standard: The role of regional and transboundary emissions. *105th Annual AMS Meeting 2025, 105*, 448980.

Huang, Y., Mubeen, S., Yang, Z., & Wang, J. (2022). Cadmium contamination in agricultural soils and crops. In *Theories and methods for minimizing cadmium pollution in crops: Case studies on water spinach*. Singapore: Springer Nature Singapore, 1–30.

Huebner, S. (2020). Tackling climate change, air pollution, and ecosystem destruction: How US-Japanese ocean industrialization and the metabolist movement's global legacy shaped environmental thought (circa 1950s–Present). *Environmental History*.

Husen, A. (2021). Morpho-anatomical, physiological, biochemical and molecular responses of plants to air pollution. *Harsh Environment and Plant Resilience: Molecular and Functional Aspects*, 203–234.

Jalota, S. K., Vashisht, B. B., Sharma, S., & Kaur, S. (2018). *Understanding climate change impacts on crop productivity and water balance*. Academic Press.

Institute for Health Metrics and Evaluation (2013). *The ase: licyThe global burden of disease: Generating evidence, guiding policy*. Seattle, WA: IHME.

Jha, S., & Yadav, A. (2023). Plants response to SO_2 or acid deposition. In *Plants and their interaction to environmental pollution*. Elsevier, 99–108.

Jiang, X., Li, G., & Fu, W. (2021). Government environmental governance, structural adjustment and air quality: A quasi-natural experiment based on the three-year Action Plan to Win the blue sky defense war. *Journal of Environmental Management, 277*, 111470.

Jiechen, W., Yue, W., Huihui, Z., Dandan, G., & Guangyu, S. (2021). Atmospheric nitrogen dioxide at different concentrations levels regulates growth and photosynthesis of tobacco plants. *Journal of Plant Interactions, 16*(1), 422–431.

Khan, R., Noorpoor, A., & Ebadi, A. G. (2022). Effects of air contamination on agriculture. In *Sustainable plant nutrition under contaminated environments*. Cham: Springer International Publishing, 1–16.

Kousehlar, M., & Widom, E. (2020). Identifying the sources of air pollution in an urban-industrial setting by lichen biomonitoring-A multi-tracer approach. *Applied Geochemistry, 121*, 104695.

Kumar, P., Zavala-Reyes, J. C., Tomson, M., & Kalaiarasan, G. (2022). Understanding the effects of roadside hedges on the horizontal and vertical distributions of air pollutants in street canyons. *Environment International, 158*, 106883.

Lahieb Faisal, M., Ali, S. A. K., Al-Sharify, Z. T., & Ali, A. (2020). The effect of smoke from factories, electricity generator and vehicles on human health and environment. A review. *Solid State Technology, 63*(6), 21390–21398.

Lee, J. K., Kwak, M. J., Jeong, S. G., & Woo, S. Y. (2022). Individual and interactive effects of elevated ozone and temperature on plant responses. *Horticulturae, 8*(3), 211.

Li, P., Xin, J., Wang, Y., Li, G., Pan, X., Wang, S., Cheng, M., Wen, T., Wang, G., & Liu, Z. (2015). Association between particulate matter and its chemical constituents of urban air pollution and daily mortality or morbidity in Beijing City. *Environmental Science and Pollution Research, 22*, 358–368.

Li, X. B., Yuan, B., Wang, S., Wang, C., Lan, J., Liu, Z., Song, Y., He, X., Huangfu, Y., Pei, C., & Cheng, P. (2022). Variations and sources of volatile organic compounds (VOCs) in urban region: Insights from measurements on a tall tower. *Atmospheric Chemistry and Physics, 22*(16), 10567–10587.

Liao, J., Jin, A. Z., Chafe, Z. A., Pillarisetti, A., Yu, T., Shan, M., Yang, X., Li, H., Liu, G., & Smith, K. R. (2017). The impact of household cooking and heating with solid fuels on ambient PM2. 5 in peri-urban Beijing. *Atmospheric Environment, 165*, 62–72.

Lin, D. Y., Waller, S. T., & Lin, M. Y. (2024). A review of urban planning approaches to reduce air pollution exposures. *Current Environmental Health Reports,* 1–10.

Liu, H., Liu, S., Xue, B., Lv, Z., Meng, Z., Yang, X., Xue, T., Yu, Q., & He, K. (2018). Ground-level ozone pollution and its health impacts in China. *Atmospheric Environment, 173*, 223–230.

Losada, H., Martínez, H., Vieyra, J., Pealing, R., Rivera, J., Zavala, R., & Cortés, J. (2018). Urban agriculture in the metropolitan zone of Mexico City: Changes over time in urban, suburban and peri-urban areas. In *The earthscan reader in rural-urban linkages*. Routledge, 247–264.

Mahajan, A., Ghosh, S., Dhawan, P., & Krishna, B. (2023). *The state of India's pollution control boards*.

Mahida, D. K., Makwana, V. M., Sankhla, M. S., Patel, A., & Dodia, P. (2023). Accumulation of heavy metals in roadside plants and their role in phytoremediation. In *Anthropogenic environmental hazards: Compensation and mitigation*. Cham: Springer Nature Switzerland, 119–141.

Mallik, C. (2019). Anthropogenic sources of air pollution. In *Air pollution: Sources, impacts and controls*. Wallingford UK: CAB International, 6–25.

Marson, F., Eagly, P., & Russell, O. M. (2023). Impact of social assistance programs on family economic welfare in urban environments: An analysis of equity and community engagement. *Law and Economics, 17*(3), 186–202.

Masutomi, Y., Kinose, Y., Takimoto, T., Yonekura, T., Oue, H., & Kobayashi, K. (2019). Ozone changes the linear relationship between photosynthesis and stomatal conductance and decreases water use efficiency in rice. *Science of the Total Environment, 655*, 1009–1016.

Mills, G., Sharps, K., Simpson, D., Pleijel, H., Frei, M., Burkey, K., Emberson, L., Uddling, J., Broberg, M., Feng, Z., & Kobayashi, K. (2018). Closing the global ozone yield gap: Quantification and cobenefits for multistress tolerance. *Global Change Biology, 24*(10), 4869–4893.

Mohammed, A., & Devi, P. (2021). Hydrogen sulfide: Risk assessment, environmental, and health hazard. In *Hazardous Gases*. Academic Press, 209–223.

Moretti, C. L., Mattos, L. M., Calbo, A. G., & Sargent, S. A. (2010). Climate changes and potential impacts on postharvest quality of fruit and vegetable crops: A review. *Food Research International, 43*(7), 1824–1832.

Mori, J., Fini, A., Galimberti, M., Ginepro, M., Burchi, G., Massa, D., & Ferrini, F. (2018). Air pollution deposition on a roadside vegetation barrier in a Mediterranean environment: Combined effect of evergreen shrub species and planting density. *Science of the Total Environment, 643*, 725–737.

Mrabet, R. (2023). Sustainable agriculture for food and nutritional security. In *Sustainable agriculture and the environment*. Academic Press, 25–90.

Nabulo, G., Young, S. D., & Black, C. R. (2010). Assessing risk to human health from tropical leafy vegetables grown on contaminated urban soils. *Science of the Total Environment, 408*(22), 5338–5351.

Naiker, Y., Diab, R. D., Zunckel, M., & Hayes, E. T. (2012). Introduction of local air quality management in south Africa: Overview and challenges. *Environmental science & policy, 17*, 62–71.

Naz, M., Dai, Z., Hussain, S., Tariq, M., Danish, S., Khan, I. U., Qi, S., & Du, D. (2022). The soil pH and heavy metals revealed their impact on soil microbial community. *Journal of Environmental Management, 321*, 115770.

Noe, S. M., Kimmel, V., Hüve, K., Copolovici, L., Portillo-Estrada, M., Püttsepp, Ü., Jõgiste, K., Niinemets, Ü., Hörtnagl, L., & Wohlfahrt, G. (2011). Ecosystem-scale biosphere–atmosphere interactions of a hemiboreal mixed forest stand at Järvselja, Estonia. *Forest Ecology and Management, 262*(2), 71–81.

Nowak, D. J., Hirabayashi, S., Bodine, A., & Greenfield, E. (2014). Tree and forest effects on air quality and human health in the United States. *Environmental Pollution, 193*, 119–129.

Odubo, T. C., & Kosoe, E. A. (2024). Sources of air pollutants: Impacts and solutions. In *Air pollutants in the context of one health: Fundamentals, sources, and impacts*. Cham: Springer Nature Switzerland, 75–121.

Oksanen, E., & Kontunen-Soppela, S. (2021). Plants have different strategies to defend against air pollutants. *Current Opinion in Environmental Science & Health, 19*, 100222.
Onivefu, A. P., & Imarhiagbe, O. (2024). Types of air pollutants. In *Air pollutants in the context of one health: Fundamentals, sources, and impacts*. Cham: Springer Nature Switzerland, 123–160.
Ottosen, T. B., & Kumar, P. (2020). The influence of the vegetation cycle on the mitigation of air pollution by a deciduous roadside hedge. *Sustainable Cities and Society, 53*, 101919.
Ozougwu, J. C. (2016). The role of reactive oxygen species and antioxidants in oxidative stress. *International Journal of Research, 1*(8), 1–8.
Panda, D. (2015). Fluoride toxicity stress: Physiological and biochemical consequences on plants. *International Journal of Bioresearch and Environmental Agricultural Science, 1*, 70–84.
Phuntsho, S., & Kanitpong, K. (2010). Vehicle emission control strategies and public opinion in Bhutan. *Transportation Research Record, 2157*(1), 103–110.
Pinto, D. M., Blande, J. D., Souza, S. R., Nerg, A. M., & Holopainen, J. K. (2010). Plant volatile organic compounds (VOCs) in ozone (O_3) polluted atmospheres: The ecological effects. *Journal of Chemical Ecology, 36*, 22–34.
Prigioniero, A., Zuzolo, D., Niinemets, Ü., & Guarino, C. (2021). Nature-based solutions as tools for air phytoremediation: A review of the current knowledge and gaps. *Environmental Pollution, 277*, 116817.
Qiu, Z., Li, X., Hao, Y., & Deng, S. (2016). Potential of diesel emissions reduction strategies in Xi'an, China. *Clean Technologies and Environmental Policy, 18*, 2717–2724.
Rafie-Rad, Z., Raza, T., Eash, N. S., Moradi-Khajevand, M., & Moradkhani, M. (2024). Effects of outdoor air pollutants on plants and agricultural productivity. In *Health and environmental effects of ambient air pollution*. Academic Press, 71–90.
Rai, P. K. (2016). Impacts of particulate matter pollution on plants: Implications for environmental biomonitoring. *Ecotoxicology and Environmental Safety, 129*, 120–136.
Rai, P. K., Kim, K. H., Lee, S. S., & Lee, J. H. (2020). Molecular mechanisms in phytoremediation of environmental contaminants and prospects of engineered transgenic plants/microbes. *Science of the Total Environment, 705*, 135858.
Rani, M., Vikas, Kumar, R., Lathwal, M., & Kamboj, A. (2024). Effect and responses of lead toxicity in plants. In *Lead toxicity mitigation: Sustainable nexus approaches*. Cham: Springer Nature Switzerland, 211–241.
Reche, C., Viana, M., Pandolfi, M., Alastuey, A., Moreno, T., Amato, F., Ripoll, A., & Querol, X. (2012). Urban NH3 levels and sources in a Mediterranean environment. *Atmospheric Environment, 57*, 153–164.
Robert, H., Brennan, E., & Leone, I. A. (2018). Air pollution, plant response, and productivity. In *Handbook of agricultural productivity*. CRC Press, 375–399.
Ryalls, J. M., Langford, B., Mullinger, N. J., Bromfield, L. M., Nemitz, E., Pfrang, C., & Girling, R. D. (2022). Anthropogenic air pollutants reduce insect-mediated pollination services. *Environmental Pollution, 297*, 118847.
Sadiq, A. A. (2022). *Effect of particulate emissions from road transportation vehicles on health of communities in urban and rural areas, Kano State, Nigeria (Doctoral dissertation, Université Claude Bernard-Lyon I)*.
Schauberger, B., Rolinski, S., Schaphoff, S., & Müller, C. (2019). Global historical soybean and wheat yield loss estimates from ozone pollution considering water and temperature as modifying effects. *Agricultural and Forest Meteorology, 265*, 1–15.
Shahid, M., Dumat, C., Khalid, S., Schreck, E., Xiong, T., & Niazi, N. K. (2017). Foliar heavy metal uptake, toxicity and detoxification in plants: A comparison of foliar and root metal uptake. *Journal of Hazardous Materials, 325*, 36–58.
Sharps, K., Hayes, F., Harmens, H., & Mills, G. (2021). Ozone-induced effects on leaves in African crop species. *Environmental Pollution, 268*, 115789.

Sheng, Q., & Zhu, Z. (2019). Effects of nitrogen dioxide on biochemical responses in 41 garden plants. *Plants, 8*(2), 45.

Shi, Z. (2019). Urban air pollution. *Atmospheric Science for Environmental Scientists,* 331.

Singh, D., & Verma, S. K. (2023). Impacts of particulate matter pollution on plants. In *New frontiers in plant-environment interactions: Innovative technologies and developments*. Cham: Springer Nature Switzerland, 483–500.

Sivakumar, M. V. (2011). Climate and land degradation. *Sustaining Soil Productivity in Response to Global Climate Change: Science, Policy, and Ethics,* 141–154.

Suner, M. (2024). Analysis of air pollution from three main transportation vehicles: A case study. *Energy Sources, Part A: Recovery, Utilization, and Environmental Effects, 46*(1), 1890–1906.

Tai, A. P., Martin, M. V., & Heald, C. L. (2014). Threat to future global food security from climate change and ozone air pollution. *Nature Climate Change, 4*(9), 817–821.

Tiwari, S., Agrawal, M., Tiwari, S., & Agrawal, M. (2018). Effect of ozone on physiological and biochemical processes of plants. *Tropospheric Ozone and its Impacts On Crop Plants: A Threat to Future Global Food Security,* 65–113.

Tomson, M., Kumar, P., Barwise, Y., Perez, P., Forehead, H., French, K., Morawska, L., & Watts, J. F. (2021). Green infrastructure for air quality improvement in street canyons. *Environment International, 146,* 106288.

Tsai, W. T. (2016). Current status of air toxics management and its strategies for controlling emissions in Taiwan. *Toxics, 4*(2), 8.

Uzoho, C. (2025). The public health impact of airborne particulate matter: Risks, mechanisms, and mitigation strategies. *Communication In Physical Sciences, 12*(2), 600–619.

Valavanidis, A., Fiotakis, K., Vlahogianni, T., Bakeas, E. B., Triantafillaki, S., Paraskevopoulou, V., & Dassenakis, M. (2006). Characterization of atmospheric particulates, particle-bound transition metals and polycyclic aromatic hydrocarbons of urban air in the centre of Athens (Greece). *Chemosphere, 65*(5), 760–768.

Verma, P., Kumari, T., & Raghubanshi, A. S. (2021). Energy emissions, consumption and impact of urban households: A review. *Renewable and Sustainable Energy Reviews, 147,* 111210.

Vitaliano, S., Cascone, S., & D'Urso, P. R. (2024). Mitigating built environment air pollution by green systems: An in-depth review. *Applied Sciences, 14*(15), 6487.

Xu, C., Dong, L., Yu, C., Zhang, Y., & Cheng, B. (2020). Can forest city construction affect urban air quality? The evidence from the Beijing-Tianjin-Hebei urban agglomeration of China. *Journal of Cleaner Production, 264,* 121607.

Yang, H., Zhou, J., & Zhou, J. (2024). Interactive effects of ammonium sulfate and lead on alfalfa in rare earth tailings: Physiological responses and toxicity thresholds. *Science of the Total Environment, 947,* 174439.

Yang, W., Yu, C., Yuan, W., Wu, X., Zhang, W., & Wang, X. (2018). High-resolution vehicle emission inventory and emission control policy scenario analysis, a case in the Beijing-Tianjin-Hebei (BTH) region, China. *Journal of Cleaner Production, 203,* 530–539.

Yang, Y., Hassan, M. F., Ali, W., Zou, H., Liu, Z., & Ma, Y. (2025). Effects of cadmium pollution on human health: A narrative review. *Atmosphere, 16*(2), 225.

Ye, J., Li, J., Wang, P., Ning, Y., Liu, J., Yu, Q., & Bi, X. (2022). Inputs and sources of Pb and other metals in urban area in the post leaded gasoline era. *Environmental Pollution, 306,* 119389.

CHAPTER SIX

Plant's role in producing volatile organic compounds and their impacts on air pollution

Rachna Tyagi
Department of Botany, D.A.V. College, Muzaffarnagar, Uttar Pradesh, India

6.1 Introduction

Plants constitute one of the primary natural sources of volatile organic compounds (VOCs) in the atmosphere. Available estimates indicate that vegetation emits more than 10^9 metric tons of these compounds each year. A notable share of the carbon fixed through photosynthesis, sometimes approaching 40%, is utilized in the synthesis of VOCs. Rather than being released individually, these compounds are emitted as complex blends. Such emissions are closely linked with plant communication, enabling interactions not only with neighboring plants but also with the surrounding biotic and abiotic environment. In the troposphere, VOCs of plant origin exhibit high reactivity and relatively short lifetimes, typically ranging from a few minutes to a few hours. Their direct contribution to radiative forcing may be limited; however, their influence on atmospheric chemistry is considerable. These compounds participate actively in oxidation–reduction reactions and thereby help maintain the chemical balance of the lower atmosphere. In several regions, biogenic hydrocarbons account for a substantial portion of reactive atmospheric constituents, at times surpassing contributions from anthropogenic as well as other natural nonbiological sources. Although VOCs are produced in both terrestrial and aquatic systems, terrestrial vegetation is responsible for the dominant share—nearly 90%—of total emissions. On a global scale, the annual release is estimated to be around 400–800 teragrams of carbon in the form of hydrocarbons, which is broadly comparable to methane emissions from combined natural and human-driven processes. Under certain conditions, a substantial proportion of assimilated carbon—reported to be as high as 36%—may be returned to the atmosphere as VOCs.

Urban Air Pollution and Plant Responses
DOI: https://doi.org/10.1016/B978-0-443-40320-0.00008-5

Emission of VOCs occurs from various plant organs, including leaves, flowers, fruits, and roots. Compounds released from roots enter the soil matrix, where they can influence below-ground interactions. The nature and quantity of VOC emissions differ widely across plant species and functional groups. Even within a single species, variation is commonly observed, often linked to genetic factors, environmental conditions, and geographical location. Ecologically, VOCs fulfill multiple roles. They are involved in defense responses, for instance, by deterring herbivores or limiting pathogen attack. At the same time, these compounds contribute to the attraction of pollinators, seed dispersers, and beneficial microorganisms. VOC-mediated signaling between plants, as well as between plants and insects, forms an important component of ecological interactions. In certain situations, VOCs are also associated with wound sealing and protection of damaged tissues. From the perspective of atmospheric processes, plant VOCs are significant because they act as precursors to a range of secondary pollutants. Through a series of chemical transformations, they give rise to compounds that may adversely affect plant health. When VOCs interact with nitrogen oxides (NO_x) in the presence of sunlight, photochemical reactions are initiated, leading to the formation of smog. One important outcome of these reactions is the production of tropospheric ozone (O_3), which is known to be harmful to both vegetation and human health. In addition, VOCs influence the oxidative capacity of the atmosphere and play a role in regulating concentrations of gases such as carbon monoxide, O_3, and secondary organic aerosols. From a chemical standpoint, plant VOCs comprise a highly diverse group of compounds. Major categories include terpenoids, fatty acid-derived C_6 compounds, and phenylpropanoid-based aromatics. In addition, several other classes—such as alkanes, alkenes, alcohols, esters, aldehydes, and ketones—are also represented (Holopainen, 2004;). To date, more than 1700 volatile compounds have been identified across over 90 plant families, accounting for roughly 1% of the known plant secondary metabolites.

6.2 Historical perspective of plant volatile organic compounds in atmospheric science

The study of VOCs released by plants started in the 1920s and 1930s. Some of the earliest research was carried out in the former Soviet Union. A major development came when Haagen-Smit et al. (1952) reported that plants growing in chaparral regions release large amounts of monoterpenes

($C_{10}H_{16}$) from their leaves. Later, Sanadze (1957), working under a Soviet research program on plant defense, discovered that some plants also release high amounts of isoprene (C_5H_8). After this, a study suggested that these plant-produced hydrocarbons could play an important role in atmospheric chemical reactions. However, even after this important idea, research in this field progressed slowly for about 10–15 years, and only a few scientists continued to study how and why plants release VOCs.

A major breakthrough came with the work of Jones and Rasmussen (1975). They carried out detailed studies on plant physiology and proposed that isoprene emission is linked to photorespiration and may be produced as a byproduct of plant metabolism. Further important findings came from experiments conducted at the NASA Ames Research Centre in California. In these studies, researchers worked on *Salvia mellifera* and found that the release of monoterpenes increases rapidly with rising temperature. They also observed that this process is not strongly affected by immediate photosynthesis or light intensity. Based on these results, scientists suggested that monoterpene emission mainly happens due to the evaporation of compounds already stored inside plant tissues, rather than being directly produced during photosynthesis at that moment. Later, improvements in micrometeorological techniques, especially the use of instruments placed on towers above plant canopies, helped scientists clearly show that VOCs released by plants move from vegetation into the lower atmosphere (troposphere). Further atmospheric studies and models showed that isoprene emissions have a strong effect on the formation of O_3 and carbon monoxide in both rural and urban areas.

6.3 Site of biosynthesis and emission of volatile organic compounds

VOCs are released from both above-ground and below-ground parts of plants (Fig. 6.1). The aerial parts, like leaves, stems, and flowers, release VOCs directly into the atmosphere, while roots release them into the surrounding soil. In above-ground plant parts, VOCs can be released very quickly when the plant tissue is damaged, such as during physical injury or herbivore attack. In such cases, the cells break open, and the compounds escape directly into the air. However, in healthy and undamaged tissues, the release of VOCs is more controlled. These compounds have to pass through several layers inside the plant, including cell organelles, cytoplasm,

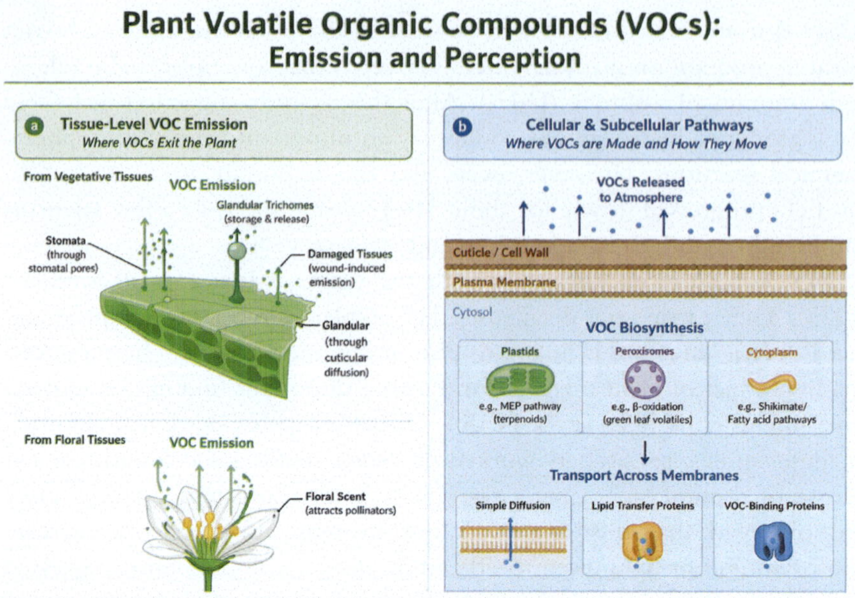

Figure 6.1 Overview of plant VOCs-emission. (This figure was generated with the assistance of AI-based tools and subsequently refined by the author to ensure scientific accuracy and clarity.) (*VOCs*, volatile organic compounds).

plasma membrane, cell wall, and finally the cuticle, which is the outer protective layer. Because of this, the release of VOCs depends on both the structure of the plant cells and the chemical nature of the compounds.

In flowers, VOCs are usually released from both upper (adaxial) and lower (abaxial) epidermal cells, which are covered by a cuticle. In leaves, the cuticle also covers the stomata and sometimes lines the inner surfaces of air spaces within the leaf. This structure affects how VOCs move out of the leaf. Some VOCs, like isoprene, are mainly produced in mesophyll cells inside the leaf. These cells do not have a cuticle barrier, so the compounds can easily move into the air spaces between cells. From there, they are released into the atmosphere mainly through the stomata. Roots are different from aerial plant parts because they do not have a cuticle or stomata. Due to this, the release of VOCs from roots follows a different pattern. The absence of a cuticle means there is less storage of these compounds and less protection from their toxic effects. At the same time, the lack of stomata slows down their movement, so VOCs are released more slowly into the soil environment around the roots.

6.4 Biosynthesis and storage of plant volatile organic compounds

The biosynthesis of VOCs in plants is closely linked to photosynthesis. During photosynthesis, plants produce carbohydrates, which are mainly used in primary metabolism to make important molecules, such as proteins, structural carbohydrates, and lipids needed for growth and survival. However, some part of this carbon is also used to produce secondary metabolites, including VOCs. These VOCs are formed from both basic metabolic pathways and specialized secondary pathways. The main types of plant VOCs include isoprenoids (like isoprene and terpenoids), fatty acid-derived compounds, phenylpropanoids and benzenoids, and compounds derived from amino acids. These compounds play important roles in plant defense, communication, and interaction with the environment. Plants produce many secondary metabolites as a defense against harmful organisms like herbivores and pathogens. Some of these compounds are produced continuously and may either be released outside the plant or stored inside special compartments within the plant. In many plants, VOCs are stored in special structures such as resin ducts and glandular trichomes. These stored compounds can be quickly released when the plant is damaged by insects or physical injury.

Since many VOCs can be harmful to plant cells if present in high amounts, plants have developed ways to safely produce and store them. One common method is to store these compounds in an inactive or less harmful form, such as bound with sugar molecules (glycosides). These can later be activated by enzymes when needed. Another method is to keep them stored in special compartments or secretory structures, which prevents damage to normal plant cells. Glandular trichomes are one of the most important structures for storing and releasing VOCs. These are small hair-like structures present on the surface of plants. They are found in many plant families, including Lamiaceae, Asteraceae, Geraniaceae, Solanaceae, and Cannabaceae. These structures develop from the outer cell layer (protoderm) and are commonly found on leaves, flowers, and sometimes seeds. When these structures are damaged, they release VOCs into the environment, which can help in repelling herbivores, stopping microbial growth, and supporting plant interactions.

6.5 Induced production of volatile organic compounds in plants

Unlike floral scents, which are mainly controlled by the plant's developmental stages, the type and amount of VOCs released from vegetative parts, such as leaves, can change a lot depending on environmental stress. VOCs released from leaves, flowers, fruits, and roots perform many important roles. They help plants defend against herbivores and pathogens and also attract pollinators, which supports successful reproduction. Apart from these direct roles, some VOCs act as airborne signals. They can travel through the air and activate defense responses in other parts of the same plant or even in nearby plants. This type of chemical communication helps plants strengthen both direct and indirect defense systems. The induced defense response in plants usually occurs in three main steps: signal perception, signal transmission, and production of defense compounds. First, the plant detects danger through specific signals released during herbivore attack or damage. Then, these signals are passed through complex pathways inside the plant. Finally, this leads to the activation of genes that produce defensive substances, including VOCs. The production of VOCs can be triggered by both biotic factors, such as insect feeding and pathogen attack, as well as abiotic stresses such as pollution. Among these, O_3 is very important. O_3 can change both the amount and type of VOCs released by plants, which in turn affects how plants interact with insects.

Different studies have shown mixed effects of O_3 on plant VOCs. For example, a study found that O_3 exposure increased egg-laying (oviposition) by the insect *Manduca sexta* on tobacco plants. This was not due to changes in leaf chemistry but because of changes in VOC emissions. On the other hand, another studyobserved that VOCs released from O_3-treated red spruce needles acted as repellents for spruce budworm larvae. O_3 can also disturb plant communication by masking important VOC signals released during herbivore attack. It can change the balance of compounds in VOC mixtures and affect how these signals are received by other organisms. At the same time, O_3 can increase the release of certain signaling compounds such as methyl salicylate and 4,8-dimethyl-1,3,7-nonatriene. These compounds are important signals for higher-level organisms like predators and parasitoids, helping them locate their prey.

6.6 Biological aspects of volatile organic compound emissions

Among the VOCs released by plants, isoprene and monoterpenes are the most common. These belong to a group called isoprenoids, which are mainly produced through two important pathways in plants: the mevalonic acid (MVA) pathway and the methylerythritol phosphate pathway. The release of VOCs from leaves into the atmosphere mainly happens through diffusion. This means that the movement of these compounds depends on the difference in their concentration between the inside of the leaf and the outside air. The rate at which VOCs are released also depends on how easily they can move through plant tissues and the resistance they face along the pathway. The release of VOCs is also closely related to where they are produced and stored in the plant. For example, isoprene is produced in chloroplasts and is not stored in plant tissues. Because of this, it is released very quickly after it is formed and mainly escapes through the stomata into the atmosphere. On the other hand, monoterpenes are produced in special secretory cells and are often stored in structures like resin ducts or secretory cavities. These storage structures help control their release and prevent them from escaping immediately.

Research studies have shown clear differences in how these compounds are released. A study found that monoterpenes can be detected on both sides of leaves that have stomata only on one surface. This shows that monoterpenes can move not only through stomata but also through the cuticle. In contrast, isoprene is released mainly through the stomata, as it is found only on the stomatal side of the leaf. From a biological and evolutionary point of view, VOCs are very important in plant–pollinator interactions. Plants produce different types of VOCs that act as scent signals to attract pollinators. Many of these compounds are also used by humans as essential oils. Pollinators depend a lot on these chemical signals, which often work together with visual signals such as flower color. The role of VOCs becomes even more important in low-light or night conditions, where visual signals are not very effective. In such cases, VOCs help pollinators find flowers and locate resources like nectar and pollen. Therefore VOC emissions are very important for plant reproduction as well as for maintaining ecological interactions.

6.7 Isoprene biosynthesis and emission

Isoprene is one of the most commonly released VOCs in land ecosystems. It is produced by many biological sources, such as leaves, flowers, stems, trunks, roots, decomposing leaf litter, and even soil microorganisms. Among these, leaves are the main source and contribute the highest amount of isoprene to the atmosphere. The production and release of isoprene are controlled by many plant and environmental factors, making it a complex process. In some plants, isoprene or similar compounds may be temporarily stored inside plant tissues and are released mainly when the tissue is damaged. However, in many other plants, isoprene is produced in cells that are in direct contact with the atmosphere. In such cases, it is released immediately after it is formed. This shows that isoprene emission is very dynamic and changes quickly depending on environmental conditions and the plant's physiological state. In addition to regular emissions, isoprene release can also increase under stress conditions such as high temperature or drought. The pathway for isoprene production is well understood, and the enzyme responsible for its final formation, called isoprene synthase, has been identified and studied. Scientists have also made progress in studying the gene responsible for this enzyme, which helps in understanding how isoprene production is controlled in plants. Research shows that isoprene helps plants tolerate environmental stress. It protects plant cells, especially during sudden increases in temperature and strong sunlight, by stabilizing cell structures. This makes isoprene an important compound for plant survival under stress conditions. Isoprene is produced inside chloroplasts through metabolic pathways related to isoprenoid synthesis (Fig. 6.2). Important precursor molecules like isopentenyl pyrophosphate and dimethylallyl pyrophosphate (DMAPP) are formed from carbon compounds produced during photosynthesis. Since plants do not have special structures to store isoprene, it is released almost immediately after it is formed. Therefore the rate of isoprene emission is closely linked to its rate of production.

Several factors affect isoprene production, including light intensity, temperature, plant species, stage of growth, and availability of resources. Light plays a very important role because the carbon fixed during photosynthesis is the main source for isoprene production. When light decreases, such as during night or low-light conditions, isoprene emission also decreases. However, isoprene production is not completely dependent on current photosynthesis. When photosynthesis is temporarily reduced,

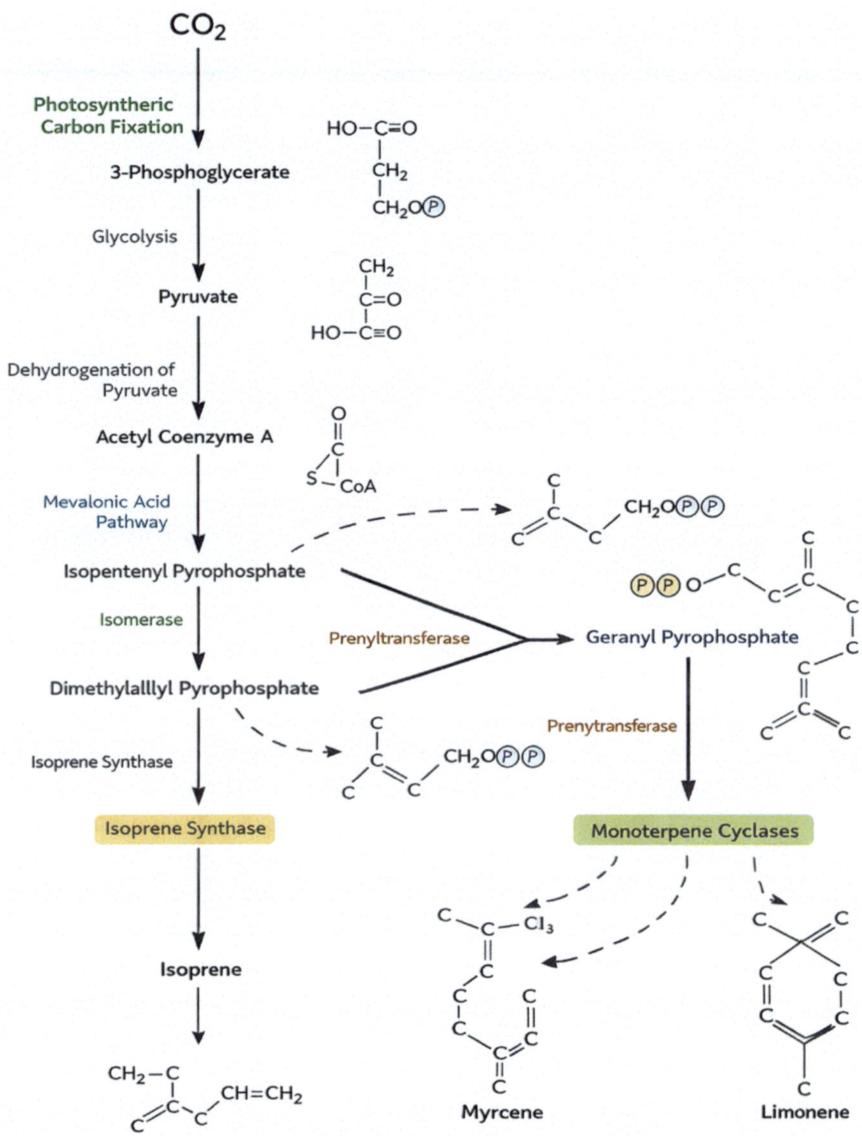

Figure 6.2 Biosynthetic pathway of isoprene and monoterpenes via the mevalonate (MVA) pathway in plants. (This figure was generated with the assistance of AI-based tools and subsequently refined by the author to ensure scientific accuracy and clarity.)

plants can use stored carbon to continue producing isoprene. For example, during drought conditions, when stomata close and carbon dioxide intake is reduced, isoprene emission can still continue for some time using stored resources. The second important role of light is related to the control of the enzyme called isoprene synthase. This enzyme helps in converting DMAPP into isoprene. It is believed that this enzyme becomes active in the presence of light. This activation may happen due to changes inside the chloroplast, such as an increase in pH and magnesium ions (Mg^{2+}), which are caused by photosynthetic electron transport.

A consistent relationship between light intensity and isoprene emission is observed across diverse plant species and varying environmental conditions. This suggests that the basic mechanism controlling isoprene production is common across plants. Isoprene emission is also strongly affected by temperature, which shows that enzymes play an important role in its production. In most plants, especially those found in tropical and temperate regions, the highest isoprene emission is observed around 40°C. When the temperature rises above this level, the emission decreases sharply because the enzymes involved in isoprene production start to lose their function due to high heat. Isoprene emission also varies between different plant groups, especially at the genus level. For example, in the Fagaceae family, many species of the genus Quercus (oak trees) release high amounts of isoprene. In contrast, related genera like *Fagus* (beeches) and *Castanea* (sweet chestnut) usually do not release isoprene. Similarly, in the Pinaceae family, species of the genus *Picea* (spruce) are known to emit isoprene, while other genera in the same family show very little or no emission. These differences suggest that the ability to produce isoprene has either evolved independently in different plant groups or has been lost in some groups during evolution. The stage of leaf development also affects isoprene emission. Young leaves usually release very small amounts of isoprene, while fully grown leaves show much higher emission rates. In addition, the basic emission rate (measured under standard light and temperature conditions) can increase if the plant has been exposed to high temperatures (above 32°C), even for a short time, such as 24 hours.

Nutrient availability also plays an important role. Plants with higher nitrogen levels tend to produce more isoprene under the same environmental conditions. Isoprene emission is also linked to the carbon status of the leaf, especially the amount of stored starch. This shows how plants use and store carbon, which affects how much isoprene they release. Even after many studies, the exact function of isoprene in plants is still not fully clear. However, one widely accepted idea is that isoprene helps plants tolerate high temperatures. Experiments support this idea. For example, when leaves of isoprene-producing plants like *Pueraria alba* or *Quercus rubra* are kept in an environment without isoprene (such as nitrogen gas), they get damaged at lower temperatures. But when isoprene is supplied again, the plants can tolerate higher temperatures. This shows that isoprene helps protect plants from heat stress.

6.8 Monoterpene biosynthesis and emission

Monoterpenes are a group of VOCs made of 10 carbon atoms (C_{10}) and belong to the isoprenoid group. They are produced through pathways related to the MVA pathway. Unlike isoprene, monoterpenes are usually stored inside special structures in plants. These structures differ in different plants. For example, they are stored in glandular trichomes in plants like mint and basil, in resin ducts in pine needles, in resin blisters in fir (*Abies*), and in secretory cavities in eucalyptus leaves. Monoterpene emission is generally not strongly dependent on light, unlike isoprene. In most plants, the cells that produce monoterpenes are located close to these storage structures. This allows the compounds to accumulate before being released. Because of this storage system, monoterpenes are mainly released through simple evaporation (volatilization) of already formed compounds, rather than being directly linked to immediate metabolic activity. The rate at which monoterpenes are released depends on their volatility and their concentration inside plant tissues. From an ecological point of view, monoterpenes play an important role in plant defense. Even though they may not always be highly toxic to herbivores, they help protect plants in both physical and chemical ways. For example, in conifer plants, resin containing monoterpenes can block the mouthparts of insects like bark beetles, preventing them from feeding further. Monoterpenes can also stop or slow down the growth of harmful fungi, thus protecting plants from infections.

The type of monoterpenes produced by a plant is mainly controlled by its genes, so different species and even different populations of the same species can show variation. Because of this, monoterpene profiles are often used in plant classification (chemotaxonomy). However, the total amount of monoterpenes depends on environmental conditions, especially the availability of nutrients. Monoterpenes are considered carbon-based defense compounds. Their production is often linked to the balance between carbon and nutrients in the plant. According to ecological theories like the carbon–nutrient balance hypothesis and growth differentiation balance theory, when plant growth is limited due to a lack of nutrients, but photosynthesis continues, extra carbon is used to produce secondary metabolites like monoterpenes. This helps the plant improve its defense system.

When nutrients are low, but photosynthesis is still active, plants may store extra carbon and use it to make more monoterpenes and other defensive compounds. In fast-growing annual plants, there is usually a clear trade-off between growth and defense when nutrients are limited; more energy is used for defense. However, in long-living plants like conifers, this trade-off is not always seen. In some cases, higher nitrogen levels may even increase monoterpene production. Although monoterpenes help in defense, they cannot fully protect plants from herbivores. When plant tissues containing stored monoterpenes are damaged by insects or physical injury, large amounts of these compounds are quickly released into the air. In such cases, the release mainly depends on diffusion, which is controlled by the concentration difference between inside the plant and the outside air, as well as the chemical properties of the compounds. Apart from this passive release, herbivore attack can also trigger active responses in the plant. The enzymes involved in monoterpene production, especially monoterpene cyclases (which convert geranyl pyrophosphate (GPP) into different monoterpenes), increase after tissue damage. Studies have shown that after herbivore attack, the activity of these enzymes increases, leading to higher production of monoterpenes. This shows that plants can actively respond to damage by increasing their defense capacity.

6.9 Diversity of plant volatile organic compounds

Plants release a very large number of different VOCs into the atmosphere. Classification, sources and functional characteristics of major plant VOCs are discussed in Table 6.1. It is estimated that there are tens of thousands of such compounds, showing the great chemical diversity present in plants. These compounds usually have low molecular weight and can be grouped into different categories. The main groups include isoprenoids (such as isoprene, monoterpenes, and sesquiterpenes), oxygenated VOCs like methanol, ethanol, acetone, and acetaldehyde, and green leaf volatiles such as hexenal and other C_6 compounds. Among all these, isoprene makes up the largest share of total VOC emissions, contributing about 44%. Monoterpenes and methanol each contribute around 10%–11%. Other compounds, such as sesquiterpenes and many minor VOCs, are released in smaller amounts, but they still play important roles in plant ecology and atmospheric processes.

Table 6.1 Classification, sources and functional characteristics of major plant VOCs.

VOC category	Representative compounds	Contribution to total emissions (%)	Biosynthetic origin/source	Key functional roles
Isoprenoids	Isoprene, Monoterpenes, Sesquiterpenes	Isoprene (~44%); Monoterpenes (~10%–11%)	Derived from the MEP pathway in plastids and the MVA pathway in the cytosol	Involved in protection against heat stress, defense responses, and participation in atmospheric chemical reactions
Oxygenated VOCs (OVOCs)	Methanol, Ethanol, Acetone, Acetaldehyde	Methanol (~10%–11%)	Produced during cell wall modification (especially pectin breakdown) and other metabolic activities	Associated with active growth phases and contribute to atmospheric chemical processes
Green Leaf Volatiles (GLVs)	Hexenal, Hexenol, C_6 compounds	Relatively low contribution	Generated through the lipoxygenase pathway following tissue injury	Play a role in defense signaling and interactions with herbivores
Methyl Butenol (MBO)	2-Methyl-3-buten-2-ol	Region-specific emissions	Released directly from foliage (e.g., Pinus taeda)	Reacts rapidly with hydroxyl radicals, thereby influencing atmospheric oxidation processes
Other minor VOCs	Carboxylic acids, Organic sulfur compounds	Minor contribution	Formed via diverse metabolic routes	Their ecological and atmospheric significance is still under investigation

MEP, methylerythritol phosphate; *MVA*, mevalonic acid; *VOCs*, volatile organic compounds.

Methanol is one of the most commonly released oxygenated VOCs. It is mainly emitted during periods of active plant growth, especially when leaves are expanding, such as in spring. This happens because methanol is produced during the breakdown of pectin in the cell wall as it expands. Similarly, some plants release acetone, especially from growing parts like buds. Even though oxygenated VOCs are widely released by plants, the exact biological processes controlling their production and release are still not fully understood. Another important compound is methyl butanol, which has been found in high amounts in the atmosphere, especially in areas like the Colorado Rockies. At first, scientists were not sure about its source, but later studies showed that it is directly released from plant leaves, such as in *Pinus taeda*. Methyl butenol is highly reactive with hydroxyl radicals (•OH), which means it can affect atmospheric chemistry by reducing the amount of these radicals and changing the oxidative capacity of the atmosphere. Apart from these, plants also release other compounds, such as carboxylic acids and organic sulfur compounds. However, their exact roles in plant ecology and the atmosphere are still not completely known and need further study.

6.10 Role of plant volatile organic compounds in biotic stress responses

Plants increase the release of VOCs when they face biotic stress, especially when they are attacked by herbivores. Research shows that plants damaged by herbivores can release up to 2.5 times more VOCs compared to healthy plants. This indicates that these stress-induced VOCs may also affect atmospheric chemistry at a larger scale (Holopainen, 2004).

VOCs act as important signaling molecules in plant systems. They help in many ecological interactions. For example:

- They help herbivores find suitable host plants.
- They support indirect plant defense by attracting predators or parasitoids that feed on herbivores.
- They act as signals to nearby plants, warning them about possible danger.
- They can also activate defense responses in other parts of the same plant.

In this way, VOCs play a key role in shaping interactions between plants and insects, both above the ground and below the ground. However, in the rhizosphere (root zone), VOC behavior is different. Because these compounds are highly reactive, especially volatile isoprenoids, they quickly undergo physical, chemical, and biological changes in the soil. This reduces their movement into the atmosphere. Plant pathogens also trigger the production of VOCs. Many of these VOCs have antimicrobial properties, which help the plant defend itself by slowing down or preventing the spread of pathogens within plant tissues.

6.11 Role of plant volatile organic compounds in abiotic stress responses

Apart from biotic stress, plants also change their VOC emissions in response to different abiotic (environmental) stresses. Even when there is no physical damage, plants can release VOCs due to changes in conditions such as light intensity, temperature, flooding, and drought. The control of VOC emissions under these conditions depends on both physical and physiological processes inside the plant. For example, when the temperature increases, VOCs evaporate more easily and diffuse faster. Higher temperatures can also affect how much VOC is stored inside plant tissues. Water availability is another very important factor. When plants face moderate water stress, VOC production may decrease because metabolic activities slow down. Under severe drought conditions, VOC emissions can be reduced even more due to the strong suppression of plant metabolism. Increased levels of atmospheric carbon dioxide (CO_2) can also influence VOC emissions indirectly. Higher CO_2 levels contribute to global warming, which may increase VOC release. However, the overall effect of CO_2 is quite complex because it interacts with other factors like temperature, light, and herbivore activity. Due to these combined effects, it is sometimes difficult to clearly understand the direct role of CO_2 in VOC emissions.

6.12 Response of plant-produced volatile organic compounds to atmospheric ozone

Atmospheric O_3 has complex and sometimes opposite effects on the release of VOCs from plants. Depending on how much O_3 is present and how long plants are exposed to it, O_3 can either increase or decrease VOC emissions in different plant species (Calfapietra et al., 2013). Because of this, scientists are very interested in understanding how O_3 changes VOC emissions and how this affects both ecology and the atmosphere. One important point in plant–O_3 interaction is the presence of a positive feedback loop. When plants are exposed to O_3, they may release more VOCs (Llusia et al., 2002). These VOCs then take part in chemical reactions in the atmosphere that can lead to the formation of more O_3 (Atkinson & Arey, 2003). So, VOCs and O_3 are closely connected and influence each other. O_3 affects VOC emissions in several ways. It can change plant biochemical and physiological processes, which affects how VOCs are produced (Long & Naidu, 2002; Rao et al., 2000). O_3 can also react directly with plant surfaces, causing changes in the type and amount of VOCs released (Fruekilde et al., 1998). These changes can influence plant interactions with pollinators, herbivores, and natural enemies (Gate et al.,1995). The interaction between O_3 and isoprene (a major VOC) is especially important in the context of climate change. Both O_3 levels and isoprene emissions are expected to increase in the future (Fares et al.,2006). Models suggest that O_3 levels in the lower atmosphere may increase by up to 50% by the end of this century due to higher emissions of NO_X, VOCs, and greenhouse gases (Fowler et al.,1999).

Experimental studies show mixed results regarding isoprene emissions under O_3 stress. In some cases, O_3 increases isoprene release. For example, higher emissions have been observed in plants like *Phragmites australis*, *Populus alba*, and *Quercus pubescens* under controlled conditions (Velikova et al., 2005; Fares et al., 2006; Pinelli & Tricoli, 2008). However, the level of increase depends on factors, such as leaf damage and reduction in photosynthesis caused by O_3 stress (Yuan et al., 2016). On the other hand, many studies report that O_3 can reduce VOC emissions. For example, exposure to moderate O_3 levels (around 80 ppb for two weeks) reduced isoprene and other VOC emissions in *Populus nigra* (Fares et al., 2010). Similarly, higher O_3 levels (about 91 ppb) caused a 21% decrease in isoprene emission, while affecting other VOCs less (Feng et al., 2019). Long-term exposure (around 69 ppb) caused an even greater reduction (about 25%), while short-term exposure to very high O_3 levels (around 250 ppb) showed only small or inconsistent effects (Feng et al., 2019). Overall, these studies show that the effect of O_3 on plant VOC emissions depends on many factors, such as O_3 concentration, duration of exposure, plant species, and plant health. This makes the relationship quite complex. Therefore it is important to study plant responses along with atmospheric processes to better understand the role of VOCs in climate and ecosystem interactions.

6.13 Global change and ecology of volatile organic compound emissions

Global environmental changes are expected to strongly affect how plants produce and release VOCs. Increasing levels of atmospheric carbon dioxide (CO_2) can change plant communities. For example, C_3 plants are likely to increase compared to C_4 plants because C_3 plants show better photosynthesis under high CO_2 conditions (Bazzaz, 1990). This is important because many C_3 plants release large amounts of VOCs like isoprene and monoterpenes, while C_4 plants generally release much less. So, changes in vegetation can directly affect total VOC emissions at the ecosystem level. Along with rising CO_2, greenhouse gases are also increasing global temperatures and changing rainfall patterns. Many regions, especially inland areas, may become drier (Schneider et al., 1993). These changes, along with human activities like deforestation and land-use change, can shift entire ecosystems (biomes). As a result, both the amount of plant biomass and the rate of VOC emission per plant may change. Temperature plays a very important role in VOC emissions. Both isoprene and monoterpene emissions increase with temperature. Monoterpenes increase rapidly because higher temperatures increase their evaporation. Isoprene emissions follow a specific pattern (Arrhenius-type), with the highest emission usually between 36°C and 40°C, depending on the plant species (Guenther et al., 1993; Lerdau & Keller, 1997). Since future temperatures are expected to rise, VOC emissions are also likely to increase. Water availability adds more complexity. In some plants, such as pines, drought can increase monoterpene levels (Lorio et al., 1995). However, long-term drought can reduce plant growth and overall vegetation cover, leading to lower total VOC emissions. For isoprene, water stress usually reduces emissions from leaves, and combined with reduced plant biomass, it may decrease total emissions at the ecosystem level.

Changes in plant species due to climate change and land use are also very important. Different plant species release different types and amounts of VOCs. For example, oak trees release large amounts of isoprene, while maple trees release very little. So, if one species replaces another, VOC emissions in that area will change. Similarly, cutting down tropical forests (which are major VOC sources) and converting them into agricultural land can reduce VOC emissions. On the other hand, in some dry areas, the spread of shrubs (which emit more VOCs than grasses) may increase emissions. It is difficult to predict the overall effect of these changes on the atmosphere because VOCs interact with other pollutants, especially NO_x from vehicles and industries. In areas with low NO_x, higher VOC levels may reduce the atmosphere's ability to clean itself. But in areas with high NO_x, VOCs can increase the formation of harmful pollutants like ground-level O_3 (Fehsenfeld et al., 1992). After being released, VOCs undergo different processes in the atmosphere. First, they enter the layer of air around the plant (boundary layer). Their movement depends on factors like evaporation rate and concentration differences. Some VOCs may stay near the leaf surface, some may be reabsorbed by the plant through stomata, and others may move into the atmosphere. These processes depend on properties like solubility, fat-solubility (lipophilicity), stomatal opening, and environmental conditions (Riederer et al., 2002). Environmental factors also affect these processes. For example, humidity can influence how O_3 interacts with plant surfaces, which can change VOC reactions. Temperature also affects how quickly VOCs evaporate and react in the atmosphere (Copolovici & Niinemets, 2005). Overall, global environmental change is expected to significantly influence VOC emissions, but the exact outcomes are complex and depend on multiple interacting factors.

6.14 Atmospheric impact of plant volatile organic compound emissions

VOCs are very important for plants because they help in defense, communication, and survival. However, when these compounds are released into the atmosphere, they can also affect the environment. Different plant species release different types and amounts of VOCs, and under certain conditions, these natural compounds can behave like air pollutants. Among all plant VOCs, isoprene is the most commonly released compound and plays a major role in atmospheric chemistry. Because it is highly reactive, isoprene takes part in many chemical reactions in the air, leading to the formation of secondary pollutants like O_3 and other oxidants. So, even though VOCs are useful for plants, they also have a strong impact on air quality and climate. When VOCs enter the atmosphere, they react with NO_x, especially in the presence of sunlight, and form tropospheric O_3. O_3 has both useful and harmful roles. In the upper atmosphere, that is, the stratosphere, it protects life by blocking harmful ultraviolet rays. But at ground level, O_3 acts as a pollutant. It contributes to global warming, causes breathing problems in humans, and damages plant tissues. VOCs also react with other atmospheric chemicals like hydroxyl radicals (•OH). These reactions can reduce the amount of these radicals, which normally help breakdown methane, a strong greenhouse gas. As a result, methane stays longer in the atmosphere, increasing its harmful effects. In addition, VOCs undergo reactions such as ozonolysis and oxidation, which lead to the formation of secondary organic aerosols (SOAs) (Laothawornkitkul et al., 2009; Seinfeld & Pandis, 2006). These aerosols are tiny particles in the air that affect both climate and air quality. They can also form secondary organic material (SOM), which may settle on plant surfaces and soil.

From an ecological point of view, both the increase in VOC emissions and their breakdown in the atmosphere can change their normal functions in plants. Changes in the type and amount of VOCs can affect plant defense, signaling, and interactions with other organisms (Holopainen, 2004; Lerdau & Slobodkin, 2002; Yuan et al., 2009). Many of the particles formed from VOCs (SOAs) remain in the atmosphere for a longer time than the original VOCs. Because they are less volatile, they can accumulate on plant surfaces and in the soil, forming SOM. VOCs also play an important role in cloud formation and climate regulation. SOAs formed from VOCs act as cloud condensation nuclei, which are necessary for cloud formation at both local and global levels (Hallquist et al., 2009; Kirkby et al., 2016; Tröstl et al., 2016). The emission of VOCs is strongly affected by environmental conditions. Both biotic and abiotic stresses can change how much VOCs are released and what type of VOCs are produced. Stress can either increase or decrease total emissions and can also activate different biochemical pathways in plants (Kleist et al., 2012; Peñuelas & Staudt, 2010). Changes in VOC composition also affect the properties of SOAs, such as particle size and their ability to absorb water (hygroscopicity). These properties are important for cloud formation. So, any change in VOC emissions can directly influence climate processes (Paasonen et al.,2013). With ongoing climate change, extreme conditions such as heat waves, droughts, and pest attacks are expected to increase. This will likely increase both the amount and variability of VOC emissions. As a result, a feedback loop may develop between plants and climate, where each influences the other.

6.15 Atmospheric chemistry and fate plant volatile organic compounds

Isoprene and other plant-produced hydrocarbons (VOCs) play a very important role in the lower atmosphere, that is, the troposphere.

Because they are highly reactive, they influence many processes such as O_3 formation, production of carbon monoxide (CO), and breakdown of methane (Baldocchi et al., 1995; Crutzen, 1979; Fehsenfeld et al., 1992; Logan et al., 1981). In the atmosphere, isoprene is quickly broken down through reactions with different oxidants. During the daytime, it mainly reacts with hydroxyl radicals (•OH). This reaction is faster than its reaction with O_3, so it is the most important pathway during the day. At night, isoprene reacts mainly with nitrate radicals (NO_3), but this process is slower because NO_3 levels are low. Overall, reaction with hydroxyl radicals is the main pathway for isoprene breakdown.

One of the main products formed during isoprene oxidation is CO. In fact, oxidation of nonmethane hydrocarbons contributes up to 35% of total atmospheric CO, which is almost equal to emissions from fossil fuels (Brasseur & Chatfield, 1991). Isoprene itself stays in the atmosphere only for a few hours and usually spreads over a short distance of about 10 km. However, its products, such as CO, remain in the atmosphere for several months and can travel long distances, affecting air quality on a larger scale. Isoprene oxidation also produces organic nitrogen compounds, which can store reactive nitrogen and influence atmospheric chemistry for a longer time.

The effect of isoprene on O_3 formation depends on NO_x levels:

- High NO_x conditions: Isoprene increases the formation of ground-level O_3. This is often seen in cities surrounded by vegetation, where O_3 remains high even if human-made pollution decreases (Chameides et al., 1988).
- Low NO_x conditions: Isoprene can actually reduce O_3 levels, helping to decrease pollution.

Isoprene does not directly trap heat (it does not absorb infrared radiation), but it still affects the climate indirectly. By reacting with atmospheric chemicals, it can increase the lifetime of methane, a strong greenhouse gas, thus influencing global warming (Wuebbles et al., 1989). Also, since isoprene emission increases with temperature, global warming may lead to more isoprene release, creating a positive feedback effect. Compared to isoprene, the behavior of monoterpenes and other VOCs is less clearly understood. Many monoterpenes react quickly with O_3 and can be removed from the atmosphere by dissolving in water droplets. These processes may lead to the deposition of partially oxidized compounds into ecosystems and may also contribute to acid rain in some regions (Crutzen, 1979).

From a plant point of view, isoprene emission is considered useful in some species. It helps plants protect themselves from oxidative stress by neutralizing harmful molecules, such as reactive oxygen species and reactive nitrogen species. This is due to its chemical structure, which makes it highly reactive and effective as an antioxidant. However, the exact reasons why plants produce isoprene are still not fully understood. Because of their high emission and reactivity, VOCs such as isoprene, monoterpenes, and sesquiterpenes play a major role in forming SOAs (Carslaw et al.,2010; Tasoglou & Pandis, 2015). These aerosols are important components of air pollution. VOCs are also involved in the formation of photochemical smog, contribute to O_3 depletion in the upper atmosphere, and help form organic acids that can lower the pH of rainwater, leading to acidification (Lee et al.,2002; Montzka et al.,2011; Zhang et al., 2019).

6.16 Conclusion

Plants play a crucial role in Earth's ecosystems. They help maintain biodiversity, regulate climate, and improve air quality. However, their role in the atmosphere is more complex than we usually think. While plants are generally considered beneficial, especially in cities, they also release different types of biogenic volatile organic compounds (BVOCs) such as isoprene, monoterpenes, sesquiterpenes, and green leaf volatiles. These BVOCs are very useful for plants. They help in defense against herbivores and pathogens, support communication between plants, and allow plants to survive under stress conditions. For example, isoprene helps plants tolerate high temperatures, while monoterpenes can protect plants by repelling herbivores. At the same time, once these compounds are released into the air, they can affect atmospheric chemistry. In areas where NO_x are high, especially in urban and industrial regions, BVOCs react in sunlight and form harmful substances like ground-level O_3 and SOAs. These are major air pollutants. O_3 can cause breathing problems and damage plants, while fine particles ($PM_{2.5}$) can affect human respiratory and heart health. Therefore BVOCs have a dual role. They are essential for plant life and ecosystem functioning, but they also influence air pollution and climate. Understanding this balance is very important for managing air quality and dealing with climate change in the future. Future research should emphasize integrated, interdisciplinary approaches to better elucidate the complex roles of VOCs within the Earth system.

References

Atkinson, R., & Arey, J. (2003). Gas-phase tropospheric chemistry of biogenic volatile organic compounds: A review. *Atmospheric Environment, 37*(2), 197–219. https://doi.org/10.1016/s1352-2310(03)00391-1.

Baldocchi, D., Guenther, A., Harley, P., Klinger, L., Zimmerman, P., Lamb, B., & Westberg, H. (1995). The fluxes and air chem istry of isoprene above a deciduous hardwood forest. *Philosophical Transactions of the Royal Society of London. Series A: Physical and Engineering Sciences. 351*(1696), 279–296. https://doi.org/10.1098/rsta.1995.0034.

Bazzaz, F. (1990). The response of natural ecosystems to the rising global CO_2 Levels. *Annual Review of Ecology and Systematics, 21*(1), 167–196. https://doi.org/10.1146/annurev.ecolsys.21.1.167.

Brasseur, G. P., & Chatfield, R. B. (1991). The fate of biogenic trace gases in the atmosphere. *Journal of Geophysical Research, 96*(D4), 7425–7437.

Calfapietra, C., Pallozzi, E., Lusini, I., & Velikova, V. (2013). Modification of BVOC emissions by changes in atmospheric CO_2 and air pollution. In Ü. Niinemets, & R. K. Monson (Eds.). *Biology, controls and models of tree volatile organic compound emissions* (pp. 253–284). Springer.

Carslaw, K. S., et al. (2010). A review of natural aerosol interactions and climate. *Nature, 465*, 627–635.

Chameides, W. L., Lindsay, R. W., Richardson, J., & Kiang, C. S. (1988). The role of biogenic hydrocarbons in urban photochemical smog. *Journal of Geophysical Research, 93*, 1589–1599.

Copolovici, L., & Niinemets, Ü. (2005). Temperature dependencies of Henry's law constants and partition coefficients for plant volatiles. *Chemosphere, 61*(9), 1390–1400.

Crutzen, P. J. (1979). The role of NO and NO_2 in the chemistry of the troposphere and stratosphere. *Annual Review of Earth and Planetary Sciences,* 7, 443–472. https://doi.org/10.1146/annurev.ea.07.050179.002303.

Fares, S., et al. (2006). Ozone exposure induces isoprene emission in plants. *Plant Physiology, 142*, 123–131.

Fares, S., et al. (2010). Ozone inhibits isoprene emission from poplar. *Plant, Cell & Environment, 33*, 123–134.

Fehsenfeld, F., Calvert, J., Fall, R., Goldan, P., Guenther, A. B., Hewitt, C. N., Lamb, B., Liu, S., Trainer, M., Westberg, H., & Zimmerman, P. (1992). Emissions of volatile organic compounds from vegetation and the implications for atmospheric chemistry. *Global Biogeochemical Cycles, 6*(4), 389–430. https://doi.org/10.1029/92gb02125.

Feng, Z., et al. (2019). Effects of elevated ozone on isoprene emission. *Environmental Pollution, 252*, 158–166.

Fowler, D., et al. (1999). Atmospheric composition change and its impact on ecosystems. *Environmental Pollution, 100*, 43–55.

Fruekilde, P., Hjorth, J., Jensen, N. R., Kotzias, D., & Larsen, B. (1998). Ozonolysis at vegetation surfaces. *Atmospheric Environment, 32*(11), 1893–1902. https://doi.org/10.1016/s1352-2310(97)00485-8.

Gate, I. M., McNeill, S., & Ashmore, M. R. (1995). Effects of air pollution on insect–plant interactions. *Water, Air, and Soil Pollution, 85*, 1425–1430.

Guenther, A. B., Zimmerman, P. R., Harley, P. C., Monson, R. K., & Fall, R. (1993). Isoprene and monoterpene emission rate variability: model evaluations and sensitivity analyses. *Journal of Geophysical Research, 98*(7), 12–617. https://doi.org/10.1029/93jd00527.

Hallquist, M., Wenger, J. C., Baltensperger, U., et al. (2009). The formation, properties and impact of secondary organic aerosol. *Atmospheric Chemistry and Physics, 9*, 5155–5236.

Holopainen, J. K. (2004). Multiple functions of plant volatiles under air pollution. *Environmental Pollution, 131*, 305–311.

Kirkby, J., Duplissy, J., Sengupta, K., Frege, C., Gordon, H., Williamson, C., Heinritzi, M., Simon, M., Yan, C., Almeida, J., Tröstl, J., Nieminen, T., Ortega, I. K., Wagner, R., Adamov, A., Amorim, A., Bernhammer, A.-K., Bianchi, F., Breitenlechner, M., ... Curtius, J. (2016). Ion-induced nucleation of pure biogenic particles. *Nature, 533*(7604), 521–526. https://doi.org/10.1038/nature17953.

Kleist, E., et al. (2012). Irreversible impacts of heat on BVOC emissions. *Biogeosciences, 9*, 5111–5123.

Laothawornkitkul, J., Taylor, J. E., Paul, N. D., & Hewitt, C. N. (2009). Biogenic volatile organic compounds in the earth system. *New Phytologist, 183*(1), 27–51. https://doi.org/10.1111/j.1469-8137.2009.02859.x.

Lee, A., et al. (2002). Atmospheric oxidation and acid formation. *Atmospheric Environment, 36*, 403–413.

Lerdau, M., & Keller, M. (1997). Controls on isoprene emission. *Ecology, 78*, 123–130.

Lerdau, M., & Slobodkin, L. (2002). Trace gas emissions and ecological interactions. *Trends in Ecology & Evolution, 17*, 126–132.

Llusia, J., Peñuelas, J., & Gimeno, B. (2002). Seasonal and species-specific responses of VOC emissions to ozone. *Atmospheric Environment, 36*, 3931–3938.

Logan, J. A., Prather, M. J., Wofsy, S. C., & McElroy, M. B. (1981). Tropospheric chemistry: a global perspective. *Journal of Geophysical Research, 86*(8), 7210–7254. https://doi.org/10.1029/JC086iC08p07210.

Long, S. P., & Naidu, S. L. (2002). Effects of oxidants on plant metabolism. *Plant Biology, 4*, 15–23.

Lorio, P. L., Stephen, F. M., & Paine, T. D. (1995). Environmental stress and plant responses. *Forest Ecology and Management, 73*, 97–110.

Montzka, S. A., et al. (2011). Ozone-depleting substances and atmospheric chemistry. *Nature, 476*, 43–49.

Paasonen, P., et al. (2013). Warming-induced aerosol increases. *Nature Geoscience, 6*, 438–442.

Peñuelas, J., & Staudt, M. (2010). BVOCs and global change. *Trends in Plant Science, 15*(3), 133–144. https://doi.org/10.1016/j.tplants.2009.12.005.

Pinelli, P., & Tricoli, D. (2008). Ozone-induced VOC emission in oak species. *Environmental Pollution, 152*, 60–65.

Rao, M. V., Koch, J. R., & Davis, K. R. (2000). Ozone as a tool to study plant stress. *Plant Molecular Biology, 44*, 345–358.

Riederer, M., Müller, C., & Markstädter, C. (2002). Transport of VOCs through plant surfaces. *Plant Biology, 4*, 66–71.

Schneider, S. H. (1993). Scenarios of global warming. In P. Kareiva, (Ed.). *Biotic interactions and global change* (pp. 9–23). Sinauer.

Seinfeld, J. H., & Pandis, S. N. (2006). *Atmospheric chemistry and physics* (2nd ed.). Wiley.

Tasoglou, A., & Pandis, S. N. (2015). Secondary organic aerosol formation. *Atmospheric Chemistry and Physics, 15*, 11055–11072.

Tröstl, J., et al. (2016). The role of organic vapours in aerosol formation. *Nature, 533*, 527–531.

Velikova, V., Pinelli, P., Pasqualini, S., Reale, L., Ferranti, F., & Loreto, F. (2005). Isoprene emission induced by ozone. *Plant Physiology, 137*, 859–869.

Wuebbles, D. J., Grant, K. E., Connell, P. S., & Penner, J. E. (1989). Atmospheric chemistry and climate change. *Journal of the Air Pollution Control Association, 39*, 22–28.

Yuan, J. S., Himanen, S. J., Holopainen, J. K., Chen, F., & Stewart, C. N. (2009). Smelling global climate change: Mitigation of BVOC emissions. *Trends in Ecology & Evolution, 24*, 323–331.

Zhang, Y., et al. (2019). Secondary organic aerosol formation and environmental impacts. *Atmospheric Chemistry and Physics, 19*, 12345–12360.

CHAPTER SEVEN

Evaluating plant health and tolerance to tropospheric ozone pollution

Vineet Upadhyay[1], Harshita Singh[2], Priyanka Singh[2], and Krishna Kumar Choudhary[1]
[1]Department of Botany, MMV, Banaras Hindu University, Varanasi, Uttar Pradesh, India,
[2]Department of Botany, Institute of Science, Banaras Hindu University, Varanasi, Uttar Pradesh, India

7.1 Introduction

In the backdrop of global warming and climate change, air pollution remains one of the foremost environmental threats, with profound implications for health, ecological stability, and food security (Vilcins et al., 2024). While diverse number of pollutants including carbon monoxide (CO), particulate matter ($PM_{2.5}$ and PM_5), oxides of sulfur (SO_x) and nitrogen (NO_x), carbon monoxide (CO), volatile organic carbon (VOC_s) and ozone (O_3) contribute to the aggravating air pollution, ground-level O_3has emerged as a particularly dangerous phytotoxicant, generating large-scale losses in crop productivity (Pei et al., 2024). Interestingly, O_3-generation is closely intertwined with climate change-driven factors, such as extreme heat, longer durations of sunshine, forest fires, and rising emissions of O_3 precursors. Climate change-induced events such as forest fires release 3.5 Tg yr^{-1} VOC_s annually (Le Roy et al., 2025) while O_3 itself exacerbates climate change by generating 0.35–0.37 W m^{-2} of radioactive forcing and decreasing carbon sequestration via plant damage (Janni et al., 2024). The increase in O_3 formation stems from a surge in emissions of NO_x, VOC_s, often as a by-product of rapid urbanization and developmental activities under higher temperatures and solar radiation (Upadhyay et al., 2025). Global mean NO_2 concentrations showed an increase in 71% cities, with particular concerns for South Asia and sub-Saharan Africa (Sicard et al., 2023). Unlike primary pollutants, O_3 is a secondary pollutant, and its precursor gases can travel long distances, contributing to O_3 formation at sites away from the source of emissions (Sahu et al., 2021). This transboundary air pollution adversely impacts crop development and productivity, threatening food

Urban Air Pollution and Plant Responses
DOI: https://doi.org/10.1016/B978-0-443-40320-0.00013-9

security. It is estimated that O_3 pollution will only increase in the future due to increasing anthropogenic and climate change-induced emissions of O_3 precursors and rising global temperatures (Pei et al., 2024). Tai et al. (2021) estimated present O_3-induced yield losses in soybean (6.7%), maize (3.6%), rice (2.6%), and wheat (7.2%), signifying an impending threat to food security by ground-level O_3.

O_3 action on plants varies from visible manifestations in the form of lesions, chlorotic, necrotic patches, and reduction in growth, biomass fixation, and reproductive success (Upadhyay et al., 2025). O_3 enters primarily through stomatal pores in plants, and once inside leaf tissues, it rapidly generates reactive oxygen species (ROS), triggering oxidative stress (Grulke & Heath, 2020). ROS damages cellular membranes via lipid peroxidation and increased electrolyte leakage rate, impacting physiological functions (Nowroz et al., 2024). Photosynthetic apparatus, particularly PSII are highly susceptible to ROS-induced reduction in light energy conversion efficiency, adversely impacting photosynthetic parameters (Basara & Gorzelany, 2025). Redox imbalance in plants increases metabolic cost, thus effectively lowering resource ceiling and reducing photosynthetic assimilates for yield productivity (Singh et al., 2023a, 2023b; Singh, Ansari, et al., 2023; Singh, Singh, et al., 2023). O_3-based crop models project progressive decline in simulated crop yields with daily O_3 concentrations beyond 25 ppb (Guarin et al., 2024). The estimated yield losses per ppb increment in O_3 were cultivar-dependent, ranging from 0.16% to 0.82% for maize, 0.05% to 0.63% for rice, 0.36% to 0.96% for soybean, and 0.26% to 1.23% for wheat (Guarin et al., 2024). Hence, studying plant responses to O_3 is critical to address agroeconomic concerns and evaluate plant health. Addressing this challenge requires an interdisciplinary approach that integrates research on O_3 formation dynamics, mechanistic plant- O_3 interactions, methodologies for studying O_3-induced phytotoxicity, and plant responses to O_3 stress. This chapter provides a comprehensive evaluation of plant health under O_3 stress, integrating methodologies for assessing damages, detailing physiological, biochemical, and molecular responses, and identifying adaptive strategies.

7.2 Methodologies for evaluating ozone impacts

Assessment of impacts of O_3 on plants requires precise experimental approaches that simulate real-world exposure conditions while allowing controlled measurements of plant responses. Two primary methodologies—open-top chambers (OTCs) and free-air concentration enrichment (FACE) systems have been widely employed to study O_3-plant interactions under near-natural conditions. OTCs are enclosed structures that allow controlled O_3 exposure while maintaining ambient environmental conditions (Bell & Ashmore, 1986). It houses an O_3 delivery system with O_3 generated from ambient air using UV lamps or corona discharge, mixed with filtered air and subsequently released into the chamber (Bell & Ashmore, 1986). They offer precise control over O_3 concentrations supplied, reduce wind speed, and intercept rainfall whilst allowing natural sunlight penetration and diurnal temperature changes. FACE represents ecologically most realistic method for studying O_3 effects, as it exposes plants to elevated O_3 in open-field conditions without physical enclosures. It ensures a uniform exposure zone and real-time monitoring and adjustments of O_3 levels via a series of vertical pipes or horizontal jets around the experimental plot in a circular or octagonal pattern (Paoletti et al., 2017). FACE ensures near-natural conditions and mimics realistic scenarios for current and future ground-level O_3 stress (Hoshika et al., 2024). Both OTCs and FACE are an intricate part of controlled exposure methodologies, while field monitoring approaches include passive O_3 samplers, continuous monitors, and remote sensing. Passive samplers such as Ogawa and Radiello samplers rely on diffusion and chemical absorption to collect O_3 over a defined exposure period. Ogawa samplers use nitrite-coated filters in which O_3 reacts to form nitrate and is later analyzed via ion chromatography, while Radiello samplers utilize potassium iodide (KI) substrate, where O_3 oxidizes iodide (I^-) to iodine (I_2), subsequently measured spectrophotometrically. These passive samplers suffer from limitations of low temporal resolution and potential interference from humidity and other oxidants. Large-scale spatial measurements of O_3 data often employ satellite remote sensing methodology, integrating satellite-based observations with ground-based measurements such as Differential Optical Absorption Spectroscopy and Light Detection and Ranging (Jian et al., 2024; Xu et al., 2024). Remote sensing-based O_3 simulation is significant for mapping high-risk agricultural zones, especially if coupled with crop growth models to predict yield losses and map vulnerable agricultural zones (Xu et al., 2024). However, it suffers from limitations of cloud cover, surface reflectance and aerosols interference. Thus, O_3 monitoring methodologies vary in scale, precision and cost. Ground-based monitoring techniques provide localized, precise data while remote sensing offers global perspectives on O_3 data. Emerging technologies (AI, drones, and low-cost sensors) are revolutionizing O_3 assessment, enabling cost-effective and precise real-time monitoring approaches (Javan et al., 2025).

7.3 Emission and sources of tropospheric ozone

Tropospheric O_3, a highly reactive and phytotoxic pollutant, is a secondary pollutant that is not directly released into the atmosphere. Instead, it is produced through a series of intricate photochemical reactions involving primary pollutants, specifically nitrogen oxides (NO_x) and volatile organic compounds (VOCs), under the influence of sunlight. These reactions typically occur during the daytime when the sunlight is most intense, triggering the transformation of precursor gases into O_3 molecules.

The origin of O_3 precursors is diverse and includes both natural and anthropogenic sources. Natural sources encompass emissions from vegetation (biogenic VOCs), wildfires, lightning, and soil microbial activity. On the other hand, anthropogenic emissions are predominantly associated with industrial activities, fossil fuel combustion (from vehicles and power plants), agricultural operations, and solvent usage. Both natural and anthropogenic contributions can significantly affect the local and regional concentrations of ground-level O_3, especially when compounded by meteorological conditions such as high temperatures, stagnant air masses, and intense solar radiation.

7.3.1 Natural sources

Natural processes play a significant role in the emission of O_3 precursor compounds, particularly VOCs and NO_x. One of the most prominent sources of biogenic VOCs is vegetation, especially forest ecosystems. Trees emit a range of reactive hydrocarbons such as isoprene and monoterpenes, which are highly influential in O_3 chemistry. Tropical forests, in particular, are major contributors, releasing large volumes of isoprene that readily react with NO_x in the presence of sunlight, accelerating tropospheric O_3 formation. Another notable natural source of NO_x is lightning, which generates NO_x through the thermal dissociation of nitrogen (N_2) and oxygen (O_2) molecules in the atmosphere during high-energy electrical discharges. These reactions form nitrogen oxides that can contribute significantly to the regional O_3 burden. In addition to lightning, soil microbial processes such as nitrification and denitrification also produce NO_x under specific environmental conditions. Meanwhile, oceanic emissions and other soil-related processes contribute to both VOC and NO_x fluxes, although their overall contribution is smaller in comparison to forests and lightning. Recent reviews have emphasized the need for deeper investigation into the biogeochemical mechanisms governing NO_x and VOC emissions from soil, as understanding these natural processes is essential for improving O_3 pollution modeling and mitigation strategies (Hui et al., 2023).

Rising global temperatures may lead to increased VOC emissions from plants and soils, more frequent lightning events, and intensified stratosphere-to-troposphere O_3 transport, all of which could elevate ambient O_3 levels (Dewan & Lakhani, 2022). Additionally, warmer and more stagnant atmospheric conditions, as well as increased emissions of methane and CO_2, may further enhance photochemical O_3 production.

7.3.2 Anthropogenic sources

Human activities are a dominant source of NO_x and VOCs, which are the primary precursors of tropospheric O_3. These emissions are particularly concentrated in urban and industrialized regions, where dense populations and economic activities intensify air pollution levels. Among these, motor vehicle emissions from diesel and gasoline-powered engines are a major contributor to ambient NO_x and VOC concentrations (Fig. 7.1). High traffic volumes in metropolitan areas release substantial amounts of these pollutants, creating conditions conducive to rapid photochemical O_3 formation under strong sunlight (Singh et al., 2023a) (Fig. 7.1).

According to the findings (Fig. 7.1), the major contributors include industrial processes, biogenic emissions, coal combustion, solvent usage,

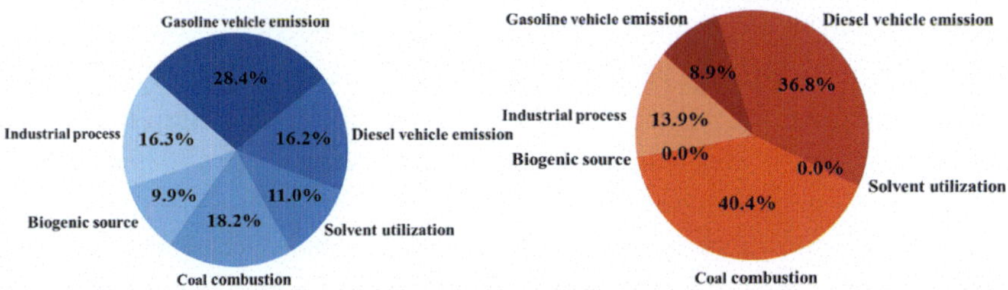

Figure 7.1 Anthropogenic sources contribution of VOCs and NOx- A: VOC contribution by sources; B: NOx contribution by sources (Donzelli & Suarez-Varela, 2024). *NOx*, nitrogen oxide; *VOC*, volatile organic compounds.

and emissions from diesel and gasoline vehicles. Of these, vehicular emissions were identified as the most significant, accounting for 44.60% of VOCs and 45.70% of NO_x emissions (Niu et al., 2024). Beyond the transport sector, industrial operations and energy production also release significant quantities of NO_x, primarily through the combustion of fossil fuels (coal, oil, and natural gas) (Berezina et al., 2020). Additionally, various industrial processes such as oil refining, chemical manufacturing, and metal processing emit VOCs either as by-products or through evaporation during handling and use of organic chemicals (Rajabi et al., 2020).

In residential and commercial zones, other anthropogenic sources also add to the VOC burden. These include space heating in colder climates, use of household solvents (paints, cleaning products, adhesives), and emissions from cooking and waste burning (Baudic et al., 2016). While individually smaller in scale, the widespread and routine nature of these activities contributes significantly to urban air quality deterioration, particularly when compounded by stagnant meteorological conditions that hinder dispersion.

7.3.3 Mechanism of tropospheric ozone formation: role of chemistry and meteorology

The generation of O_3 in the troposphere (ground-level O_3) is driven by a complex chain of photochemical reactions. The process begins when nitrogen dioxide (NO_2) absorbs sunlight, particularly ultraviolet radiation, triggering a reaction known as photolysis. This reaction breaks NO_2 into nitric oxide (NO) and a highly reactive atomic oxygen (O). The free oxygen atom combines with molecular oxygen (O_2) already present in the atmosphere, forming O_3. While nitrogen oxides (NO_x=NO + NO_2) are central to initiating O_3 formation, VOCs play a critical amplifying role. VOCs react with hydroxyl radicals (OH) and other reactive species in the atmosphere to produce organic peroxy radicals (RO_2). These radicals convert NO back into NO_2 without consuming O_3, allowing the cycle of NO_2 photolysis and O_3 formation to continue repeatedly under sunlight. This mechanism results in a self-sustaining cycle that can lead to the accumulation of O_3, particularly during sunny, stagnant weather conditions.

The formation of tropospheric O_3 involves complex chemical reactions that are closely modulated by meteorological conditions. These chemical–meteorological interactions are effectively summarized in Fig. 7.1, which illustrates that O_3 formation chemistry is intricately linked to atmospheric parameters such as temperature, solar radiation, and humidity. Ground-level

O_3 (O_3), also known as tropospheric O_3, can originate from two primary sources. One pathway involves the downward transport of O_3 from the stratosphere into the troposphere, known as stratospheric intrusion (Li et al., 2024; Wang et al., 2022). However, the predominant source of ground-level O_3 is in-situ photochemical production within the troposphere. This process occurs through complex reactions involving NO_x and VOCs in the presence of sunlight.

A key aspect of O_3 formation lies in its dependency on its precursor species, primarily NO_x and VOCs, and their interplay under daytime and nighttime conditions. The diurnal variation in O_3 levels is largely driven by the photochemical activity during the day and the absence of sunlight at night, leading to significant shifts in chemical pathways. One of the central chemical processes in O_3 formation is the cycling of reactive radicals, collectively referred to as, which includes hydroxyl radicals (OH), hydroperoxyl radicals (HO_2), and organic peroxy radicals (RO_2). The interplay between RO_x and NO_x ($NO + NO_2$) is critical, though their contributions to O_3 formation differ in mechanism and effectiveness.

Studies have shown that the RO_x radical cycle plays a more dominant role in sustaining O_3 production compared to the NO_x cycle (Wang et al., 2022). In the RO_x pathway, peroxy radicals (RO_2) efficiently oxidize nitric oxide (NO) to nitrogen dioxide (NO_2) without consuming O_3, allowing further photolysis of NO_2 to generate O_3 continuously. This leads to a net accumulation of O_3 in the atmosphere. In contrast, the NO_x cycle, particularly under high NO conditions, can suppress O_3 buildup due to the NO titration effect, where NO reacts directly with O_3, reducing its concentration rather than contributing to its formation. A major pathway for net O_3 production is the reaction between RO_2 and NO, which forms NO_2 and alkoxy radicals, thereby feeding the O_3 production loop. The formation of RO_x radicals is tightly linked to the oxidation of VOCs, a process initiated by the reaction of VOCs with OH radicals. This reaction generates RO_2 radicals, which are further transformed into RO and subsequently to HO_2. The HO_2 radicals then regenerate OH via further reactions, establishing a self-sustaining RO_x cycle (Wang et al., 2017). This regenerative mechanism ensures a persistent supply of oxidizing agents essential for sustained O_3 formation.

Moreover, the photolysis of oxygenated volatile organic compounds, such as formaldehyde and acetaldehyde, represents another significant source of RO_x radicals in the atmosphere. In recent years, additional reservoir species have been identified as important contributors to radical generation. Compounds like nitrous acid and nitryl chloride can release OH and Cl radicals, respectively, upon photolysis, further enhancing the radical budget and indirectly promoting O_3 formation (Dai and Wang, 2021; Zhang et al., 2017; Zhang et al., 2017).

Meteorological factors play a critical role in governing the formation, accumulation, dispersion, deposition, and removal of ground-level O_3 across spatial and temporal scales (Fig. 7.2). The behavior of O_3 in the atmosphere is not solely a function of precursor emissions; it is significantly influenced by prevailing weather conditions, which modulate chemical reactivity and atmospheric transport processes. Specific weather patterns have been found to favor elevated O_3 levels. These conditions typically include: Intense solar radiation, which accelerates the photochemical reactions responsible for O_3 formation.

High ambient temperatures, which enhance the emission and reactivity of VOCs and increase the rate of photochemical reactions. Low cloud cover, allowing for maximum sunlight penetration (Fig. 7.2). Low relative humidity, which can reduce the likelihood of wet deposition and cloud scavenging. Atmospheric stability, which suppresses vertical mixing and allows pollutants to accumulate near the surface. These meteorological setups create a conducive environment for local O_3 buildup and can also facilitate its regional transport, especially under stable wind patterns. In contrast, water vapor plays a dual role in O_3 chemistry. On one hand, it

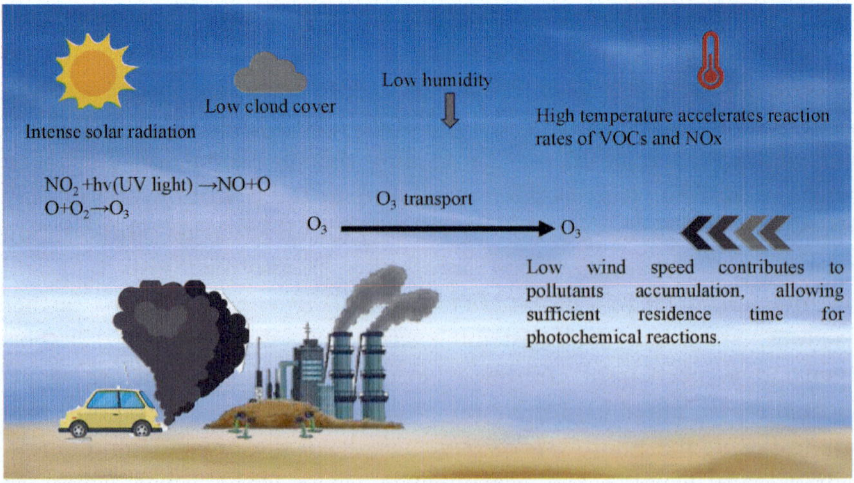

Figure 7.2 Meteorological modulation of O_3 formation.

contributes to the loss of O_3 by participating in reactions involving photolysis products of O_3, which react with H_2O to produce OH radicals. This increase in OH production enhances the oxidative capacity of the atmosphere, which can, in turn, lead to the degradation of VOCs and impact O_3 dynamics. On the other hand, wet scavenging, the removal of O_3 via precipitation, significantly reduces O_3 concentrations, particularly in humid regions. This process helps explain the spatial variation in O_3 concentrations. Inland regions, which are generally warmer and drier, tend to exhibit higher O_3 levels due to less efficient removal by water vapor and precipitation. In contrast, coastal regions, characterized by higher humidity and greater availability of water vapor, often experience more effective O_3 removal through wet deposition and chemical scavenging. Thus meteorology not only influences the chemical pathways of O_3 formation but also determines its atmospheric residence time and spatial distribution, highlighting its integral role in air quality dynamics.

7.4 Responses of plants to ozone pollution

O_3 stress causes a range of physiological and morphological symptoms in plants, primarily due to the oxidative damage induced by ROS.

7.4.1 Visual symptoms (chlorosis, necrosis, bronzing of leaves)

Plants exposed to O_3 (O_3) often exhibit striking visual injuries that serve as an early warning sign of stress. Several physiological and biochemical factors altered multiple metabolic pathways, resulting in evident foliar damage (Shahzadi et al., 2022). Visible leaf injury is a primary indicator of plant damage caused by O_3 exposure (Fig. 7.3). Therefore visible damage is regarded as a thumb rule when foliar O_3 injury appears, indicating damage to plants' metabolism. O_3 also influences other major factors such as stomatal conductance (g_s) and photosynthesis, along with visible foliar injuries (Marchica et al., 2022).

Leaf damage varied based on O_3 concentration and species tolerance (Feng et al., 2008). Visible O_3 injury, an easily detectable trait, reflects atmospheric O_3 levels (Leung et al., 2020). Heagle et al. (1991) noted that initial O_3 injury appears as small pale yellow and brown specks in the interveinal areas of the upper leaf surface, progressing to bifacial necrosis in later growth stages. Certain medicinal plants also exhibited foliar damage when exposed to elevated O_3 levels. Pellegrini et al. (2018b) observed

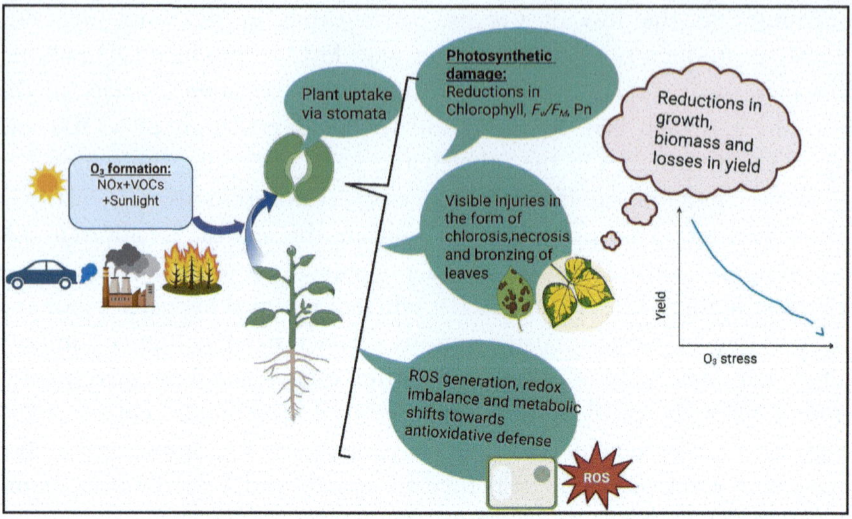

Figure 7.3 O_3 uptake and symptomatic manifestations on plants.

significantly small, dark-blackish necrotic spots in the interveinal regions of the upper leaf surface in *Liriodendron tulipifera* and Melissa *officinalis* after 45 days of O_3 exposure. Dolker et al. (2019) reported foliar damage in grass species (*Panicum maximum* and *Cenchrus ciliaris*) under similar O_3 treatments. In Mungbean and potatoes, foliar injuries like chlorosis and necrosis appeared on the upper leaf surface, progressing to more severe conditions (Kumari & Agrawal, 2014; Mina et al., 2010; Upadhyay et al., 2025) (Table 7.1). Feng et al. (2019) noted light-brown stippling and white flecking in the interveinal areas of older leaves on the adaxial surface of *Hibiscus syriacus*. Numerous studies have documented the detrimental effects of O_3 exposure on the foliage of various plant species, highlighting the sensitivity of different crops to elevated atmospheric O_3 levels. Research has identified foliar injuries in several agriculturally and economically important plants, including wheat (*Triticum aestivum*), where studies by Mishra et al. (2013), Picchi et al. (2017), and Singh et al. (2018) reported visible damage such as chlorosis, necrosis, and stippling on leaves, which can impair photosynthesis and reduce crop yield. Similarly, bean (*Vigna radiata*) plants exhibited foliar symptoms under O_3 stress, as noted by Chaudhary et al. (2013) and Singh et al. (2010), with characteristic leaf damage that included dark stipples and reduced leaf vitality, affecting plant growth and productivity. Okra (*Abelmoschus esculentus*) also showed

Table 7.1 Observed responses of O_3 on various parameters of plant species.

Parameter	Observed responses	Plant species	Key studies
Visual symptoms			
Chlorosis/Necrosis	Interveinal chlorosis, brown/black necrotic lesions along veins	Rice, *Leucaena*, *Phoebe* spp., *Vigna radiata* L.	Chen et al. (2018), Singh, Kannaujia, et al. (2021), Upadhyay et al. (2025)
Leaf bronzing	Bronzing of leaves due to oxidative cell damage	Rice, *Leucaena*, *Phoebe zhennan*	Chen et al. (2018), Rajonandraina et al. (2023), Singh, Kannaujia, et al. (2021)
Premature senescence	Accelerated aging of leaves and abscission	*Phoebe bournei*, Rice, *Arabidopsis sp.*	Chen et al. (2018), Burkart et al. (2013), Boublin et al. (2022)
Biochemical markers			
Lipid peroxidation (Malondialdehyde)	30%–150% increase in leaves/roots (dose-dependent)	*Leucaena*, Rice, *Phoebe* spp., Bean, and tobacco plants	Chen et al. (2018), Ramya et al. (2021), Singh, Kannaujia, et al. (2021), Vougeleka et al. (2024)
Electrolyte leakage	25%–80% increase (membrane integrity loss)	Rice, *Pinus mugo, Conocarpus erectus*	Bičárová et al. (2023), Ueda et al. (2013), El-Mahrouk et al. (2024)
Antioxidant enzymes			
SOD (Superoxide Dismutase) CAT (Catalase) APX (Ascorbate Peroxidase)	Differential responses under acute and chronic exposure, depending on test cultivars; however, a generally elevated level of enzymatic-antioxidant activities was observed; organelle-specific isoforms responsive	*Arachis hypogea* L., Rice, *Sida cordifolia* L., *Vigna mungo*	Rathore and Chaudhary (2021), Ansari et al. (2023), Dhevagi et al. (2023), Nowroz et al. (2024)

(*Continued*)

Table 7.1 (Continued)

Parameter	Observed responses	Plant species	Key studies
Nonenzymatic antioxidants			
Ascorbic acid	A general increase was noted across the test plants. Ascorbic acid serves as the first antioxidant barrier in the apoplast.	*Vigna radiata* L., crop plants, Soybean (*Glycine max* L.)	Shahzadi et al. (2022), Chaudhary et al. (2023), Guo et al. (2024)
Phenolics/ Isoprenoids	A general increase was noted in the production of total phenolics due to upregulation of the phenylpropanoid pathway.	Crop plants, *Salvia officinalis; Melissa officinalis; Lactuca sativa* L.	Bachheti et al. (2022), Guo et al. (2024), Lee et al. (2022), Marchica et al. (2021)
Photosynthetic parameters			
Total Chlorophyll	↓ (dose-dependent reductions; drop in chlorophyll a > b).	Rice, Urban trees, *Pinus roxburghii*	Kittipornkul et al. (2020), Khan et al. (2024), Ramya et al. (2021)
Carotenoids	A general decrease was noted due to direct oxidative degradation and disruption of chloroplast membranes.	*Brassica campestris* L. *ssp. chinensis, Phillyrea angustifolia, Poplar clones, Rice*	Han et al. (2023), Pellegrini et al. (2021), Pisuttu et al. (2024), He et al. (2024)
F_v/F_m (PSII efficiency)	Reduction in key fluorescence matrices mainly due to disruption of thylakoid membranes and photoinhibition leading to reduced quantum yield of PSII.	*Lonicera caerulea* varieties, various crop plants	Basara and Gorzelany (2025), Guo et al. (2024), Bhattacharjee and Bhattacharjee (2019)
Net photosynthesis (Pn)	Reduction in electron transport rate, photochemical efficiencies of PSII, and stomatal conductance significantly decreased Pn.	Soybean, Wheat, *Quercus sp., Fagus crenata*	Cheng et al. (2023), Hoshika et al. (2022), Xu et al. (2023), Yamaguchi et al. (2024)

susceptibility to O_3, with Bulbovas et al. (2008) observing foliar lesions and discoloration that compromised plant health. In grapevine (*Vitis vinifera*), Feng et al. (2014) documented O_3-induced foliar injuries, such as interveinal necrosis and premature leaf senescence, which can impact fruit quality and vineyard sustainability. Likewise, watermelon (*Citrullus lanatus*) displayed similar foliar damage under O_3 exposure, as reported by Hayes et al. (2007), with symptoms including light-brown stippling and leaf necrosis that could reduce fruit yield and quality. Prolonged O_3 exposure typically results in leaf yellowing and premature senescence, while acute exposure causes distinct dark brown necrotic spots (Kangasjärvi et al., 2005). Common visible symptoms include stippling, flecking, surface bleaching, necrosis, reddening, bronzing, and chlorosis, each reflecting varying levels of cellular damage (Feng et al., 2014). Due to these clear and measurable responses, plant sensitivity to O_3 makes them powerful and reliable bioindicators of atmospheric pollution.

7.4.2 Biochemical markers

Biochemical markers of O_3 stress in plants provide insights into the physiological and metabolic responses that occur upon exposure to elevated O_3 levels. These markers are crucial for the early detection of oxidative damage and for understanding plant defense mechanisms. Key biochemical markers include:

7.4.2.1 Lipid peroxidation (malondialdehyde content), solute leakage

Lipid peroxidation (LPO) is a naturally occurring but harmful process in all living organisms, exacerbated under conditions of oxidative stress such as O_3 exposure. Excessive ROS generated during O_3 stress initiate LPO by abstracting hydrogen atoms from the unsaturated acyl chains of polyunsaturated fatty acids in cellular and organelle membranes (Fig. 7.3). This chain reaction results in the formation of toxic lipid-derived compounds, including lipid alkenes, aldehydes, and hydroxyalkenals such as 4-hydroxy-2-nonenal and malondialdehyde (MDA), which serve as critical biochemical markers of membrane damage. The accumulation of these compounds disrupts normal physiological activities by degrading membrane proteins, increasing membrane permeability (leakiness), and reducing membrane fluidity, thereby impairing membrane integrity and function (Gill & Tuteja, 2010). Several studies have documented elevated MDA levels under O_3 stress, indicating enhanced lipid peroxidation in plants like *Solanum lycopersicum* (Gupta et al., 2023a), *Costus pictus* (Ansari et al., 2021),

Solanum melongena (Li et al., 2019), and *V. radiata* (Shahzadi et al., 2022). However, contrasting findings were reported by Ghosh et al. (2021), who observed no significant changes in MDA content in *T. aestivum* (HD2967 cultivar) under elevated O_3. This resilience was attributed to a robust antioxidant defense system that effectively scavenged ROS, preventing extensive lipid peroxidation and subsequent membrane damage.

The impact of O_3-induced oxidative stress on membrane integrity can be further evaluated through electrolyte leakage assays, which serve as a dynamic measure of membrane permeability changes. Elevated O_3 concentrations lead to ROS accumulation beyond the cellular antioxidant capacity, resulting in increased lipid peroxidation and loss of membrane stability. This manifests as enhanced leakage of solutes and electrolytes from foliar tissues, signaling compromised membrane function (Matłok et al., 2022). The disruption in membrane integrity not only accelerates leaf senescence but also impairs photosynthetic efficiency by affecting chloroplast membranes, ultimately reducing plant productivity (Gupta et al., 2023a). Supporting this, Ansari et al. (2023) reported an increase in hydrogen peroxide (H_2O_2) concentration in *Sida cordifolia L.* exposed to elevated O_3 (ambient +20 ppb), highlighting the role of ROS intermediates in exacerbating oxidative damage. Collectively, these findings underscore the complex interplay between ROS generation, lipid peroxidation, membrane integrity, and antioxidant defenses in modulating plant responses to O_3 stress

7.4.2.2 Antioxidant enzyme activities (superoxide dismutase, catalase, ascorbate peroxidase)

Plants combat excessive ROS generated under O_3 stress by employing both enzymatic and nonenzymatic antioxidants, as outlined in Table 7.1. Key enzymatic antioxidants include superoxide dismutase (SOD), catalase (CAT), ascorbate peroxidase (APX), peroxidase (POD), and glutathione reductase (GR). SOD plays a critical role in neutralizing stress-induced free radicals (O_2^-) (You & Chan, 2015), while CAT and POD work synergistically with SOD to mitigate harmful ROS production. In *S. lycopersicum* (tomato), elevated O_3 exposure enhanced SOD activity across all growth stages in most cultivars, except H-86, and increased CAT activity except in Kashi Chayan and VRT 02 (Gupta et al., 2023a). Similarly, in *Solanum melongena* (eggplant) exposed to 110 nmol mol^{-1} O_3 for 25 days, Li et al. (2019) observed heightened SOD and CAT activities after 15 and 10 days, respectively, highlighting the coordinated antioxidant response to O_3 stress across plant species. APX, also known as ascorbate (AsA)-dependent peroxidase, is a class I hemoperoxidase and a vital component of the Foyer-Halliwell-Asada Pathway. It acts as a key scavenger of hydrogen peroxide (H_2O_2) within plant cells, using ascorbate as a specific electron donor to convert H_2O_2 into water, while also serving as a redox sensor. In contrast, POD, classified as Class III peroxidases, operate in the extracellular environment to neutralize H_2O_2 (Rajput et al., 2021). Studies show that *S. cordifolia* under elevated O_3exposure exhibited increased APX and POD activities, whereas *Costus pictus* showed enhanced APX but reduced POD activity (Ansari et al., 2021, 2023). Ma et al. (2024) reported consistently elevated APX activity in *T, aestivum* cultivars (Yangmai16, Yangmai23, Zhenmai12, Yangmai29, and Yangmai30) exposed to high O_3 levels over two consecutive years. GR, an NAD(P) H-dependent flavoprotein oxidoreductase, plays a critical role in the ascorbate-glutathione pathway by converting oxidized glutathione (GSSG) to reduced glutathione (GSH) (Sharma, 2012). Pandey et al. (2018) found upregulated GR activity in *T. aestivum* cultivars (HD2967 and Sonalika) under elevated O_3 (ambient + 20 ppb), while *Hypericum perforatum* showed reduced GR activity at 110 ppb O_3 (Pellegrini et al., 2018). Gupta et al. (2023a) observed varied responses in *S. lycopersicum* cultivars under elevated O_3 (ambient + 20 ppb), with increased GR activity in some cultivars (EC-620438) and decreased activity in others (HR-86).

7.4.2.3 Nonenzymatic antioxidants (ascorbic acid, phenolics, flavonoids)

AsA, widely known as vitamin C, serves as a potent nonenzymatic antioxidant and a critical substrate for detoxifying ROS in plants (Pirnia et al., 2022; Smirnoff & Wheeler, 2000). AsA, found in the apoplast, plays a key role in perceiving environmental stress, facilitating downstream stress signaling and responses (Akram et al., 2017) (Fig. 7.3). It significantly contributes to plant development, hormone signaling, and stress response, particularly in fine-tuning ROS homeostasis to enhance tolerance to O_3 stress (Ansari et al., 2021, 2023; Boubakri, 2018; Gupta et al., 2023a; Li et al., 2019; Vidal-Meireles et al., 2017; Viviani et al., 2021) (Table 7.1). However, Rathore and Chaudhary (2021) reported a decrease in AsA content across five *Arachis hypogaea* cultivars under acute O_3 exposure (75 ppb). Alongside AsA, phenolic compounds, the most abundant plant secondary metabolites derived from the shikimate/phenylpropanoid pathway, play a crucial role in mitigating O_3-induced oxidative stress (Cheynier et al., 2013; Kangasjärvi et al., 2005). These compounds, ranging from simple aromatic molecules to complex structures, neutralize ROS directly or reduce ROS production, protecting chloroplasts from photo-oxidative damage (Brunetti et al., 2015; Zhang et al., 2018). Plants capable of synthesizing phenolics, particularly flavonoids, exhibit greater resistance to O_3 stress compared to mutants deficient in the flavonoid pathway (Rao & Zheng, 2025). Studies on Abelmoschus cytotypes showed increased total phenol concentrations under elevated O_3 (Singh et al., 2023a,b; Singh, Ansari, et al., 2023; Singh, Singh, et al., 2023), while Pellegrini et al. (2018) observed a reduction in phenolic content under similar conditions. However, multiple studies consistently report elevated phenolic content in response to high O_3 levels, underscoring their protective role (Ansari et al., 2021, 2023; Chen et al., 2020; Gupta et al., 2023a; Marchica et al., 2020; Rathore & Chaudhary, 2021). Flavonoids are the most abundant group of polyphenolic compounds present in plants and are widely recognized for their potent antioxidant properties and potential health benefits, particularly in disease prevention. Beyond their antioxidant role, flavonoids also contribute to plant reproduction by imparting pigmentation to petals, aiding in signal transduction, guiding pollinators through nectar cues, and facilitating processes such as allelopathy, microbial interactions, and the formation of root nodules and mycorrhizal associations (Dudek et al., 2016; Thepphakhun & Intanon, 2020). Among phenolic compounds, flavonoids play a major role in scavenging ROS, owing to their extensive presence across various plant organs, tissues, and cellular compartments (Gill & Tuteja, 2010). Pellegrini et al. (2018) demonstrated that O_3 (O_3) exposure can activate phenylalanine ammonia lyase (PAL), a key enzyme in the phenylpropanoid pathway, thereby promoting flavonoid biosynthesis. Supporting this, Gupta et al. (2023a) observed that elevated PAL activity in *S. lycopersicum* corresponded with increased flavonoid accumulation. Similar trends were reported by Marchica et al. (2020) in *Salvia officinalis* and Ansari et al. (2021) in *Costus pictus*.

7.4.3 Impacts of ozone on photosynthetic, chlorophyll parameters (Total chlorophyll, a/b, carotenoids, Fv/Fm) and physiology in Plants

Photosynthesis (P_s), a fundamental and complex physiological process in plants, is highly sensitive to environmental stress, which often impairs plant physiological functions. Photosynthesis serves as a critical indicator for assessing plant health under stress conditions. Thwe et al. (2014) observed that elevated air pollution, particularly O_3 stress, leads to a decline in the net photosynthetic rate, accelerating plant aging and reducing overall productivity. Joffe et al. (2022) noted that the primary cause of photosynthetic breakdown is a reduced carboxylation rate, which alters light reactions and disrupts electron transport between photosystems I and II. Marchica et al. (2022) reported that exposure to 120 ppb O_3 in *S. officinalis* L. decreased the photosynthetic rate. Additionally, O_3 stress triggers ROS formation, linked to stomatal closure, which influences ion channel activity and protein kinase function in stomatal guard cells (Joffe et al., 2022; Vainonen & Kangasjärvi, 2015). Gupta et al. (2023b) identified a positive correlation between photosynthetic rate (P_s) and g_s, with g_s significantly contributing to P_s reduction in eight tomato cultivars (Superbug, Kashi chayan, VRT02, Sel7, H86, VRT06, EC-620438, and EC-620421) under elevated O_3 levels. Declines in Ps linked to reduced g_s under high O_3 conditions were also observed in *S. cordifolia* (Ansari et al., 2023), *V. radiata* (Dhevagi et al., 2022), *Phillyrea angustifolia* (Pellegrini et al., 2021), Fraxinus excelsior (Cotrozzi et al., 2018), and *S. officinalis* (Pellegrini et al., 2021). Fig. 7.4 shows the maximum reduction in Pn of plants due to O_3-induced stress. Furthermore, responses to internal CO_2 concentration (C_i) varied across species and growth stages. Gupta et al. (2023b) reported that in tomato cultivars (Sel7, H86, and VRT06), a decrease in C_i during early growth stages likely resulted from reduced CO_2 uptake due to lower g_s under elevated O_3. Conversely, at later growth stages, an increase in C_i

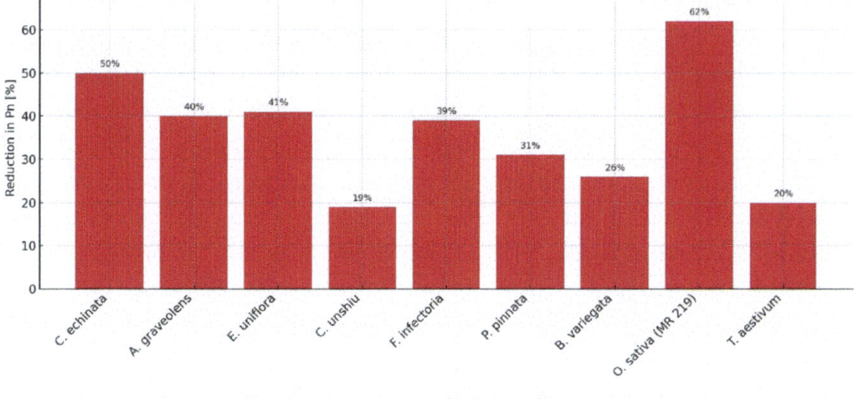

Figure 7.4 Maximum reduction in the Pn of plants due to O_3 stress.

suggested CO_2 accumulation due to the inactivity of the CO_2-fixing enzyme Rubisco under high O_3 conditions. Chlorophyll Fluorescence (F_v/F_m) is a ratio that is essential to the photosynthetic system and measures the stability and efficacy of PSII. F_v/F_m acts as a useful indicator of photoinhibition of photosynthesis since fluctuations in it reflect variations in PSII's photochemical conversion efficiency (Ranjbarfordoei et al., 2011). Ansari et al. (2023) found that the F_v/F_m value of *S. cordifolia* varied from 0.79 to 0.82 under no stress conditions; however, it ranged from 0.70 to 0.75 when exposed to elevated levels of O_3. In addition, *Abelmoschus* cytotypes exposed to ambient+20 ppb O_3 showed nonsignificant changes in F_v/F_m at initial growth stages but exhibited a decline at the latter growth stage (Singh, Ansari, et al., 2024; Singh, Gupta, et al., 2024).

Photosynthetic pigments are also key stress indicators in plants. Chlorophyll is required for photosynthesis and acts as a marker of leaf senescence. Reduced levels of photosynthetic pigments after elevated O_3 exposure could be attributed to chlorophyll molecule breakage or pigment suppression. According to Joffe et al. (2022), O_3-induced oxidative stress damages chlorophyll molecules by encouraging lipid peroxidation in the chloroplast membrane. Several plant species, including *Abelmoschus* species (Singh, Ansari, et al., 2024; Singh, Gupta, et al., 2024), *S. cordifolia* (Ansari et al., 2023), *V. radiata* (Dhevagi et al., 2022), and *S. lycopersicum* (Gupta et al., 2023b), showed reduced chlorophyll content when exposed to increased O_3 levels. Li et al. (2019) found that prolonged exposure to elevated O_3 levels in *S. melongena* resulted in a considerable decrease in chlorophyll content, possibly due to inactivation of chlorophyll production or degradation. Dhevagi et al. (2022) found that exposing *V. radaiata* to 50 and 100 ppb of O_3 reduced chlorophyll content considerably. On the other hand, Bindi et al. (2002) observed an increase in chlorophyll content in *Solanum tuberosum* before the tuber initiation stage under elevated O_3. From this observation, Bindi et al. (2002) suggested that, as chlorophyll is essential for antioxidant defense in stressful situations, *S. tuberosum* may have evolved to exhibit this rise in chlorophyll under elevated O_3 exposure. Carotenoids also play an important function in protecting the photosynthetic machinery from photo-oxidative damage and dispersing surplus excitation energy through the xanthophyll cycle. Elevated O_3 levels lowered carotenoid concentration in *Abelmoschus* cytotypes (Singh, Ansari, et al., 2024; Singh, Gupta, et al., 2024), *S. cordifolia* (Ansari et al., 2023), and *S. lycopersicum* (Gupta et al., 2023b). In contrast, Chaudhary et al. (2023) in

Glycine max, Döring et al. (2020) in *M. officinalis*, and Pellegrini et al. (2019) in *Quercus* cultivars found an increase in carotenoids.

Numerous studies have consistently reported the adverse effects of tropospheric O_3 on plant photosynthesis (Rai & Agrawal, 2008; Sarkar & Agrawal, 2010; Singh et al., 2009). One of the key physiological impairments observed is the reduction in photosynthetic carbon assimilation, which is often accompanied by changes in stomatal function and water use efficiency. Both acute and chronic exposure to elevated levels of tropospheric O_3 have been shown to significantly impair various physiological functions in plants. Studies by Moraes et al. (2004) and Du et al. (2018) have documented that exposure to O_3 can cause immediate reductions in plant physiology. Similarly, sustained exposure to O_3, as reported by Pina and Moraes (2010) and Pina et al. (2017), leads to cumulative physiological stress, resulting in prolonged damage.

Quantitative data underscores the severity of O_3 responses in plants. For instance, seedlings of *Caesalpinia echinata* exposed to 68 ppb of O_3 exhibited a 50% reduction in net photosynthetic rate (Pn) and a 42% decline in stomatal conductance (gs) (Moraes et al., 2006). In another study, *Anethum graveolens* subjected to 70 ppb of O_3 showed similar trends whereby Pn and gs decreased by 40% and 36%, respectively, while respiration rate (R_D) was elevated (Moura et al., 2018). The decline in photosynthetic activity is primarily attributed to O_3-induced damage to the photosynthetic apparatus, particularly the thylakoid membranes and key photosynthetic enzymes such as Rubisco. This impairment reduces the efficiency of carbon fixation, leading to decreased Pn. At the same time, the observed increase in intercellular CO_2 concentration (Ci) is likely a consequence of reduced gs, which limits CO_2 diffusion out of the leaf, creating a buildup of internal CO_2 (Hoshika et al., 2015). Moreover, the increase in R_D under O_3 stress is believed to be a protective metabolic response. It may reflect a reprogramming of mitochondrial activity aimed at supporting cellular repair mechanisms, enhancing detoxification processes, and supplying energy for the maintenance of redox homeostasis and damage repair (Hoshika et al., 2013; Kitao et al., 2009). This upregulated respiration represents a shift in energy allocation from growth and development to defense and survival, which can ultimately impair a plant's productivity and biomass accumulation over time.

In a study by Da Silva Engela et al. (2021), *Eugenia uniflora* exhibited a 31%–41% reduction in Pn under O_3 stress. This decline in carbon assimilation was partially offset by an increase in the number of leaves, suggesting a morphological acclimation strategy that may help maintain overall carbon gain despite impaired photosynthetic efficiency per leaf. Similarly, Calatayud et al. (2006) investigated the impact of chronic O_3 exposure (over a period of one year) on *Citrus unshiu* saplings. They reported a decrease in Pn (19%), accompanied by a 41% reduction in gs and a decline in noncyclic electron transport, a key component of the light-dependent reactions of photosynthesis. In addition, an increase in Ci was observed, likely reflecting stomatal closure and impaired CO_2 assimilation under O_3 stress. Further, comparative assessments by Chapla and Kamalakar (2004) on three tropical tree species—*Ficus infectoria*, *Pongamia pinnata*, and *Bauhinia variegata* revealed varying degrees of photosynthetic inhibition under elevated O_3 levels ranging from 40 to 120 ppb. *F. infectoria* showed the highest sensitivity with Pn reductions between 16% and 39%, followed by *P. pinnata* with 7%–31% decline. *B. variegata* showed the lowest reduction in Pn, ranging from 6% to 26%. Additionally, O_3 stress adversely affected ribulose-1,5-bisphosphate carboxylase activity, a crucial enzyme in the Calvin cycle responsible for CO_2 fixation. *B. variegata* exhibited the most pronounced reduction in carboxylase activity (10%–32%), followed by *F. infectoria* (10%–23%) and *P. pinnata* (9%–15%).

In *T. aestivum* (wheat), ambient O_3 concentrations have been shown to decrease Pn, which can be inferred as an effect closely associated with reduced leaf longevity and accelerated leaf senescence, indicating that O_3 exposure shortens the photosynthetically active period of leaves (Sarkar & Agrawal, 2010). Similar detrimental effects have been observed in a Malaysian *Oryza sativa* (rice) cultivar (MR 219), where reductions were observed in Pn, gs, and transpiration rate (E) under ambient O_3 conditions (Shahadan, 2005). However, the initial chlorophyll fluorescence remained unchanged, suggesting that the light-harvesting complex and associated antenna pigments were not initially impaired by O_3. Furthermore, under elevated O_3 exposure, Pn was reduced by 62%, and this decline was attributed to stomatal closure rather than damage to the photochemical components of photosynthesis. In a comparative study by Hayes et al. (2020), on *T. aestivum* and *Phaseolus vulgaris*, it was reported that as O_3 concentrations increased, the gs was significantly affected in both. In *Eleusine coracana* (finger millet) and *Pennisetum glaucum* (pearl millet), gs remained unaffected under O_3 stress, suggesting a degree of O_3 tolerance in their stomatal regulation mechanisms.

Physiological investigations conducted by Singh et al. (2009, 2013) on different varieties of *Brassica*, specifically *Brassica campestris* and *Brassica juncea*, have reported noticeable reductions in the Pn, gs, and the maximum quantum yield of photosystem II (Fv/Fm). These parameters are critical indicators of photosynthetic performance and photochemical efficiency, suggesting that O_3 negatively impacts both stomatal function and the integrity of the photosynthetic apparatus. Supporting these findings, Thwe et al. (2014) explored the sensitivity of tomato (*Lycopersicon esculentum*, cultivar "Look Tor") to acute O_3 exposures at concentrations of 100, 175, and 250 ppb. Their results revealed that even independent of the plant's developmental stage, O_3 stress led to marked physiological impairments. Specifically, the reduction in gs, which not only restricted CO_2 uptake but also contributed to damage of photosynthetic machinery, including suppression of photosystem II (PSII) activity. Ariyaphanphitak et al. (2005) attributed the decline in Pn to O_3-induced stomatal dysfunction and the production of ROS, which likely originated from disruptions in the plant's electron transport rate. The build-up of ROS under O_3 stress contributes to oxidative damage at the cellular level, particularly affecting chloroplasts and key enzymes involved in carbon fixation. In India, multiple studies have focused on the physiological response of wheat cultivars to O_3 exposure (Ghosh et al., 2020; Rai et al., 2007; Yadav et al., 2020). These studies consistently reported declines in Pn, gs, water-use efficiency (WUE), Fv/Fm, and photosynthetic nitrogen use efficiency. Such impairments indicate that O_3 interferes with the plant's ability to use both water and nitrogen efficiently. Mishra and Agrawal (2015) examined *V. radiata* (mung bean) cultivars under an O_3 concentration of 68 ppb and found that the reduction in WUE was primarily due to a more severe decline in Pn compared to E.

7.4.4 Growth, biomass, and productivity

Studies on O_3 exposure have consistently demonstrated its negative effects on plant biomass, with multiple studies reporting declines in dry matter, growth, and overall productivity (da Silva Engela et al., 2021; Moura et al., 2021). These impacts vary depending on species, exposure levels, and duration, but the overarching trend highlights the deleterious influence of O_3 stress on plant development. A study by Cassimiro and Moraes (2016) observed that *A, graveolens* experienced an early reduction in both plant height and dry biomass when grown under ambient air conditions for 5 months, compared to plants exposed to filtered air (i.e., air free of O_3). This suggested that even naturally occurring background levels of O_3 can adversely affect sensitive species. A similar trend was observed in *E. uniflora*, where fumigation with double the ambient O_3 concentration for 75 days in an O_3-FACE system resulted in a decline in plant height and root-to-shoot ratio (da Silva Engela et al., 2021), indicating that elevated O_3 can disrupt biomass partitioning and root development, potentially affecting nutrient and water uptake. Ramya et al. (2021) evaluated fifteen cultivars of *O. sativa* under elevated O_3 conditions and found average reductions of 11.9% in root length, 2.11% in shoot length, and a striking 22.52% decline in the number of tillers, highlighting its strong inhibitory effect on early-stage growth and productive tillering. These observations are consistent with previous studies. Akhtar et al. (2010) and Shao et al. (2020) reported reduced plant height and tiller numbers in various rice subspecies, including Indica, Bangladeshi, and Japonica varieties, when grown under elevated O_3 conditions. The findings suggest a broad sensitivity of rice genotypes to O_3, especially in terms of structural development.

Moura et al. (2021) extended these findings in *Moringa oleifera*, examining plant responses to a gradient of O_3 exposures: ambient (36 ppb), moderate (53 ppb, 1.5X ambient), and high (64 ppb, 2X ambient). The study revealed a significant decrease in leaf biomass under both moderate and high O_3 concentrations, emphasizing the sensitivity of foliar tissues to oxidative stress. In contrast, certain species, such as *C. echinata* (Moraes et al., 2006) and *Ficus insipida* (Schneider et al., 2017), displayed no significant changes in leaf, stem, or root biomass or in root-to-shoot ratios, suggesting inherent tolerance mechanisms or structural barriers that mitigated O_3 damage. A notable adaptation was reported by Dias et al. (2019) in *Cordia floribunda*, where the presence of nonglandular emergences (trichome-like structures) covering the leaf surface acted as protective barriers against O_3 penetration. When the intact leaves were exposed to 100 ppb of O_3, no morphological damage was observed; however, when the trichome-like structures were removed the O_3-induced injury symptoms developed, confirming their role in defense. Chaudhary and Agrawal (2013) studied six cultivars of *Trifolium alexandrinum* under an elevated O_3 regime (ambient +10 ppb) and reported decreases in key morphological traits, including plant height, leaf number, and leaf area. Among the tested cultivars, "Bundel" and "Wardan" showed greater sensitivity, indicating genotype-specific variability in O_3 response.

Further long-term insights were provided by Singh et al. (2021), who reported that *Leucaena leucocephala*, under chronic O_3 exposure (59–79 ppb) for 2 years, exhibited a reduction in leaf size. In another recent study, Singh et al. (2023), Singh et al. (2023), Singh et al. (2023a,b) evaluated the impact of 20 ppb of elevated O_3 above ambient on two hybrid cultivars of *Mangifera indica* (mango). The plants experienced a loss in productivity and the findings suggesting that under O_3 stress, these cultivars prioritized vegetative growth over reproductive development, leading to compromised fruit production. This allocation shift reflects a stress-avoidance strategy, where plants invest in maintaining structural growth at the cost of yield.

Exposure to elevated O_3 has been reported to adversely impact various growth parameters in crop species. In *Panicum miliaceum* (proso millet), as well as in two rice cultivars (*Oryza sativa*cvs. Malviya Dhan 36 and Shivani), significant reductions were observed in root and shoot lengths, the number of tillers, leaves, and ears (Agrawal et al., 1983; Sarkar & Agrawal, 2012). Similarly, studies conducted on *V. radiata* (mung bean) demonstrated that key vegetative traits such as plant height, leaf number, and leaf area declined significantly under both ambient and elevated O_3 concentrations (Chaudhary & Agrawal, 2015). Furthermore, O_3 exposure has been observed to significantly reduce overall plant productivity. Under ambient O_3 levels, total biomass declined by approximately 50% in *T. aestivum* (wheat), 24.2% in *Pisum sativum* (pea), and 42% in *V. radiata*. Correspondingly, yield losses were substantial, with reductions of 47.5%, 56%, and 48% reported in these species, respectively (Adrees, Ibrahim, et al., 2016; Adrees, Saleem, et al., 2016). In *Vigna unguiculata* (cowpea), plant height remained unaffected by O_3 exposure across different varieties; however, elevated O_3 significantly reduced biomass accumulation in leaves, stems, and roots, along with a significant decrease in the root-to-shoot ratio, indicating a shift in biomass allocation under stress (Tetteh et al., 2015). The *L. esculentum* (tomato) cultivars exposed to O_3 concentrations of 75 ppb and 150 ppb exhibited a significant decline in shoot length (2%–10.5%), shoot biomass (44%), and root biomass (8%–65%), as reported by Mina et al. (2010). These reductions reflect the direct impairment of both above- and below-ground growth under oxidative stress. In *T. aestivum*, the impact of elevated O_3 has been assessed in both field and controlled environments. A field study conducted in highly polluted areas around Varanasi indicated that ambient O_3 pollution was a major contributor to wheat yield losses of up to 32%, accompanied by a significant decline in seed quality (Rajput & Agrawal, 2005). Supporting evidence from OTC experiments revealed yield losses ranging from 9% to 25% under elevated O_3 conditions (Rai et al., 2007; Sarkar & Agrawal, 2010). Studies by Hayes et al. (2020) demonstrated that increasing O_3

exposure led to yield reductions in *T. aestivum* and *P. vulgaris* (common bean), while millets such as *E. coracana* (finger millet) and *P. glaucum* (pearl millet) showed relative resilience, with no significant O_3-induced yield impacts. These findings highlight species-specific and varietal differences in O_3 sensitivity, which are critical for crop selection and management under changing environmental conditions.

Studies have demonstrated the detrimental effects of elevated O_3 on the growth and development of various rice cultivars. In particular, Thai rice varieties exhibited heightened sensitivity to O_3 during the tillering stage, which was found to be more vulnerable than the booting and maturing stages. Exposure to elevated O_3 levels (50 ppb and 150 ppb) significantly hampered both root and shoot elongation. Among the growth stages, the tillering phase experienced the greatest reduction in shoot dry weight (29%) compared to the maturing stage (18%). Moreover, yield-related traits were negatively impacted, with reductions in both grain quality and quantity. These effects were attributed to disrupted nutrient accumulation in seeds under high O_3 exposure (Ariyaphanphitak et al., 2003). Similarly, in Malaysian rice cultivars, elevated O_3 decreased plant height, number of tillers, and dry biomass of root and shoot (Ishii et al., 2004; Shahadan, 2005). Ismail et al. (2015) further reported substantial yield losses at O_3 concentrations of 60 ppb and 120 ppb, with more pronounced declines under the higher concentration. Beyond rice, O_3 stress has also been shown to influence the phenology of other crops. In *Cicer arietinum* (chickpea), O_3 stress (60 ppb) induced a 14-day reduction in its growth duration (Singh, Mukherjee, et al., 2021). This early senescence was linked to a decrease in chlorophyll content and essential nutrients, which accelerated leaf aging and hastened plant maturation.

Both acute (>100 ppb) and chronic (40–60 ppb) O_3 exposures have been shown to negatively impact leguminous crops, primarily by reducing root nodulation. This decline adversely affects the biological nitrogen fixation, which is essential for legume productivity and soil fertility (Hewitt et al., 2016). Acute exposure of 137 ppb O_3 resulted in a 25% reduction in below-ground biomass, a 30% decrease in nodule biomass, and a 21% decline in nitrogen fixation capacity in *Glycine max* (soybean), highlighting the substantial sensitivity of legume root systems to elevated O_3 stress (Biancari et al., 2021).

7.4.5 Molecular and genetic responses

O_3 (O_3) stress is a significant environmental challenge that negatively impacts plant growth and productivity by inducing oxidative damage.

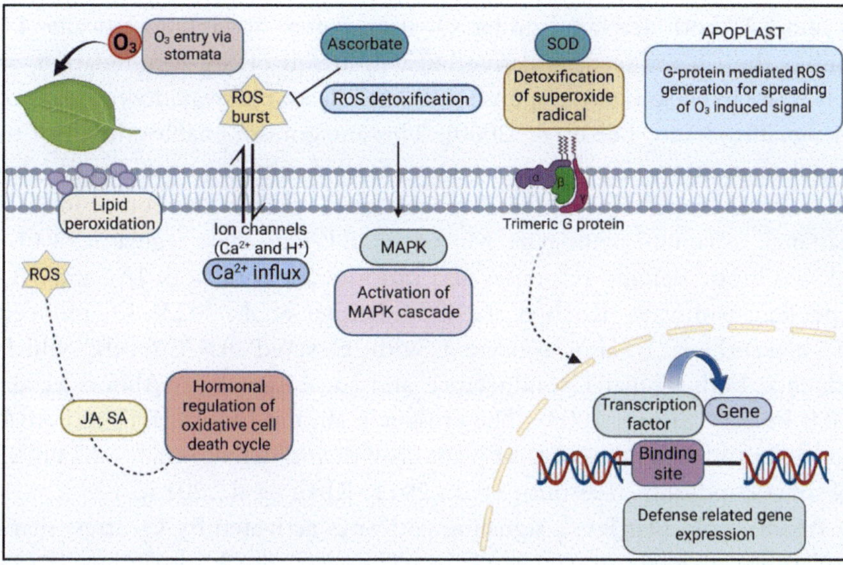

Figure 7.5 Pictorial representation of O_3 entry and trigger of hormonal and molecular signaling.

Upon exposure, plants initiate a complex network of molecular and genetic responses to mitigate their harmful effects (Fig. 7.5). These include activation of antioxidant enzymes, expression of stress-related genes, and modulation of hormonal and signaling pathways. Understanding these responses is crucial for developing O_3-resilient crop varieties.

7.4.5.1 Changes in gene expression related to stress signaling and defense

When plants are exposed to O_3 (O_3) stress, it leads to a rapid oxidative burst that generates apoplastic ROS, such as hydrogen peroxide (H_2O_2). These ROS serve as important signaling molecules, altering the redox state of the cell and initiating downstream signaling events. One of the key signaling mechanisms activated by ROS is the mitogen-activated protein kinase cascade, which subsequently induces the expression of genes involved in hormone biosynthesis specifically those for ethylene (ET), jasmonic acid (JA), salicylic acid (SA), and abscisic acid (ABA) (Ainsworth, 2017) (Fig. 7.5). These signaling pathways also increase the expression of genes related to antioxidant metabolism and respiration, while suppressing genes associated with photosynthesis. The hormone signaling network triggered by O_3 is intricate, involving both synergistic and antagonistic interactions.

JA and SA, both activated under O_3 stress, often act antagonistically. JA inhibits the O_3-induced oxidative burst upstream of SA accumulation, as well as the hypersensitive response (HR) cell death pathway downstream of SA signaling (Rao, Lee et al., 2000). This antagonism enables the plant to modulate its stress responses, particularly cell death; however, it may also suppress SA-dependent defense mechanisms, potentially compromising resistance to certain pathogens. Moreover, ET is a major regulator of O_3-induced biotic defense responses and can interact with both JA- and SA-dependent pathways (Koch et al., 2000; Leisner et al., 2023). O_3-induced ET biosynthesis is also associated with elevated ABA levels, which influence both stomatal conductance and sugar signaling (Ahlfors et al., 2004; León & Sheen, 2003). The cross-talk among JA, SA, ET, and ABA thus determines the outcome of plant growth and defense trade-offs under O_3 stress conditions (Eastburn et al., 2011; Khan et al., 2015).

At the molecular level, signaling pathways activated by O_3 stress share similarities with those triggered by pathogen attack, including ROS production and the induction of HR (Rao et al., 2002). O_3 also enhances the activity of enzymes in the phenylpropanoid, flavonoid, and lignin biosynthesis pathways, all of which are essential for structural and chemical defense (Leisner et al., 2023). Further, Yoshida et al. (2009) demonstrated that de novo biosynthesis of glutathione, a major antioxidant, constitutes a critical defense mechanism against O_3 stress and is largely regulated by SA and ET. In another study, a stimulus-specific calcium signature regulating O_3-induced transcriptional changes was identified in the Col-0 ecotype of *Arabidopsis thaliana* (Short et al., 2012). This study also highlighted the antagonistic interplay between SA and apoplastic ROS in modulating gene expression, with plants dynamically prioritizing between these two signaling mechanisms based on the O_3 exposure dose and duration. Additionally, Xu et al. (2015) investigated O_3 responses in O_3-sensitive (Te) and O_3-tolerant (C24) ecotypes of *A. thaliana*. Their findings revealed that gene expression and cell death in response to apoplastic ROS are not governed by a single signaling pathway. Instead, a network of overlapping hormonal and signaling cascades collectively modulates the plant's response to O_3-induced stress.

7.4.5.2 Role of transcription factors and reactive oxygen species-responsive genes

Plants utilize complex and multifaceted signaling networks to perceive and respond to O_3 (O_3) stress, often integrating signals from multiple pathways, including JA and redox signaling. Recent transcriptomic analyzes in *Brassica rapa* (Pak Choi) under acute O_3 exposure have revealed the upregulation of several transcription factor (TF) families known to be involved in abiotic stress responses, including AP2/ERF, Trihelix, WRKY, and MYB genes (Zhang et al., 2017; Zhang et al., 2017). Members of the AP2/ERF family are key regulators of stress and redox-responsive gene networks and have been shown to confer tolerance to multiple abiotic stresses. In *B. campestris* (Pak Choi), differentially expressed genes (DEGs) encoding AP2/ERF proteins were significantly upregulated upon elevated O_3 exposure. Notably, *Bra024953*, encoding ERF5, showed higher expression under O_3 stress. Similarly, CBF4, traditionally associated with drought tolerance, was also implicated in O_3 stress response, suggesting its broader functional role (Zhang, Li, et al., 2017; Zhang, Xu, et al., 2017).

Trihelix TFs, though traditionally linked to plant development, are emerging as potential players in abiotic stress responses. In this context, *Bra005127*, *Bra005688*, and *Bra028899* encoding Trihelix proteins were induced by O_3 in *B. campestris* (Zhang, Li, et al., 2017; Zhang, Xu, et al., 2017). WRKY transcription factors, characterized by a conserved WRKY domain and zinc finger motif, act as transcriptional activators or repressors during stress responses (Marchica et al., 2021). Their upregulation under EO_3 in *S. officinalis* and *B. campestris* suggests that WRKY TFs are involved in senescence and oxidative stress responses. Four WRKY genes, *WRKY4*, *WRKY5*, *WRKY11*, and *WRKY46*, and three WRKY genes, *WRKY28*, *WRKY48*, and *WRKY57*, were newly identified as differentially expressed, likely contributing specifically to the O_3 stress response in *S. officinalis* and *B. campestris*, respectively (Marchica et al., 2021; Zhang, Li, et al., 2017; Zhang, Xu, et al., 2017). MYB transcription factors also showed enhanced expression in response to EO_3, with one upregulated *MYB* gene linked to the accumulation of anthocyanins, a secondary metabolite involved in oxidative stress mitigation. This observation aligns with known regulatory roles of MYBs in anthocyanin biosynthesis and their interaction with the JA-mediated JAZ-DELLA-MYBL2 module, which coordinates stress-induced pigment accumulation in *Arabidopsis* (Zhang et al., 2025). In addition to TFs, several downstream functional genes were activated under O_3 stress. These include genes involved in cell wall modification (e.g., XETs), ROS detoxification (e.g., glutathione S-transferase), sulfur metabolism, and phenylpropanoid biosynthesis. Notably, the expression of *Bra038031*, encoding an enzyme catalyzing L-serine to O-acetyl-L-serine a key signaling molecule under sulfur starvation, was significantly upregulated. This enzyme, along with *Bra010645* (involved in sulfate assimilation) and *Bra015981*, likely contributes to a compensatory response to O_3-induced sulfur starvation, as supported by KEGG enrichment analysis (Zhang, Li, et al., 2017; Zhang, Xu, et al., 2017). Collectively, these findings highlight a complex transcriptional network regulated by TFs and downstream genes that modulate defense, metabolic reprogramming, and stress adaptation in plants under O_3 stress. They also underscore the integration of redox signaling, hormone pathways, and metabolic responses in coordinating plant defense.

7.4.6 Use of omics (transcriptomics, proteomics, metabolomics) in ozone stress studies

Advancements in multiomics technologies such as transcriptomics, proteomics, and metabolomics have significantly transformed the understanding of plant stress physiology by offering detailed molecular insights. These integrative approaches facilitate the identification of critical genes, proteins, and metabolic pathways involved in O_3 tolerance and damage alleviation. Transcriptomics involves the large-scale analysis of gene expression patterns under O_3 exposure using RNA sequencing (RNA-seq) or microarrays, helping in the identification of stress-responsive genes and regulatory networks. Under O_3 stress, plants reprogram their transcriptome to activate defense pathways, modulate hormone signaling, and enhance antioxidant responses. Studies on plants exposed to elevated O_3 revealed differential expression of genes involved in SA, JA, and ethylene signaling pathways (Tamaoki et al., 2003). RNA-Seq on Pak-choi plants has uncovered DEGs with most enrichments were oxidoreductase activity, response to stress, and response to stimulus (Zhang, Li, et al., 2017; Zhang, Xu, et al., 2017). Studies suggest that genes related to stomatal density and cell wall repair show increased expression under O_3 (Zhang, Li, et al., 2017; Zhang, Xu, et al., 2017). O_3 exposure rapidly increases lipoxygenase activity, leading to JA biosynthesis via the octadecanoid pathway (Singh et al., 2022). Phytohormones have been noted to be involved in stomatal regulation via guard cell signaling in plants under abiotic stress, including O_3. Key transcription factors such as WRKY, MYB, and AP2/ERF families that link hormone signaling to antioxidant biosynthesis were shown to be upregulated in response to O_3 (Antenozio et al., 2024; Zhang, Li, et al., 2017; Zhang, Xu, et al., 2017). O_3 exposure led to the upregulation of genes associated with photosystem II repair, detoxification, and ubiquitin-mediated protein degradation (Zhang et al., 2017; Zhang et al., 2017). In a transcriptomic analysis, have noted the intermediary roles of ROS in signaling of downstream hormones (JA and SA) via identification of ROS-responsive elements such as the TCA element and CGTCA-motifABRE in hormones-associated differential genes. Proteomics provides direct insights into functional molecules by analyzing changes in the abundance, modification, and interaction of proteins. Since transcript levels do not always correlate with protein abundance, proteomics offers a complementary view to transcriptomics. O_3 stress typically leads to the induction of heat shock proteins, antioxidant enzymes (e.g., catalase, peroxidases), and PR proteins (Gupta et al., 2023; Lee et al., 2022). Proteomic studies have also revealed changes in the abundance of proteins related to N2 metabolism, secondary metabolites, TCA cycle, and glycolysis (Zentella et al., 2023). Metabolomics captures the biochemical phenotype by profiling primary and secondary metabolites, providing insights into metabolic adjustments and defense activation under O_3 exposure. Metabolite profiling involves Gas Chromatography-Mass Spectrometry (GC-MS), Liquid Chromatography-Mass Spectrometry (LC-MS), High-Resolution Mass Spectrometry (HRMS), Nuclear Magnetic Resonance (NMR) Spectroscopy, and Direct Infusion Mass Spectrometry. While GC-MS offers a better approach for volatile or primary metabolite screening, LC-MS/HRMS dominates untargeted secondary metaboliteanalyzes on account of high mass accuracy. Metabolite profiling has imparted critical insights into biochemical adjustments of plants under elevated O_3. O_3 induced accumulation of organic acids (malate, citrate), amino acids (proline, GABA), which function as Osmo protectants and maintain redox homeostasis (Cho et al., 2008; Galati et al., 2019). Enhanced levels of nonenzymatic antioxidants such as ascorbate, glutathione, flavonoids, and phenolic compounds have also been confirmed in plants, suggesting adaptive measures in plants under O_3-induced oxidative stress (Zhang et al., 2017). While transcriptomics provides early indicators of O_3 response, proteomics captures the active biological processes, and metabolomics reflects the actual physiological outcomes (Galati et al., 2019). Integrated omics approaches are therefore essential for a holistic understanding of plant defense mechanisms under O_3 stress.

7.5 Mechanism of ozone entry into plant tissues and reactive oxygen species generation

O_3 actively enters plant tissues and disrupts their internal systems. The primary site for O_3 entry into a plant is through stomata on the leaf surface. The stomata are essential for gas exchange but also serve as the main entry point for O_3 (Cho et al., 2011). Upon entering through the stomata, O_3 reaches the apoplastic space of plant tissues and reacts with the cell membrane, triggering oxidative stress. Beyond the apoplastic region, O_3 can directly interact with the plasma membrane via a chemical reaction known as ozonolysis, which disrupts membrane structure and function. Alternatively, O_3 can spark a chain reaction that leads to the overproduction of ROS that cause oxidative damage to cellular components (Singh et al., 2023b; Singh, Ansari, et al., 2024; Singh, Gupta, et al., 2024). The intensity of oxidative stress and plant response to cope with it largely determines its sensitivity or tolerance to O_3. O_3-sensitive plants tend to experience higher ROS accumulation. As a result, they suffer more extensive cellular damage, leading to significant losses in yield and overall productivity. In contrast, O_3-tolerant plants rapidly activate and enhance their antioxidant defense systems, including both enzymatic (e.g., catalase, peroxidase, superoxide dismutase) and nonenzymatic antioxidants (e.g., ascorbic acid and glutathione). The regulation of these antioxidants is often localized, depending on the site of oxidative stress. In some cases, studies have reported elevated levels of ascorbic acid specifically in the apoplast. This increase suggests a direct neutralization of O_3 by apoplastic ascorbate, which helps to prevent oxidative damage to cell membranes in more resistant plant varieties. However, when the antioxidant response is insufficient (sensitive plants), the oxidative stress progresses further, leading to peroxidation of membrane lipids, degradation of photosynthetic pigments, and the activation of stress-related signaling pathways that can influence various physiological processes. Additionally, ROS can migrate into the symplast either directly or through specific channel proteins, triggering further cellular responses. Within the symplast, a strong link between GSH and ascorbate recycling systems has been observed. To maintain antioxidant balance, plants stimulate the transport and regeneration of these compounds, especially to replenish depleted apoplastic ascorbate pools (Baier & Dietz, 2005). Key enzymes, such as catalase and peroxidase, detoxify H_2O_2 by converting it into harmless water, while glutathione plays a supporting role in maintaining redox balance (Noctor and Foyer, 1998). Overall, the coordinated action of these antioxidants is vital for limiting O_3-induced damage.

However, this antioxidant response is energy-consuming, and plants often undergo a trade-off, diverting energy and resources away from growth and reproduction to prioritize defense. This shift, although protective in the short term, can compromise the plant's long-term fitness and yield potential.

7.6 Conclusions and future perspectives

The escalating threat of tropospheric O_3 to global food security and ecosystem stability demands innovative and integrated adaptive strategies. With O_3 concentrations projected to increase by 20%–60% by 2100, primarily due to industrial emissions and fossil fuel combustion, developing comprehensive approaches to mitigate phytotoxic impacts has become imperative. Future directions must bridge molecular biology, agricultural science, and policy reform, focusing on three interconnected pillars: the development of O_3-tolerant crops, strategic revisions to air quality governance, and the application of multiomics technologies for precision plant health management. Traditional breeding methods should be combined with marker-assisted selection and genomic selection to identify and incorporate O_3-tolerant traits into desired cultivars. Advanced biotechnological tools like CRISPR-Cas9 should be employed to precisely modify genes involved in O_3 detoxification, repair mechanisms, or stress signaling pathways. This could involve enhancing antioxidant defense systems, improving stomatal control to limit O_3 uptake, and strengthening cell membrane integrity. Large populations of crop plants can be screened for O_3 tolerance under controlled and field conditions by developing high-throughput phenotyping platforms. While breeding offers a solution at the plant level, broader systemic changes are essential to reduce O_3 exposure. Future efforts must integrate air quality policy, such as refining emission regulations of O_3 precursors, with sustainable agricultural practices. Regional and global collaboration should be fostered to address transboundary O_3 pollution. Sustainable agricultural practices such as optimized nutrient management, biochar, and antiozonant amendment, that enhance ecosystem resilience and potentially mitigate O_3 stress, should be increasingly adopted. Incorporating multiomics studies for a deeper understanding of plant responses to O_3 at the molecular level is crucial for informed breeding and management strategies. By synergistically pursuing these future directions, researchers, policymakers, and agricultural communities can collectively work towards safeguarding global food security against the escalating threat of tropospheric O_3.

Acknowledgment

The authors are thankful to the Head, Department of Botany and Section Incharge, Department of Botany (MMV), Banaras Hindu University, for providing the facilities required to perform a comprehensive literature review. The authors are grateful to the Coordinators, Institute of Eminence, Center of Advanced Study (CAS) in Botany, Banaras Hindu University. Vineet Upadhyay is highly appreciative of the University Grants Commission (UGC), New Delhi, India, for providing a Junior Research Fellowship (JRF). Harshita Singh acknowledges the Department of Science and Technology (INSPIRE) for financial assistance. Krishna Kumar Choudhary appreciates the ANRF (SERB), New Delhi (EEQ/2023/000403) and Institute of Eminence (IoE - Dev. Scheme no. 6031), Banaras Hindu University, Varanasi, India, for the financial assistance.

Declarations

Ethics approval and consent to participate

Not applicable.

Consent for publication

Not applicable.

Availability of data and materials

Not applicable.

Competing interests

The authors declare that they have no known competing interests that could influence the work reported in this paper.

References

Adrees, M., Ibrahim, M., Shah, A. M., Abbas, F., Saleem, F., Rizwan, M., Hina, S., Jabeen, F., & Ali, S. (2016). Gaseous pollutants from brick kiln industry decreased the growth, photosynthesis, and yield of wheat (*Triticum aestivum* L.). *Environmental Monitoring and Assessment, 188*(5), 1–11.

Adrees, M., Saleem, F., Jabeen, F., Rizwan, M., Ali, S., Khalid, S., Ibrahim, M., Iqbal, N., & Abbas, F. (2016). Effects of ambient gaseous pollutants on photosynthesis, growth, yield and grain quality of selected crops grown at different sites varying in pollution levels. *Archives of Agronomy and Soil Science, 62*(9), 1195–1207.

Agrawal, M., Nandi, P. K., & Rao, D. N. (1983). Ozone and sulphur dioxide effects on *Panicum miliaceum* plants. *Bulletin of the Torrey Botanical Club, 110*(4), 435–441. https://doi.org/10.2307/2996276.

Ahlfors, R., Lång, S., Overmyer, K., Jaspers, P., Broscheé, M., Tauriainen, A., & Kangasjärvi, J. (2004). *Arabidopsis* RADICAL-INDUCED CELL DEATH1 belongs to the WWE protein–protein interaction domain protein family and modulates abscisic acid, ethylene, and methyl jasmonate responses. *The Plant Cell, 16*(7), 1925–1937.

Ainsworth, E. A. (2017). Understanding and improving global crop response to ozone pollution. *The Plant Journal, 90*(5), 886–897.

Akhtar, N., Yamaguchi, M., Inada, H., Hoshino, D., Kondo, T., Fukami, M., et al. (2010). Effects of ozone on growth, yield and leaf gas exchange rates of four Bangladeshi cultivars of rice (Oryza sativa L.). *Environmental Pollution, 158*(9), 2970–2976.

Akram, N. A., Shafiq, F., & Ashraf, M. (2017). Ascorbic acid-a potential oxidant scavenger and its role in plant development and abiotic stress tolerance. *Frontiers in Plant Science, 8*, 613.

Ansari, N., Yadav, D. S., Agrawal, M., & Agrawal, S. B. (2021). The impact of elevated ozone on growth, secondary metabolites, production of reactive oxygen species and antioxidant response in an anti-diabetic plant *Costus pictus*. *Functional Plant Biology, 48*(6), 597–610.

Ansari, N., Yadav, D. S., Singh, P., Agrawal, M., & Agrawal, S. B. (2023). O_3 exposure response on physiological and biochemical parameters vis-a-vis secondary metabolites in a traditional medicinal plant *Sida cordifolia* L. *Industrial Crops and Products, 194*, 116267.

Antenozio, M. L., Caissutti, C., Caporusso, F. M., Marzi, D., & Brunetti, P. (2024). Urban air pollution and plant tolerance: Omics responses to O_3, nitrogen oxides, and particulate matter. *Plants, 13*(15), 2027.

Ariyaphanphitak, W., Chidthaisong, A., Sarobol, E., Bashkin, V. N., & Towprayoon, S. (2005). Effects of elevated ozone concentrations on Thai jasmine rice cultivars (*Oryza sativa* L.). *Water, Air, and Soil Pollution, 167*(1), 179–200.

Ariyaphanphitak, W., Chidthaisong, A., Sarobol, E., Towprayoon, S., & Bashkin, V. N. (2003). Responses of Thai Rice Cultivar Chainat I (Oryza sativa L.)to Enhanced Tropospheric Ozone Concentrations. In *Proceedings of the 2nd Regional Conference on Energy Technology Towards a Clean Environment 12-14 February 2003, Phuket, Thailand.*

Bachheti, A. J., Bhalla, P., Bachheti, R. K., & Husen, A. (2022). Growth and development of medicinal plants, and production of secondary metabolites under O_3 pollution. In *Environmental pollution and medicinal plants*. Boca Raton: CRC Press, 25–38.

Baier, M., & Dietz, K. J. (2005). Chloroplasts as source and target of cellular redox regulation: a discussion on chloroplast redox signals in the context of plant physiology. *Journal of Experimental Botany, 56*(416), 1449–1462.

Basara, O., & Gorzelany, J. (2025). Impact of O_3 exposure on chlorophyll fluorescence, pigment content and leaf gas exchange on *Lonicera caerulea* var. *kamtschatica* and *Lonicera caerulea* var. emphyllocalyx. *Sustainability, 17*(7), 2820.

Baudic, A., Gros, V., Sauvage, S., Locoge, N., Sanchez, O., Sarda-Estève, R., Kalogridis, C., Petit, J.-E., Bonnaire, N., Baisnée, D., et al. (2016). Seasonal variability and source apportionment of volatile organic compounds (VOCs) in the Paris megacity (France). *Atmospheric Chemistry and Physics, 16*, 11961–11989.

Bell, J. N. B., & Ashmore, M. R. (1986). Design and construction of open top chambers and methods of filtration (equipment and cost). In *Proceedings of II European open top chambers workshop*. Freiburg, Germany: CEC.

Berezina, E., Moiseenko, K., Skorokhod, A., Pankratova, N. V., Belikov, I., Belousov, V., & Elansky, N. F. (2020). Impact of VOCs and NOx on ozone formation in Moscow. *Atmosphere, 11*, 1262.

Bhattacharjee, S., & Bhattacharjee, S. (2019). ROS and antioxidants: Relationship in green cells. In *Reactive Oxygen Species in Plant Biology*. New Delhi: Springer, 33–63. https://doi.org/10.1007/978-81-322-3941-3_2.

Biancari, L., Cerrotta, C., Menéndez, A. I., Gundel, P. E., & Martínez-Ghersa, M. A. (2021). Episodes of high tropospheric ozone reduce nodulation, seed production and quality in soybean (*Glycine max* (L.) merr.) on low fertility soils. *Environmental Pollution, 269*, 116–117.

Bičárová, S., Lukasová, V., Adamčíková, K., Žatková, L., Milovský, R., Shashikumar, A., & Bilčík, D. (2023). Modified electrolyte leakage method for testing the oxidative stability of *Pinus mugo* Turra under O_3-induced stress. *Folia Oecologica, 50*(1), 1–15.

Bindi, M., Hacour, A., Vandermeiren, K., Craigon, J., Ojanperä, K.,Selldén, G., et al. (2002). Chlorophyll concentration of potatoes grown under elevated carbon dioxide and/or ozone concentrations. *European Journal of Agronomy, 17*(4), 319–335.

Boubakri, H. (2018). The role of ascorbic acid in plant–pathogen interactions. In *Ascorbic acid in plant growth, development and stress tolerance.* Cham: Springer International Publishing, 255–271.

Boublin, F., Cabassa-Hourton, C., Leymarie, J., & Leitao, L. (2022). Potential involvement of proline and flavonols in plant responses to O_3 *Environmental Research, 207*, 112214.

Brunetti, C., Guidi, L., Sebastiani, F., & Tattini, M. (2015). Isoprenoids and phenylpropanoids are key components of the antioxidant defense system of plants facing severe excess light stress. *Environmental and Experimental Botany, 119*, 54–62.

Bulbovas, P., Sant'Anna, S. M. R., Moraes, R. M. D., Lima, E. D. S., Pina, J. M., Esposito, M. P., & Domingos, M. (2008). Avaliação da sensibilidade de plantasjovens de quiabo (*Abelmoschus esculentus* (L.) Moench.-Malvaceae) aoozônio. *Hoehnea, 35*, 359–366.

Burkart, S., Bender, J., Tarkotta, B., Faust, S., Castagna, A., Ranieri, A., & Weigel, H. J. (2013). Effects of O_3 on leaf senescence, photochemical efficiency and grain yield in two winter wheat cultivars. *Journal of Agronomy and Crop Science, 199*(4), 275–285.

Calatayud, A., Iglesias, D. J., Talón, M., & Barreno, E. (2006). Effects of long-term ozone exposure on citrus: Chlorophyll a fluorescence and gas exchange. *Photosynthetica, 44*(4), 548–554.

Cassimiro, J. C., & Moraes, R. M. (2016). Responses of a tropical tree species to ozone: Visible leaf injury, growth, and lipid peroxidation. *Environmental Science and Pollution Research, 23*(8), 8085–8090.

Chapla, J., & Kamalakar, J. A. (2004). Metabolic responses of tropical trees to ozone pollution. *Journal of Environmental Biology, 25*(3), 287–290.

Chaudhary, I. J., Nigam, B., & Rathore, D. (2023). Effect of elevated ozone on soybean (*Glycine max* L.) cultivar: Role of orange juice and synthetic ascorbic acid. *Nature Environment and Pollution Technology, 22*(3), 1225–1238.

Chaudhary, N., Singh, S., Agrawal, S. B., & Agrawal, M. (2013). Assessment of six Indian cultivars of mung bean against ozone by using foliar injury index and changes in carbon assimilation, gas exchange, chlorophyll fluorescence and photosynthetic pigments. *Environmental Monitoring and Assessment, 185*(9), 7793–7807.

Chaudhary, N., & Agrawal, S. B. (2013). Intraspecific responses of six Indian clover cultivars under ambient and elevated levels of ozone. *Environmental Science and Pollution Research, 20*(8), 5318–5329.

Chaudhary, N., & Agrawal, S. B. (2015). The role of elevated ozone on growth, yield and seed quality amongst six cultivars of mung bean. *Ecotoxicology and Environmental Safety, 111*, 286–294.

Chen, C., Chen, S., Jha, R. K., Cotrozzi, L., Nali, C., Lorenzini, G., & Ma, L. (2020). Phenol metabolism of two cultivars of durum wheat (*Triticum durum*Desf.) as affected by ozone and flooding stress. *Journal of Agronomy and Crop Science, 206*(3), 338–351.

Chen, Z., Cao, J., Yu, H., & Shang, H. (2018). Effects of elevated O_3 levels on photosynthesis, biomass and non-structural carbohydrates of *Phoebe bournei* and *Phoebe zhennan* in subtropical China. *Frontiers in Plant Science, 9*, 1764.

Cheng, S., Wang, Y., Sun, M., Zheng, J., Zhu, X., & Zhao, T. (2023). Effects of elevated surface ozone concentration on photosynthetic fluorescence characteristics and yield of soybean parents and offspring. *Atmosphere, 14*(9), 1368.

Cheynier, V., Comte, G., Davies, K. M., Lattanzio, V., & Martens, S. (2013). Plant phenolics: Recent advances on their biosynthesis, genetics, and ecophysiology. *Plant Physiology and Biochemistry, 72*, 1–20.

Cho, K., Shibato, J., Agrawal, G. K., Jung, Y. H., Kubo, A., Jwa, N. S., & Rakwal, R. (2008). Integrated transcriptomics, proteomics, and metabolomics analyses to survey O_3 responses in the leaves of rice seedling. *Journal of Proteome Research, 7*(7), 2980–2998.

Cho, K., Tiwari, S., Agrawal, S. B., Torres, N. L., Agrawal, M., Sarkar, A., Shibato, J., Agrawal, G. K., Kubo, A., & Rakwal, R. (2011). Tropospheric ozone and plants: Absorption, responses, and consequences. In D. Whitacre (Ed.). *Reviews of Environmental Contamination and Toxicology Volume 212. Reviews of Environmental Contamination and Toxicology, vol 212* (pp. 61–111). New York, NY: Springer. https://doi.org/10.1007/978-1-4419-8453-1_3.

Cotrozzi, L., Campanella, A., Pellegrini, E., Lorenzini, G., Nali, C., & Paoletti, E. (2018). Phenylpropanoids are key players in the antioxidant defense to ozone of European ash, *Fraxinus excelsior. Environmental Science and Pollution Research, 25*(9), 8137–8147.

Dai, J., & Wang, T. (2021). Impact of international shipping emissions on ozone and PM2.5 in East Asia during summer: The important role of HONO and $ClNO_2$. *Atmospheric Chemistry and Physics, 21*, 8747–8759. https://doi.org/10.5194/acp-21-8747-2021.

Dewan, S., & Lakhani, A. (2022). Tropospheric ozone and its natural precursors impacted by climatic changes in emission and dynamics. *Frontiers in Environmental Science, 10*, 1007942.

Dhevagi, P., Ramya, A., Poornima, R., & Chandrakumar, K. (2023). Effectiveness of ethylene diurea in ameliorating O_3 stress in blackgram varieties *(Vigna mungo* L.). *Archives of Agronomy and Soil Science, 69*(9), 1503–1518.

Dhevagi, P., Ramya, A., Priyatharshini, S., & Poornima, R. (2022). Effect of elevated tropospheric ozone on *Vigna Mungo* L. varieties. *Ozone: Science & Engineering, 44*(6), 566–586.

Dias, M. G., Moura, B. B., Pedrosa, G.da S., de Souza, S. R., & Cardoso-Gustavson, P. (2019). The role of non-glandular emergences in *Croton floribundus* (Euphorbiaceae) upon elevated ozone exposures. *Nordic Journal of Botany, 37*(11). https://doi.org/10.1111/njb.02338.

Dolker, T., Mukherjee, A., Bhushan Agrawal, S., & Agrawal, M. (2019). Ozone phytotoxicity to *Panicum maximum* and *Cenchrus ciliaris* at indo-gangetic plains: An assessment of antioxidative defense and growth responses. *Ecotoxicology (London, England), 28*(8), 853–868.

Donzelli, G., & Suarez-Varela, M. M. (2024). "Tropospheric Ozone: A critical review of the literature on emissions, exposure, and health effects. *Atmosphere, 15*(7), 779.

Du, B., Kreuzwieser, J., Winkler, J. B., Ghirardo, A., Schnitzler, J.-P., Ache, P., Alfarraj, S., Hedrich, R., White, P., & Rennenberg, H. (2018). Physiological responses of date palm (Phoenix dactylifera) seedlings to acute ozone exposure at high temperature. *Environmental Pollution, 242*, 905–913.

Dudek, B., Warskulat, A. C., & Schneider, B. (2016). The occurrence of flavonoids and related compounds in flower sections of Papaver nudicaule. *Plants, 5*(2), 28.

Döring, A. S., Cotrozzi, L., Lorenzini, G., Nali, C., Petersen, M., & Pellegrini, E. (2020). Deciphering the role of low molecular weight antioxidants in the sensitivity of *Melissa officinalis* L. to realistic ozone concentrations. *Industrial Crops and Products, 150*, 112369.

Eastburn, D. M., McElrone, A. J., & Bilgin, D. D. (2011). Influence of atmospheric and climatic change on plant–pathogen interactions. *Plant Pathology, 60*(1), 54–69.

El-Mahrouk, E. S. M., Eldawansy, S. M., El-Tarawy, A. M., Ebrahim, H. M. A., Eisa, E. A., Tilly-Mándy, A., & Honfi, P. (2024). Evaluation of the growth, enzymatic activity, electrolyte leakage, and phytoremediation efficiency of *Conocarpus erectus* under Cadmium and lead stress. *Frontiers in Plant Science, 15*, 1466697.

Engela, da S., Gonçalves, M. R., Furlan, C. M., Esposito, M. P., Fernandes, F. F., Carrari, E., Domingos, M., Paoletti, E., & Hoshika, Y. (2021). Metabolic and physiological alterations indicate that the tropical broadleaf tree *Eugenia uniflora* L. is sensitive to ozone. *Science of the Total Environment, 769*, 145080.

Feng, Z., Kobayashi, K., & Ainsworth, E. A. (2008). Impact of elevated ozone concentration on growth, physiology, and yield of wheat (*Triticum aestivum* L.): A meta-analysis. *Global Change Biology, 14*(11), 2696–2708.

Feng, Z., De Marco, A., Anav, A., Gualtieri, M., Sicard, P., Tian, H., & Paoletti, E. (2019). Economic losses due to ozone impacts on human health, forest productivity and crop yield across China. *Environment International, 131*, 104966.

Feng, Z., Sun, J., Wan, W., Hu, E., & Calatayud, V. (2014). Evidence of widespread ozone-induced visible injury on plants in Beijing, China. *Environmental Pollution, 193*, 296–301.

Galati, G., Gandin, A., Jolivet, Y., Larbat, R., & Hehn, A. (2019). Untargeted metabolomics approach reveals diverse responses of *Pastinaca sativa* to O_3 and wounding stresses. *Metabolites, 9*(7), 153.

Ghosh, A., Pandey, A. K., Agrawal, M., & Agrawal, S. B. (2020). Assessment of growth, physiological, and yield attributes of wheat cultivar HD 2967 under elevated ozone exposure adopting timely and delayed sowing conditions. *Environmental Science and Pollution Research, 27*(14), 17205–17220.

Ghosh, A., Agrawal, M., & Agrawal, S. B. (2021). Examining the effectiveness of biomass-derived biochar for the amelioration of tropospheric ozone-induced phytotoxicity in the Indian wheat cultivar HD 2967. *Journal of Hazardous Materials, 408*, 124968. https://doi.org/10.1016/j.jhazmat.2020.124968.

Gill, S. S., & Tuteja, N. (2010). Reactive oxygen species and antioxidant machinery in abiotic stress tolerance in crop plants. *Plant Physiology and Biochemistry, 48*(12), 909–930.

Grulke, N. E., & Heath, R. L. (2020). O3 effects on plants in natural ecosystems. *Plant Biology, 22*, 12–37. https://doi.org/10.1111/plb.12971.

Guarin, J. R., Jägermeyr, J., Ainsworth, E. A., Oliveira, F. A., Asseng, S., Boote, K., & Sharps, K. (2024). Modeling the effects of tropospheric O_3 on the growth and yield of global staple crops with DSSAT v4. 8.0. *Geoscientific Model Development, 17*(7), 2547–2567.

Guo, C., Wang, X., Wang, Q., Zhao, Z., Xie, B., Xu, L., & Zhang, R. (2024). Plant defense mechanisms against O_3 stress: Insights from secondary metabolism. *Environmental and Experimental Botany, 217*, 105553.

Gupta, A., Agrawal, S. B., & Agrawal, M. (2023a). Evaluation of toxicity of tropospheric ozone on tomato (*Solanum lycopersicum* L.) cultivars: ROS production, defense strategies and intraspecific sensitivity. *Journal of Plant Growth Regulation, 42*(10), 6444–6460.

Gupta, A., Agrawal, S. B., & Agrawal, M. (2023b). Responses of eight differentially heat sensitive tomato cultivars against chronic ozone exposure in the indo-gangetic plain: Growth, physiology, and yield. *Agronomy, 13*(3), 717.

Gupta, A., Yadav, D. S., Agrawal, S. B., & Agrawal, M. (2023). Individual effects of high temperature and tropospheric O_3 on tomato: A review. *Journal of Plant Growth Regulation, 42*(3), 1421–1443.

Han, Y. J., Beck, W., Mewis, I., Förster, N., & Ulrichs, C. (2023). Effect of O_3 stresses on growth and secondary plant metabolism of *Brassica campestris* L. ssp. chinensis. *Horticulturae, 9*(9), 966.

Hayes, F., Jones, M. L. M., Mills, G., & Ashmore, M. (2007). Meta-analysis of the relative sensitivity of semi-natural vegetation species to ozone. *Environmental Pollution, 146*(3), 754–762.

Hayes, F., Harmens, H., Sharps, K., & Radbourne, A. (2020). Ozone dose-response relationships for tropical crops reveal potential threat to legume and wheat production, but not to millets. *Scientific African, 9*, e00482.

He, L., Bao, M., Li, Y., Xu, Y., Shao, Z., Ma, Y., & Feng, Z. (2024). Leaf biochemical and physiological responses to elevated atmospheric O_3 concentration in eight modern rice cultivars. *Ecosystem Health and Sustainability, 10*, 0269.

Heagle, A. S., McLUGHLIN, M. R., Miller, J. E., Joyner, R. L., & Spruill, S. E. (1991). Adaptation of a white clover population to ozone stress. *New Phytologist, 119*(1), 61–68.

Hewitt, D. K. L., Mills, G., Hayes, F., Norris, D., Coyle, M., Wilkinson, S., & Davies, W. (2016). N-fixation in legumes–An assessment of the potential threat posed by ozone pollution. *Environmental Pollution, 208*, 909–918.

Hoshika, Y., Agathokleous, E., Moura, B. B., & Paoletti, E. (2024). O_3 risk assessment with free-air controlled exposure (FACE) experiments: A critical revisit. *Environmental Research, 255*, 119215.

Hoshika, Y., Paoletti, E., Centritto, M., Gomes, M. T. G., Puértolas, J., & Haworth, M. (2022). Species-specific variation of photosynthesis and mesophyll conductance to ozone and drought in three Mediterranean oaks. *Physiologia Plantarum, 174*(1), e13639.

Hoshika, Y., Watanabe, M., Inada, N., & Koike, T. (2013). Model-based analysis of avoidance of ozone stress by stomatal closure in Siebold's beech (*Fagus crenata*). *Annals of Botany, 112*(6), 1149–1158.

Hoshika, Y., Watanabe, M., Kitao, M., Häberle, K. H., Grams, T. E. E., Koike, T., & Matyssek, R. (2015). Ozone induces stomatal narrowing in European and Siebold's beeches: A comparison between two experiments of free-air ozone exposure. *Environmental Pollution, 196*, 527–533.

Hui, K., Yuan, Y., Xi, B., & Tan, W. (2023). A review of the factors affecting the emission of the ozone chemical precursors VOCs and NOx from the soil. *Environment International, 172*, 107799.

Ishii, S., Marshall, F. M., & Bell, J. N. B. (2004). Physiological and morphological responses of locally grown Malaysian rice cultivars (*Oryza sativa* L.) to different ozone concentrations. *Water, Air, and Soil Pollution, 155*(1), 205–221.

Ismail, M., Suroto, A., & Abdullah, S. (2015). Response of Malaysian local rice cultivars induced by elevated ozone stress. *Environment Asia, 8*(1), 86–93. https://doi.org/10.14456/ea.2015.11.

Janni, M., Maestri, E., Gullì, M., Marmiroli, M., & Marmiroli, N. (2024). Plant responses to climate change, how global warming may impact on food security: A critical review. *Frontiers in Plant Science, 14*, 1297569.

Javan, F. D., Samadzadegan, F., & Toosi, A. (2025). Air pollution observation—bridging spaceborne to unmanned airborne remote sensing: A systematic review and meta-analysis. *Air Quality, Atmosphere & Health, 18*, 2481–2549. https://doi.org/10.1007/s11869-025-01771-y.

Jian, X. U., Zhuo, Z. H. A. N. G., Lanlan, R. A. O., Yapeng, W. A. N. G., Huanhuan, Y. A. N., Letu Husi, C. S., & Jian Cheng, S. H. I. (2024). Tropospheric O_3 retrieval from satellite remote sensing—A review. *Advances in Earth Science, 39*(1), 56.

Joffe, R., Berthe, A., Jolivet, Y., & Gandin, A. (2022). The response of mesophyll conductance to ozone-induced oxidative stress is genotype-dependent in poplar. *Journal of Experimental Botany, 73*(14), 4850–4866.

Kangasjärvi, J., Jaspers, P., & Kollist, H. (2005). Signalling and cell death in ozone-exposed plants. *Plant, Cell & Environment, 28*(8), 1021–1036.

Khan, M. I. R., Fatma, M., Per, T. S., Anjum, N. A., & Khan, N. A. (2015). Salicylic acid-induced abiotic stress tolerance and underlying mechanisms in plants. *Frontiers in Plant Science, 6*, 462.

Khan, S., Alam, O., Jamil, H. M. A., Khan, W. U., Ullah, A., Khan, K. N., & Ahmad, R. (2024). Impact of high tropospheric O_3 and nitrogen concentration on total protein and chlorophyll content of '*Pinusroxburghii*'. *Plant Omics, 16*(1), 11–18.

Kitao, M., Löw, M., Heerdt, C., Grams, T. E. E., Häberle, K. H., & Matyssek, R. (2009). Effects of chronic elevated ozone exposure on gas exchange responses of adult beech trees (*Fagus sylvatica*) as related to the within-canopy light gradient. *Environmental Pollution, 157*(2), 537–544.

Kittipornkul, P., Treesubsuntorn, C., & Thiravetyan, P. (2020). Effect of exogenous catechin and salicylic acid on rice productivity under ozone stress: the role of chlorophyll contents, lipid peroxidation, and antioxidant enzymes. *Environmental Science and Pollution Research, 27*(20), 25774–25784.

Koch, J. R., Creelman, R. A., Eshita, S. M., Seskar, M., Mullet, J. E., & Davis, K. R. (2000). Ozone sensitivity in hybrid poplar correlates with insensitivity to both salicylic acid and jasmonic acid. The role of programmed cell death in lesion formation. *Plant Physiology, 123*(2), 487–496.

Kumari, S., & Agrawal, M. (2014). Growth, yield and quality attributes of a tropical potato variety (*Solanum tuberosum* L. cv Kufri chandramukhi) under ambient and elevated carbon dioxide and ozone and their interactions. *Ecotoxicology and Environmental Safety, 101*, 146–156.

Lee, J. K., Kwak, M. J., Jeong, S. G., & Woo, S. Y. (2022). Individual and interactive effects of elevated O_3 and temperature on plant responses. *Horticulturae, 8*(3), 211.

Leisner, C. P., Potnis, N., & Sanz-Saez, A. (2023). Crosstalk and trade-offs: Plant responses to climate change-associated abiotic and biotic stresses. *Plant, Cell & Environment, 46*(10), 2946–2963.

Le Roy, E. J., Wong, A. Y., Eastham, S. D., Fiore, A. M., & Selin, N. E. (2025). Impact of climate variability and change on the surface O_3 response to NO x emissions reductions. *Environmental Science & Technology, 59*(21), 10422–10433. https://doi.org/10.1021/acs.est.5c01347.

Leung, F., Pang, J. Y., Tai, A. P., Lam, T., Tao, D. K., & Sharps, K. (2020). Evidence of ozone-induced visible foliar injury in Hong Kong using *Phaseolus vulgaris* as a bioindicator. *Atmosphere, 11*(3), 266.

León, P., & Sheen, J. (2003). Sugar and hormone connections. *Trends in Plant Science, 8*(3), 110–116.

Li, Y., Liu, M., Lv, L., Liang, J., Ma, M., Liu, M., & Fu, P. (2024). Quantitative analysis of spatiotemporal patterns and factor contributions of surface ozone in the North China Plain. *Applied Sciences, 14*(12), 5026. https://doi.org/10.3390/app14125026.

Li, Y., Woo, S. Y., Lee, J. K., Kwak, M. J., Khaine, I., Jang, J. H., & Lim, Y. J. (2019). Effects of elevated ozone on physiological, biochemical and morphological characteristics of eggplant. *Horticulture, Environment, and Biotechnology, 60*(6), 809–820.

Ma, Y., Agathokleous, E., Xu, Y., Cao, R., He, L., & Feng, Z. (2024). Cultivar-specific regulation of antioxidant enzyme activity and stomatal closure confer tolerance of wheat to elevated ozone: A two-year open-field study with five cultivars. *Plant Stress, 12*, 100479.

Marchica, A., Ascrizzi, R., Flamini, G., Cotrozzi, L., Tonelli, M., Lorenzini, G., & Pellegrini, E. (2021). O_3 as eustress for enhancing secondary metabolites and bioactive properties in *Salvia officinalis. Industrial Crops and Products, 170*, 113730.

Marchica, A., Cotrozzi, L., Detti, R., Lorenzini, G., Pellegrini, E., Petersen, M., & Nali, C. (2020). The biosynthesis of phenolic compounds is an integrated defense mechanism to prevent ozone injury in *Salvia officinalis. Antioxidants, 9*(12), 1274.

Marchica, A., Cotrozzi, L., Lorenzini, G., Nali, C., & Pellegrini, E. (2022). Antioxidants and phytohormones act in coordination to regulate sage response to long term ozone exposure. *Plants, 11*(7), 904.

Matłok, N., Piechowiak, T., Zardzewiały, M., & Balawejder, M. (2022). Effect of ozone treatment on the contents of selected bioactive phytochemicals in leaves of alligator plant *Kalanchoe daigremontiana. Applied Sciences, 12*(18), 8934.

Mina, U., Kumar, P., & Varshney, C. (2010). Effect of ozone exposure on growth, yield and isoprene emission from tomato (*Lycopersicon esculentum* L.) plants. *Vegetable Crops Research Bulletin, 72*, 35.

Mishra, A. K., Rai, R., & Agrawal, S. B. (2013). Differential response of dwarf and tall tropical wheat cultivars to elevated ozone with and without carbon dioxide enrichment: Growth, yield and grain quality. *Field Crops Research, 145*, 21–32.

Mishra, A. K., & Agrawal, S. B. (2015). Biochemical and physiological characteristics of tropical mung bean (*Vigna radiata* L.) cultivars against chronic ozone stress: An insight to cultivar-specific response. *Protoplasma, 252*(3), 797–811.

Moraes, R. M., Bulbovas, P., Furlan, C. M., Domingos, M., Meirelles, S. T., Delitti, W. B. C., & Sanz, M. J. (2006). Physiological responses of saplings of *Caesalpinia echinata* Lam., a Brazilian tree species, under ozone fumigation. *Ecotoxicology and Environmental Safety, 63*(2), 306–312.

Moraes, R. M., Furlan, C. M., Bulbovas, P., Domingos, M., Meirelles, S. T., Salatino, A., Delitti, W. B. C., & Sanz, M. J. (2004). Photosynthetic responses of tropical trees to short-term exposure to ozone. *Photosynthetica, 42*(2), 291–293.

Moura, B. B., Alves, E. S., Marabesi, M. A., de Souza, S. R., Schaub, M., & Vollenweider, P. (2018). Ozone affects leaf physiology and causes injury to foliage of native tree species from the tropical Atlantic Forest of southern Brazil. *Science of the Total Environment, 610*, 912–925.

Moura, B. B., Brunetti, C., da Silva Engela, M. R. G., Hoshika, Y., Paoletti, E., & Ferrini, F. (2021). Experimental assessment of ozone risk on ecotypes of the tropical tree Moringa oleifera. *Environmental Research, 201*, 111475.

Niu, Y., Yan, Y., Xing, Y., Duan, X., Yue, K., Dong, J., Hu, D., Wang, Y., & Peng, L. (2024). Analyzing ozone formation sensitivity in a typical industrial city in China: Implications for effective source control in the chemical transition regime. *The Science of the Total Environment, 919*, 170559.

Noctor, G., & Foyer, C.H. (1998). Ascorbate and glutathione: keeping active oxygen under control. *Annual Review of Plant Biology, 49*(1), 249–279.

Nowroz, F., Hasanuzzaman, M., Siddika, A., Parvin, K., Caparros, P. G., Nahar, K., & Prasad, P. V. (2024). Elevated tropospheric O_3 and crop production: Potential negative effects and plant defense mechanisms. *Frontiers in Plant Science, 14*, 1244515.

Pandey, A. K., Ghosh, A., Agrawal, M., & Agrawal, S. B. (2018). Effect of elevated ozone and varying levels of soil nitrogen in two wheat (*Triticum aestivum* L.) cultivars: Growth, gas-exchange, antioxidant status, grain yield and quality. *Ecotoxicology and Environmental Safety, 158*, 59–68.

Paoletti, E., Materassi, A., Fasano, G., Hoshika, Y., Carriero, G., Silaghi, D., & Badea, O. (2017). A new-generation 3D O_3 FACE (free air controlled exposure). *Science of the Total Environment, 575*, 1407–1414. https://doi.org/10.1016/j.scitotenv.2016.09.217.

Pei, J., Liu, P., Feng, Z., Chang, M., Wang, J., Fang, H., & Huang, B. (2024). Long-term trajectory of O_3 impact on maize and soybean yields in the United States: A 40-year spatial-temporal analysis. *Environmental Pollution, 344*, 123407.

Pellegrini, E., Cotrozzi, L., Neri, L., Baraldi, R., Carrari, E., Nali, C., & Hoshika, Y. (2021). Stress markers and physiochemical responses of the Mediterranean shrub *Phillyrea angustifolia* under current and future drought and ozone scenarios. *Environmental Research, 201*, 111615.

Pellegrini, E., Hoshika, Y., Dusart, N., Cotrozzi, L., Gérard, J., Nali, C., & Paoletti, E. (2019). Antioxidative responses of three oak species under ozone and water stress conditions. *Science of the Total Environment, 647*, 390–399.

Pellegrini, M., Lucas-Gonzalez, R., Sayas-Barberá, E., Fernández-López, J., Pérez-Álvarez, J. A., & Viuda-Martos, M. (2018). Bioaccessibility of phenolic compounds and antioxidant capacity of chia (*Salvia hispanica* L.) seeds. *Plant Foods for Human Nutrition, 73*(1), 47–53.

Picchi, V., Monga, R., Marzuoli, R., Gerosa, G., & Faoro, F. (2017). The ozone-like syndrome in durum wheat (*Triticum durum*Desf.): Mechanisms underlying the different symptomatic responses of two sensitive cultivars. *Plant Physiology and Biochemistry, 112*, 261–269.

Pina, J. M., & Moraes, R. M. (2010). Gas exchange, antioxidants and foliar injuries in saplings of a tropical woody species exposed to ozone. *Ecotoxicology and Environmental Safety, 73*(4), 685–691.

Pina, J. M., Souza, S. R., Sérgio, Meirelles, T., & Moraes, R. M. (2017). Psidium guajava Paluma responses to environmental conditions and ozone concentrations in the urban forest of São Paulo, SE-Brazil. *Ecological Indicators, 77*, 1–7.

Pirnia, M., Shirani, K., Yazdi, F. T., Moratazavi, S. A., & Mohebbi, M. (2022). Characterization of antioxidant active biopolymer bilayer film based on gelatin-frankincense incorporated with ascorbic acid and *Hyssopus officinalis* essential oil. *Food Chemistry: X, 14*, 100300.

Pisuttu, C., Risoli, S., Cotrozzi, L., Nali, C., Pellegrini, E., Hoshika, Y., & Paoletti, E. (2024). Untangling the role of leaf age specific osmoprotectant and antioxidant responses of two poplar clones under increasing O_3 concentrations. *Plant Physiology and Biochemistry, 208*, 108450.

Rai, R., & Agrawal, M. (2008). Evaluation of physiological and biochemical responses of two rice (*Oryza sativa* L.) cultivars to ambient air pollution using open top chambers at a rural site in India. *Science of the Total Environment, 407*(1), 679–691.

Rai, R., Agrawal, M., & Agrawal, S. B. (2007). Assessment of yield losses in tropical wheat using open top chambers. *Atmospheric Environment, 41*(40), 9543–9554.

Rajabi, H., Hadi Mosleh, M., Mandal, P., Lea-Langton, A., & Sedighi, M. (2020). Emissions of volatile organic compounds from crude oil processing—Global emission inventory and environmental release. *The Science of the Total Environment, 727*, 138654.

Rajonandraina, T., Ueda, Y., Wissuwa, M., Kirk, G. J. D., Rakotoson, T., Manwaring, H., Andriamananjara, A., & Razafimbelo, T. (2023). Magnesium supply alleviates iron toxicity-induced leaf bronzing in rice through exclusion and tissue-tolerance mechanisms. *Frontiers in Plant Science, 14*, 1213456. https://doi.org/10.3389/fpls.2023.1213456.

Rajput, Harish, V. D., Singh, R. K., Verma, K. K., Sharma, L., Quiroz-Figueroa, F. R., & Mandzhieva, S. (2021). Recent developments in enzymatic antioxidant defense mechanism in plants with special reference to abiotic stress. *Biology, 10*(4), 267.

Rajput, M., & Agrawal, M. (2005). Biomonitoring of air pollution in a seasonally dry tropical suburban area using wheat transplants. *Environmental Monitoring and Assessment, 101*(1), 39–53.

Ramya, A., Dhevagi, P., Priyatharshini, S., Saraswathi, R., Avudainayagam, S., & Venkataramani, S. (2021). Response of rice (*Oryza sativa* L.) cultivars to elevated O_3 stress. *Environmental Monitoring and Assessment, 193*(12), 808. https://doi.org/10.1007/s10661-021-09595-w.

Ranjbarfordoei, A., Samson, R., & Van Damme, P. (2011). Photosynthesis performance in sweet almond [*Prunus dulcis* (Mill) D. Webb] exposed to supplemental UV-B radiation. *Photosynthetica, 49*(1), 107–111.

Rao, M. J., & Zheng, B. (2025). The role of polyphenols in abiotic stress tolerance and their antioxidant properties to scavenge reactive oxygen species and free radicals. *Antioxidants, 14*(1), 74.

Rao, M. V., Lee, H. I., Creelman, R. A., Mullet, J. E., & Davis, K. R. (2000). Jasmonic acid signaling modulates ozone-induced hypersensitive cell death. *The Plant Cell, 12*(9), 1633–1646.

Rao, M. V., Lee, H. I., & Davis, K. R. (2002). Ozone-induced ethylene production is dependent on salicylic acid, and both salicylic acid and ethylene act in concert to regulate ozone-induced cell death. *The Plant Journal, 32*(4), 447–456.

Rathore, D., & Chaudhary, I. J. (2021). Effects of tropospheric ozone on groundnut (*Arachis hypogea* L.) cultivars: Role of plant age and antioxidative potential. *Atmospheric Pollution Research, 12*(3), 381–395.

Sahu, S. K., Liu, S., Liu, S., Ding, D., & Xing, J. (2021). O_3 pollution in China: Background and transboundary contributions to O_3 concentration & related health effects across the country. *Science of the Total Environment, 761*, 144131.

Sarkar, A., & Agrawal, S. B. (2010). Elevated ozone and two modern wheat cultivars: An assessment of dose dependent sensitivity with respect to growth, reproductive and yield parameters. *Environmental and Experimental Botany, 69*(3), 328–337.

Sarkar, A., & Agrawal, S. B. (2012). Evaluating the response of two high yielding Indian rice cultivars against ambient and elevated levels of ozone by using open top chambers. *Journal of Environmental Management, 95*, S19–S24.

Schneider, G. F., Cheesman, A. W., Winter, K., Turner, B. L., Sitch, S., & Kursar, T. A. (2017). Current ambient concentrations of ozone in *Panama modulate* the leaf chemistry of the tropical tree Ficus insipida. *Chemosphere, 172*, 363–372.

Shahadan, N. A. (2005). *Measurement of growth, photosynthesis and transpiration rates of a Malaysian local rice cultivar (MR 219) for assessing ambient ozone stress* (Doctoral dissertation, Universiti Putra Malaysia).

Shahzadi, E., Nawaz, M., Adrees, M., Asghar, M. J., & Iqbal, N. (2022). Elevated ozone phytotoxicity ameliorations in mung bean {*Vigna radiata* (L.) Wilczek} by foliar nebulization of silicic acid and ascorbic acid. *Environmental Science and Pollution Research, 29*(46), 69680–69690.

Shao, Z., Zhang, Y., Mu, H., Wang, Y., Wang, Y., & Yang, L. (2020). Ozone-induced reduction in rice yield is closely related to the response of spikelet density under ozone stress. *The Science of the Total Environment, 712*, 136560. https://doi.org/10.1016/j.scitotenv.2020 136560.

Sharma, I. (2012). Arsenic induced oxidative stress in plants. *Biologia (Lahore, Pakistan), 67*(3), 447–453.

Short, E. F., North, K. A., Roberts, M. R., Hetherington, A. M., Shirras, A. D., & McAinsh, M. R. (2012). A stress-specific calcium signature regulating an ozone-responsive gene expression network in *Arabidopsis*. *The Plant Journal, 71*(6), 948–961.

Sicard, P., Agathokleous, E., Anenberg, S. C., De Marco, A., Paoletti, E., & Calatayud, V. (2023). Trends in urban air pollution over the last two decades: A global perspective. *Science of the Total Environment, 858*, 160064.

Singh, A. A., Fatima, A., Mishra, A. K., Chaudhary, N., Mukherjee, A., Agrawal, M., & Agrawal, S. B. (2018). Assessment of ozone toxicity among 14 Indian wheat cultivars under field conditions: Growth and productivity. *Environmental Monitoring and Assessment, 190*(4), 190.

Singh, E., Tiwari, S., & Agrawal, M. (2009). Effects of elevated ozone on photosynthesis and stomatal conductance of two soybean varieties: A case study to assess impacts of one component of predicted global climate change. *Plant Biology, 11*, 101–108.

Singh, H., Gupta, A., Yadav, D. S., Singh, P., Singh, P., Agrawal, S. B., & Agrawal, M. (2024). Secluding the vegetation of India in retaliation to tropospheric ozone: A mechanistic approach. *Discover Agriculture, 2*(1), 27.

Singh, P., Ansari, N., Mishra, A. K., Agrawal, M., & Agrawal, S. B. (2024). Growth, ultrastructural and physiological characteristics of *Abelmoschus cytotypes* under elevated ozone stress: A study on ploidy-specific responses. *Functional Plant Biology, 51*, FP23229. https://doi.org/10.1071/FP23229.

Singh, P., Arif, Y., Miszczuk, E., Bajguz, A., & Hayat, S. (2022). Specific roles of lipoxygenases in development and responses to stress in plants. *Plants, 11*(7), 979.

Singh, H., Singh, P., Agrawal, S. B., & Agrawal, M. (2023a). Assessment of the reverberations caused by predominant air pollutants on urban vegetation: A multi-site study in Varanasi located in Indo-Gangetic Plains. *Gases, 3*(2), 57–76.

Singh, P., Singh, H., Agrawal, S. B., & Agrawal, M. (2023b). Assessment of the differential trade-off between growth, subsistence, and productivity of two popular Indian hybrid mango varieties under elevated O_3 exposure. *Science of the Total Environment, 889*, 164275.

Singh, P., Ansari, N., Rai, S. P., Agrawal, M., & Agrawal, S. B. (2023). Effect of elevated ozone on the antioxidant response, genomic stability, DNA methylation pattern and yield in three species of *Abelmoschus having* different ploidy levels. *Environmental Science and Pollution Research, 30*(21), 59401–59423.

Singh, P., Singh, H., Agrawal, S. B., & Agrawal, M. (2023). Assessment of the differential trade–off between growth, subsistence, and productivity of two popular Indian hybrid mango varieties under elevated ozone exposure. *Science of the Total Environment, 889*, 164275. https://doi.org/10.1016/j.scitotenv.2023.164275.

Singh, P., Kannaujia, R., Narayan, S., Tewari, A., Shirke, P. A., & Pandey, V. (2021). Impact of chronic elevated O_3 exposure on photosynthetic traits and anti-oxidative defense responses of *Leucaena leucocephala* (Lam.) de wit tree under field conditions. *Science of the Total Environment, 782*, 146907.

Singh, R. N., Mukherjee, J., Sehgal, V. K., Krishnan, P., Das, D. K., Dhakar, R. K., & Bhatia, A. (2021). Interactive effect of elevated tropospheric ozone and carbon dioxide on radiation utilisation, growth and yield of chickpea (*Cicer arietinum* L.). *International Journal of Biometeorology, 65*(11), 1939–1952.

Singh, S., Agrawal, S. B., Singh, P., & Agrawal, M. (2010). Screening three cultivars of *Vigna mungo* L. against ozone by application of ethylenediurea (EDU). *Ecotoxicology and Environmental Safety, 73*(7), 1765–1775.

Singh, S., Bhatia, A., Tomer, R., Kumar, V., Singh, B., & Singh, S. D. (2013). Synergistic action of tropospheric ozone and carbon dioxide on yield and nutritional quality of Indian mustard (*Brassica juncea* (L.) Czern.). *Environmental Monitoring and Assessment, 185*(8), 6517–6529.

Smirnoff, N., & Wheeler, G. L. (2000). Ascorbic acid in plants: biosynthesis and function. *Critical Reviews in Plant Sciences, 19*(4), 267–290.

Tai, A. P., Sadiq, M., Pang, J. Y., Yung, D. H., & Feng, Z. (2021). Impacts of surface O_3 pollution on global crop yields: Comparing different O_3 exposure metrics and incorporating co-effects of CO2. *Frontiers in Sustainable Food Systems, 5*, 534616.

Tamaoki, M., Nakajima, N., Kubo, A., Aono, M., Matsuyama, T., & Saji, H. (2003). Transcriptome analysis of O3-exposed Arabidopsis reveals that multiple signal pathways act mutually antagonistically to induce gene expression. *Plant Molecular Biology, 53*(4), 443–456.

Tetteh, R., Yamaguchi, M., Wada, Y., Funada, R., & Izuta, T. (2015). Effects of ozone on growth, net photosynthesis and yield of two African varieties of Vigna unguiculata. *Environmental Pollution, 196*, 230–238.

Thepphakhun, T., & &Intanon, S. (2020). Total phenolics, flavonoids, antioxidant activity, and allelopathic potential of praxelis. *Journal of Current Science and Technology, 10*(1), 59–65.

Thwe, A. A., Vercambre, G., Gautier, H., Gay, F., Phattaralerphong, J., & &Kasemsap, P. (2014). Response of photosynthesis and chlorophyll fluorescence to acute ozone stress in tomato (*Solanum lycopersicum* Mill.). *Photosynthetica, 52*(1), 105–116.

Ueda, Y., Uehara, N., Sasaki, H., Kobayashi, K., & Yamakawa, T. (2013). Impacts of acute ozone stress on superoxide dismutase (SOD) expression and reactive oxygen species (ROS) formation in rice leaves. *Plant Physiology and Biochemistry, 70*, 396–402.

Upadhyay, V., Singh, P., Choudhary, K. K., Agrawal, M., & Agrawal, S. B. (2025). Navigating biomass trade-offs: Earmarking sustainable food security through biochar interventions in mung bean cultivars under high ozone atmosphere. *Water, Air, & Soil Pollution, 236*(5), 301.

Vainonen, J. P., & &Kangasjärvi, J. (2015). Plant signalling in acute ozone exposure. *Plant, Cell & Environment, 38*(2), 240–252.

Vidal-Meireles, A., Neupert, J., Zsigmond, L., Rosado-Souza, L., Kovács, L., Nagy, V., & Tóth, S. Z. (2017). Regulation of ascorbate biosynthesis in green algae has evolved to enable rapid stress-induced response via the VTC_2 gene encoding GDP-l-galactose phosphorylase. *New Phytologist, 214*(2), 668–681.

Vilcins, D., Christofferson, R. C., Yoon, J. H., Nazli, S. N., Sly, P. D., Cormier, S. A., & Shen, G. (2024). Updates in air pollution: Current research and future challenges. *Annals of Global Health, 90*(1), 9.

Viviani, A., Verma, B. C., Giordani, T., & Fambrini, M. (2021). L-Ascorbic acid in plants: From biosynthesis to its role in plant development and stress response. *Agrochimica: International Journal of Plant Chemistry, Soil Science and Plant Nutrition of the University of Pisa, 65*(2), 151–171.

Vougeleka, V., Risoli, S., Saitanis, C., Agathokleous, E., Ntatsi, G., Lorenzini, G., & Pisuttu, C. (2024). Exogenous application of melatonin protects bean and tobacco plants against O_3 damage by improving antioxidant enzyme activities, enhancing photosynthetic performance, and preventing membrane damage. *Environmental Pollution, 343*, 123180.

Wang, T., Xue, L. K., Brimblecombe, P., Lam, Y. F., Li, L., & Zhang, L. (2017). Ozone pollution in China: A review of concentrations, meteorological influences, chemical precursors, and effects. *The Science of the Total Environment, 575*, 1582–1596.

Wang, T., Xue, L. K., Feng, Z. Z., Dai, J., Zhang, Y. N., & Tan, Y. (2022). Ground-level ozone pollution in China: A synthesis of recent findings on influencing factors and impacts. *Environmental Research Letters, 17*, 063003.

Xu, E., Vaahtera, L., Horak, H., Hincha, D. K., Heyer, A. G., & Brosche, M. (2015). Quantitative trait loci mapping and transcriptome analysis reveal candidate genes regulating the response to ozone in *Arabidopsis thaliana. Plant, Cell & Environment, 38*(7), 1418–1433.

Xu, J., Zhang, Z., Rao, L., Wang, Y., Letu, H., Shi, C., & Shi, J. (2024). Remote sensing of tropospheric O_3 from space: Progress and challenges. *Journal of Remote Sensing, 4*, 0178.

Xu, J., Zhao, H., Wang, S., Zheng, Y., Mai, B., & Zhang, X. (2023). Assessment of photosynthesis and yield loss of winter wheat under ground-level ozone exposure. *Environmental Technology & Innovation, 29*, 103013.

Yadav, D. S., Mishra, A. K., Rai, R., Chaudhary, N., Mukherjee, A., Agrawal, S. B., & Agrawal, M. (2020). Responses of an old and a modern Indian wheat cultivar to future O_3 level: Physiological, yield and grain quality parameters. *Environmental Pollution, 259*, 113939.

Yamaguchi, M., Matsumoto, M., Miyaguchi, K., Li, J., Aoki, T., Ariura, R., & Izuta, T. (2024). Reduced ascorbate pool and its maintenance are important determinants of O_3 damage to net photosynthetic rate in Fagus crenata under elevated CO_2 and soil N supply. *Science of the Total Environment, 907*, 168102.

Yoshida, S., Tamaoki, M., Ioki, M., Ogawa, D., Sato, Y., Aono, M., & Nakajima, N. (2009). Ethylene and salicylic acid control glutathione biosynthesis in ozone-exposed *Arabidopsis thaliana. Physiologia Plantarum, 136*(3), 284–298.

You, J., & Chan, Z. (2015). ROS regulation during abiotic stress responses in crop plants. *Frontiers in Plant Science, 6*, 1092.

Zentella, R., Burkey, K. O., & Tisdale, R. H. (2023). Impact of tropospheric ozone on root proteomes of two soybean genotypes with contrasting sensitivity to ozone. *Environmental and Experimental Botany, 208*, 105269.

Zhang, D., Zhou, H., Zhang, Y., Zhao, Y., Zhang, Y., Feng, X., & Lin, H. (2025). Diverse roles of MYB transcription factors in plants. *Journal of Integrative Plant Biology, 67*(3), 539–562.

Zhang, L., Li, Q., Wang, T., Ahmadov, R., Zhang, Q., Li, M., & Lv, M. (2017). Combined impacts of nitrous acid and nitryl chloride on lower tropospheric ozone: New module development in WRF-Chem and application to China. *Atmospheric Chemistry and Physics, 17*, 9733–9750.

Zhang, L., Xu, B., Wu, T., Wen, M. X., Fan, L. X., Feng, Z. Z., & Paoletti, E. (2017). Transcriptomic analysis of Pak Choi under acute ozone exposure revealed regulatory mechanism against ozone stress. *BMC Plant Biology, 17*(1), 236. https://doi.org/10.1186/s12870-017-1202-4.

Zhang, Y., Dai, M., & Yuan, Z. (2018). Methods for the detection of reactive oxygen species. *Analytical Methods, 10*(38), 4625–4638.

CHAPTER EIGHT

Plant bioindicators of urban air quality

Clement Ogunkunle[1], Stephen Oyedeji[1], Abdulfatai Rufai[2], Nikita Patel[3], Ayodele Oluwatobi[4], and Paul Fatoba[1]
[1]Department of Plant Biology, University of Ilorin, Ilorin, Kwara, Nigeria,
[2]Department of Plant Biology, Osun State University, Osogbo, Osun, Nigeria,
[3]Kishorbhai Institute of Agriculture Sciences and Research Centre, Uka Tarsadia University, Bardoli, Gujarat, India,
[4]Department of Biological Sciences, Ojaja University, Eiyenkorin, Kwara, Nigeria

8.1 Introduction

Urban air quality is one of the most pressing environmental challenges faced by cities worldwide, worsening at an alarming pace due to rapid urbanization (Shiva-Nagendra et al., 2021). While urbanization has significantly improved human comfort and quality of life, it has also led to unforeseen consequences (Piracha & Chaudhary, 2022). This phenomenon has created a major global public health issue (Cohen et al., 2017), contributing to approximately 4.2 million premature, nonaccidental deaths in both urban and rural areas worldwide in 2019 alone (Murray et al., 2020). In 2016, 54% of the global population lived in urban areas, where the most harmful air pollutants to human health include nitrogen dioxide (NO_2), particulate matter smaller than 2.5 μm ($PM_{2.5}$), and volatile organic compounds (VOCs) (Anenberg et al., 2022; Manisalidis et al., 2020; Sicard et al., 2023) (Fig. 8.1). By 2050, it is expected that 70% of the global population will reside in urban areas, with outdoor air pollution projected to cause 6.6 million premature deaths annually (Sicard et al., 2023). Though the pollutants are invisible to the naked eye, they pose significant health risks, ranging from mild respiratory issues to severe cardiovascular diseases (WHO, 2022). Humans have long overlooked nature's potential to enhance our understanding of air quality. However, research has shown that certain plant species are sensitive to air pollution and can serve as both chemical and visual indicators of environmental conditions (Tiwari & Tiwari, 2006; Agbaire & Esiefarienrhe, 2009; Joshi & Swami, 2009). Utilizing plants as bioindicators offers a cost-effective and innovative

Urban Air Pollution and Plant Responses
DOI: https://doi.org/10.1016/B978-0-443-40320-0.00011-5

Figure 8.1 Exposure routes of the global urban population to air pollutants. Source: Authors.

approach to monitoring urban air quality. By observing specific changes in plant physiology—such as tissue necrosis, stunted growth, or leaf discoloration—scientists and urban planners can gather valuable data on pollution levels within urban centers. In this way, plants complement traditional air quality monitoring methods, providing a low-cost alternative for broad-scale environmental assessment (Tiwary & Kumar, 2020). Plants can absorb airborne particulate matter and gaseous pollutants, converting them into nontoxic forms through adsorption onto their leaves and branches, further highlighting their potential as bioindicators (Rai & Panda, 2014).

8.2 Urbanization and air pollution

Urbanization is a global trend characterized by the concentration of human populations in specific areas, leading to land transformation for various uses such as residential, commercial, industrial, and transportation purposes (United States Environmental Protection Agency USEPA, 2024). It is a double-edged sword; while urbanization drives economic growth, innovation, and improved living standards, it also presents significant environmental challenges, with air pollution being the most pressing (Srivastava & Rao, 2011). Over half of the world's population now resides in urban areas, a trend accelerated by the expansion of cities over the past century (Sicard et al., 2023).

However, many of these cities suffer from outdoor air quality that falls below the World Health Organization's (WHO) recommendations for healthy living (Bikis, 2023). Air pollution is the contamination of indoor or outdoor air by harmful gases and particles, which can alter the natural properties of the environment, according to the World Health Organization (WHO, 2019). It is the leading environmental health risk, affecting all life forms, including humans, animals, and plants. The severity of air pollution has prompted the United Nations to include it as a key issue to address in the Sustainable Development Goals (SDGs), such as SDG 3 (Good Health and Wellbeing), SDG 7 (Affordable and Clean Energy), SDG 11 (Sustainable Cities and Communities), and SDG 13 (Climate Action). Air pollution is also reported to be responsible for approximately 7 million deaths annually and contributes to 25% of all cardiovascular disease deaths worldwide (World Heart Foundation, 2024). The primary source of air pollution is the heavy reliance on transportation, alongside industrial, commercial, and residential activities (Jafari et al., 2021). In major cities of developing countries, 70%–80% of air pollution is attributed to emissions from outdated vehicles, coupled with poor road infrastructure, inadequate vehicle maintenance, and low-quality fuels (Badami, 2005; Anjaneyulu et al., 2006; Singh & Gupta, Gupta et al., 2007; Wang et al., 2010). According to a 2024 WHO report, 99% of the global population breathes air that does not meet acceptable quality levels. Furthermore, the United Nations Environment Programme (2021) reported that over 92% of the world's population lives in areas with unsafe air quality, and in at least 34% of countries, air quality standards are not legally enforced, leading to the unchecked release of pollutants, particularly from industrial activities. The main air pollutants include carbon monoxide (CO), sulfur dioxide (SO_2), NO_2, particulate matter ($PM_{2.5}$ and PM_{10}), and ozone (O_3) (Zhou et al., 2017). As urbanization replaces the natural landscapes with concrete, the Urban Heat Island effect intensifies, leading to significantly higher temperatures and reduced air circulation (Li et al., 2024). This increase in heat poses a serious health risk, particularly in underdeveloped areas, and with climate change worsening, the frequency of extreme heat events is expected to increase. By 2100, an estimated 950 million people could be at risk. However, cities can mitigate these effects and adapt to rising temperatures by expanding green spaces, protecting water bodies, and improving urban planning (Li et al., 2022). While the impacts of air pollution and urbanization are evident, finding solutions to these challenges is often complex. Monitoring air pollution in low- and middle-income countries has always been difficult (Pongpiachan & Iijima, 2016). As vehicle ownership increased, the use of hydrocarbons such as gasoline led to higher emissions of CO_2 and $PM_{2.5}$ into the atmosphere (Choochuay et al., 2020). These emissions are a major factor in the deterioration of urban air quality (Kumie et al., 2021). Addressing air pollution, particularly in urban areas, becomes more feasible when governments and organizations prioritize green infrastructure. Urban greenery plays a key role in providing ecological services and health benefits; plants act as biofilters, reducing air pollution by trapping particulate matter and absorbing harmful gases (Chen et al., 2017; Abhijith & Kumar, 2019). Implementing green infrastructure such as street greenery, green walls, green buildings, and integrating vegetation into the urban environment are effective strategies for combating air pollution (Berardi et al., 2014; Beatley, 2016) (Fig. 8.2).

8.3 Plants as air quality indicators

Air quality monitoring is both costly and logistically challenging. As a result, there has been significant progress in utilizing cosmopolitan organisms for monitoring and assessing air pollution over the past few decades. These organisms absorb pollutants from their environment and

Figure 8.2 Street greenery to reduce air pollutants in the environment (Tripathi & Nema, 2023). Adapted from Tripathi, D. P., & Nema, A. K. (2023). Seasonal variation of biochemical parameters and air pollution tolerance index (APTI) of selected plant species in Delhi city, and detailed meta-analysis from Indian metropolitan cities. Atmospheric Environment, 309, 119862. https://doi.org/10.1016/j.atmosenv.2023.119862.

can serve as indicators of a contaminant's bioavailability over time. This approach allows for the comparison of pollution levels across different geographic regions. In this context, the Organization for Economic Cooperation and Development has recommended broad eight programs to monitor and assess environmental impacts, including evaluating potentially hazardous substances (Giordano et al., 2021). Biomonitors, which can be classified as either accumulative or sensitive, are organisms used to quantitatively assess contaminants. Among sensitive biomonitors are optical biomonitors, which function as early warning systems and integrators of environmental stress caused by pollutants. These biomonitors rely on visual indicators, such as morphological changes or behavioral shifts related to environmental conditions, as well as on chemical and physical factors, including alterations in enzyme activity, photosynthetic or respiratory processes. Biological indicators, which include species, communities, or ecological processes, are used to assess environmental quality and its changes over time. While research on biological indicators primarily focuses on human-induced stressors, such as pollution and land-use changes, natural stressors, such as droughts or late spring frosts, are also recognized as contributing to environmental variation. The development and use of bioindicators were notably advanced in the 1960s (Marć et al., 2015). Air quality indicators include plants, algae, mosses, honeybees, mycorrhiza, and lichens, all of which act as nature's early warning system for air pollution. These organisms show visible signs of stress when exposed to pollutants, offering insights into air quality fluctuations (Table 8.1). Unlike complex air monitoring technologies, which are often limited to specific locations, these natural indicators are typically more effective at detecting pollution in particular areas (Poikolainen et al., 2000).

8.3.1 Algae

Algae are commonly found in aquatic environments and are highly adaptable to changes in their surroundings, making them excellent indicators of pollution due to their sensitivity to shifts in environmental quality. Algae species like *Spirulina*, which can produce oxygen and reduce pollutants such as carbon dioxide (CO_2), NO_2, and SO_2 in polluted air, are particularly valuable. The bioreactor used for cultivating algae consists of an air supply unit and a culture tank filled with a solution containing algae (Srivastava & Gupta, 2020). Photosynthesis, the process by which algae convert CO_2 into oxygen, occurs when light, provided by fluorescent lamps at night and sunlight during the day, radiates through the system. During this process, algae also utilize SO_2 and

Table 8.1 Sources of urban air pollutants and their effects on human and plant health.

Pollutant	Primary sources	Health effects	Plant responses	References
Particulate matter ($PM_{2.5}$)	Vehicle emissions, Industrial combustion	Lung irritation, cardiovascular diseases	Leaf surface deposition, reduced photosynthesis	Bailey-Serres et al. (2019), Dockery and Pope (2006)
Nitrogen oxides (NOx)	Vehicles, power plants, and industrial activity	Respiratory irritation, increased asthma risk	Leaf necrosis, chlorophyll reduction	Gillner et al. (2015), Vuković, Urošević, Goryainova, et al. (2015)
Sulfur dioxide (SO_2)	Fossil fuel combustion, industrial processes	Lung diseases, respiratory system irritation	Stunted growth, leaf chlorosis (yellowing)	Singh and Agrawal (2007), Ashenden and Bell (1987)
Carbon monoxide (CO)	Motor vehicles, fuel-burning appliances	Headaches, dizziness, impaired cognitive function	Stunted plant growth, reduced root respiration	Wolff and Perry (2010), Halliwell and Gutteridge (2015)
Volatile organic compounds (VOCs)	Paints, solvents, and industrial emissions	Eye, nose, and throat irritation, potential carcinogenic effects	Altered growth, leaf discoloration	Calfapietra et al. (2013), Vardoulakis et al. (2011)
Ozone (O_3)	Photochemical oxidation of VOCs and NOx	Lung and respiratory diseases.	Stunted growth, spotting, mottling, chlorosis, necrosis	Zhang et al. (2019), Sharps et al. (2021)

NO_2 as nutrients. As a result, the contaminated air passing through the system is transformed into cleaner air with higher oxygen levels (Meeranayak et al. 2020). Algae also serve as effective bioindicators for metal contamination, which can be assessed by studying water, sediments, and marine life. Aquatic organisms, including algae, are ideal markers for identifying the sources of bioavailable metal pollution. Found in a variety of aquatic environments, algae are stationary, easy to collect and identify, and accumulate trace metals, making them suitable as bioindicators. Algal cell walls contain polysaccharides and proteins with anionic groups such as carboxyl, sulfate, or phosphate, which are excellent binding sites for metals. Algae can accumulate trace metals at concentrations thousands of times higher than those found in the surrounding water (Parus & Karbowska, 2020). Green algae like Chlorella are particularly effective in detecting NO_2 contamination, especially in freshwater environments. Known for its rapid response to increased nitrogen dioxide levels, Chlorella plays a crucial role in air quality monitoring. Changes in Chlorella's chlorophyll production serve as a clear and accurate indicator of NO_2 pollution. One of the most notable signs of pollution is the change in color algae undergo in response to contaminants. Exposure to nitrogen dioxide causes Chlorella to shift from its usual green to a yellowish hue, providing an obvious visual signal that allows scientists and environmentalists to gauge the severity of pollution (Liu & Chen, 2016). The chlorophyll production in Chlorella is directly influenced by nitrogen dioxide, a common product of combustion processes. As nitrogen dioxide levels rise, Chlorella reduces its chlorophyll production, resulting in a noticeable color change. This response provides a clear, easily observable indicator of nitrogen dioxide-induced air pollution (Shing et al., 2018). Another important algal species, Spirogyra, is a key indicator of SO_2 pollution in the atmosphere. Characterized by its spiral-shaped chloroplasts, Spirogyra undergoes significant changes in its cell structure and reproductive patterns when exposed to high levels of sulfur dioxide. These alterations offer visible and tangible evidence of sulfur dioxide contamination in the surrounding air, making Spirogyra a valuable marker for assessing sulfur dioxide pollution. The physiological changes in Spirogyra in response to elevated SO_2 levels enhance our understanding of air quality in a given habitat (Mane & Bhosle, 2012).

8.3.2 Lichens

Lichens are symbiotic organisms composed of algae and fungi and are commonly recognized as primary colonizers of rock and other similar surfaces. Under favorable climatic conditions, such as brief periods of dampness, lichens can thrive not only on calcareous mountain cliffs but also on urban concrete walls. SO_2, a known phytotoxic gas, is a major byproduct of burning coal and petroleum products containing sulfur. The absence of lichens in many urban and industrial areas is likely due to a combination of high SO_2 levels and other urban pollutants (Gonzalez & Pignata, 1994). Although the role of sulfur dioxide in lichen decline in urban settings was suggested by lichenologists, the first experimental findings were not published until 1965 by Pearson and Skye (1965). Research has shown that lichens vary in their sensitivity to sulfur dioxide in real-world environments (Conti & Cecchetti, 2001). There is a strong correlation between the 12 susceptibility of lichen species to SO_2, as demonstrated in both fumigation and field studies (Yang et al., 2023). The SO_2 concentrations used in these studies reflect those typically found in industrial and urban areas. For example, 24-hour average SO_2 concentrations as high as 2.6 mg/m^3 have been recorded in locations such as London, near smelters in Sudbury, Ontario, and New York City (Mikhaylov, 2020). Certain lichens are known to be sensitive to SO_2 levels even below these maximum concentrations. Furthermore, experimental studies have shown that some lichens are as, or even more, sensitive to SO_2 than the most sensitive higher plants. Thus, wherever SO_2 levels are high enough to cause necrosis or chlorosis in higher plants, they are also likely to be toxic to lichens. Lichens are more vulnerable to air pollution than higher plants due to their lack of protective mechanisms such as leaf shedding and stomatal regulation of gas exchange (Ristić et al., 2021). Although less common than sulfur dioxide, hydrogen fluoride (HF) is another significant pollutant, released by industries such as aluminum production, rare-earth metal processing, and ceramics. Lichen communities are notably sparse near aluminum mills, with field studies showing that lichen diversity is nearly eliminated around fluoride sources, except in protected microhabitats. Transplanted lichens often show signs of chlorosis and die when placed near fluoride emissions. Gilbert (1971) identified a threshold range of 20–47 pg soluble F/g dry weight for lichen damage. While there is still insufficient data to compare the sensitivity of lichens to HF with that of higher plants, it is known that certain Gladiolus species are far more susceptible to HF than any lichen studied to date (Lounamaa, 1956). In recent years, lichens have also been explored as potential biomonitors for organic air pollution. Lichens lack stomata and a waxy cuticle but produce and release lipid compounds onto their surfaces that can function similarly to a plant's cuticle. These lipophilic surfaces make lichens highly attractive to lipophilic substances such as semivolatile organic compounds, which are easily absorbed and metabolized by the lichens (Oksanen, 2006). Organic pollutants, being hydrophobic, tend to deposit through dry deposition, while metal ions and other water-soluble substances settle on the lichen surface through different mechanisms, such as wind impaction, rainfall, or volatilization from the soil (Jones & Duarte-Davidson, 1997). Lichens can be classified as either sensitive or tolerant to air pollution. Pollution-sensitive lichens, such as *Allocetraria oakesiana, Cetrelia olivetorum, Parmelia squarrosa, Ramalina spp., Tuckermannopsis spp.*, and *Usnea spp.*, are vulnerable to various air pollutants. In contrast, pollution-tolerant lichens, such as Physcia millegrana, Parmotrema hypotropum, Candelaria concolor, Parmelia sulcata, and Punctelia rudecta, are more resilient to air pollution (Jovan, 2008).

8.3.3 Mosses

Mosses, which thrive in humid environments, are also effective bioindicators and biomonitors of air pollution, particularly biomonitors of trace elements. Unlike plants with true roots, mosses absorb nutrients from both dry (i.e., bulk) and wet atmospheric deposits through their thin cuticles, which facilitate nutrient uptake from the air (Chakrabortty & Paratkar, 2006; UNECE - ICP Vegetation et al., 2020). The high surface-to-weight ratio of mosses enhances their ability to adsorb pollutants, and numerous studies have shown that mosses are more efficient than lichens at retaining dust and pollutants (e.g., Galsomiès et al., 1999; Grodzińska & Szarek-Łukaszewska, 2001; Figueira et al., 2002; Fatoba et al., 2013; Oyedeji et al., 2016). Further, uptake or other contributions from the substrate may occur in lichens, as metal uptake by rhizines is possible. There is also the possibility of bark-lichen transfer of absorbed pollutants that may influence the actual environmental concentration in lichen tissues. Mosses have proven particularly effective in biomonitoring heavy metals (Table 8.2) because they cause fewer technical and analytical problems than other biomonitors such as tree barks and lichens. In addition, the option of using moss transplants as active biomonitors against the traditional native mosses has also provided a time-integrated measure of atmospheric depositions of air pollutants such as heavy metals and persistent aromatic hydrocarbons. A hypothesis suggests that native moss may adjust and adapt to the level of air pollution and pollutants in its natural environment (Fernández et al., 2000), which has led to the promotion of using the moss transplant technique for air quality monitoring.

Table 8.2 Some mosses that have been used as biomonitors of air pollutants (HMs and PAHs) in selected urban areas of the world.

Moss species	Air pollutants	Locality	References
Pleurozium schreberi	As, Cd, Cu, Cr, Fe, Ni, Pb, Zn, V and 25 other metallic elements	Moscow, Russia	Vergel et al. (2022)
Pleurozium schreberi, Sphagnum fallax, and *Dicranum polysetum*	Mn, Fe, Ni, Cu, Zn, Cd, and Pb	Tula, Russia	Świsłowski and Rajfur (2024)
Sphagnum girgensohnii *Hypnum cupressiforme*	Al, Ba, Cd, Co, Cr, Cu, Fe, Ni, Pb, Sr, V and Zn	Belgrade, Serbia	Vuković et al. (2015)
Moss varieties	Cd, Cr, Cu, Pb and Zn	Jos and Kaduna, Nigeria	Oyedeji et al. (2016)
Barbula lambarenensis	Pb, Cu, Cr and Zn	Shagamu, Nigeria	Ogunkunle and Fatoba (2012)
Leptodon smithii	Cr, Mn, Fe, Co, Zn, As, Se, Sr, Ba and 24 other elements	Western Cape Province, South Africa	Ndlovu et al. (2021)
Sphagnum papillosum	Al, As, Cd, Cu, Cr, Fe, Hg, Na, Ni, Pb, Ti, V and Zn,	Turku region, Finland	Salo et al. (2016)
Sphagnum palustre	Sc, Fe, Co, Zn, As, Br, Mo, Sb, Ba, La, Ce, Sm, Yb, Lu, Hf and Th	Gdansk City, Poland	Astel et al. (2008)
Hypnum amabile	Al, V, Cr, Mn, Fe, Co, Ni, Cu. Zn, As, Mo, Cd, Sb, Pb and PAHs	Mexico City, Mexico	Gómez-Arroyo et al. (2021)
Orthotrichum lyellii	As, Cd, Co, Ni , Pb and PAHs	Seattle and Portland, USA	Jovan et al. (2021, 2022)

Physcomitrium cyathicarpum; Barbula constricta	Cu, Cr, Co, Cd, Cu, Fe, Hg, Ni and Pb	Delhi region, India	Vats et al. (2010)
Haplocladium angustifolium	Ag, As, Cd, Co, Cr, Cu, Mn, Mo, Ni, V, Pb, and Zn	Wuhan City, China	Jiang et al. (2018)
Pleurozium schreberi	Cd, Cr, Cu, Fe, Ni, Pb, Zn	National survey, Poland	Grodzińska and Szarek-Łukaszewska (2001)
Pseudoscleropodium purum	Metals and Metalloids	Galicia, Spain	Varela et al. (2014)
Hylocomium splendens	Al, Co, Cr, Fe, Ni, Ni, Pb, Cd, Cu, Na, Zn	Southern Alps, northern Italy	Gerdol et al. (2002)
Pseudoscleropodium purum	Cd, Cu, Hg, Pb & Zn	Galicia, Spain	Boquette et al. (2013)

8.3.4 Shrubs and trees

Rapid industrialization and unchecked urbanization are causing significant damage to the natural landscapes across the globe. The growing population in urban areas has reduced open spaces, which were once essential in absorbing and buffering the high levels of pollution common in cities. Expansive green areas, with their abundant vegetation, served as natural filters, helping to reduce the impact of pollution. Tree canopies, in particular, act as effective filters by trapping and adhering to contaminants in the air (Koul & Thakur, 2016). However, the main challenge in cities is the lack of space required to accommodate the large trunks and wide canopies of mature trees. The limited space along busy roads and in surrounding areas has prompted the need for alternative planting strategies, such as incorporating shrubs that thrive in smaller spaces in urban environments. In addition to shrubs, planting smaller trees and tall plants can help increase ground vegetation without taking up much room. These plants can serve as multilayered traps, working in combination with shrubs to improve air quality. Shrubs play an important role in nutrient and gas cycling, including the absorption of CO_2 and O_2, helping to maintain ecological balance. They also contribute to reducing pollution by absorbing contaminants through their leaf surfaces. Furthermore, shrubs provide vertical stratification to vegetation, which helps ensure resource equity across plant layers (Escobedo et al., 2008). Ogunkunle et al. (2015), using the air pollution tolerance index (APTI), assessed the sensitivity of four tree species—*Acacia nilotica, Prosopis africana, Vitellaria paradoxa*, and *Terminalia catappa*—to air pollution. While all species were considered sensitive to air pollution, *Terminalia catappa* showed an intermediate response. Similarly, Oyedeji et al. (2019) found that *Terminalia catappa, Anacardium occidentale*, and *Tectona grandis* all exhibited intermediate responses to air pollution. Ornamental species such as Polyalthia longifolia, Thuja occidentalis, and Duranta repens have also been studied for their role in monitoring urban air pollution (Ogunkunle et al., 2019). The study indicated that the air pollution sensitivity of T. occidentalis and D. repens made them effective indicators of pollution, while the more tolerant *P. longifolia* was useful for mitigating pollution levels. Terekhina and Ufimtseva (2020) reported that tree and shrub leaves in Saint Petersburg, Russia, accumulated between 19% and 62% of air pollutants, with the accumulation varying depending on the leaf structure of each species and the surrounding ecological conditions. Species such as *Ulmus laevis, Tilia cordata, Populus sp.*, and *Rosa rugosa* demonstrated a higher ability to accumulate pollutants. While tree and shrub foliage are commonly used to monitor air pollution, studies have also shown that tree bark can be an effective tool for this purpose (Moreira et al., 2018; Rahman et al., 2022). Carneiro et al. (2011) used tree bark to track the geographical distribution of chemical elements correlated with traffic intensity and observed that the precision of this biomonitoring method extended up to 100 meters from the roadside.

8.4 Methods of assessing air quality

Air quality assessment focuses on evaluating pollution levels in a given environment. Using plants as biomonitors of air pollution offers a more cost-effective, practical, and reliable alternative to traditional methods. Plants display visible damage in response to various air pollutants, which can be used to identify the types of pollutants present. Additionally, the extent of plant damage and the duration of exposure can help determine the concentration of pollutants in the environment (Cen, 2015). Due to their natural ability to adapt to changing conditions, plants are particularly effective in tracking the spatial and temporal patterns of airborne pollutant deposition and its effects on organisms and ecosystems. This makes it possible to predict environmental changes at both local and larger regional or global scales. As a result, plant-based biomonitoring is increasingly seen as an effective method for collecting air pollution data across different locations (Giordano et al., 2021). Several methods have been developed for using plants as biomonitors of air quality. These include the index of purity, physiological response methods (such as the air pollution tolerance index and anticipated performance), stomatal conductance analysis, and annual ring analysis, all of which use various plant categories like lichens, mosses, and higher plants.

8.4.1 Index of purity in evaluating air quality

The Index of Purity (or Atmospheric Purity) is a method used to assess air quality based on qualitative or quantitative measures that indicate the extent to which a particular area is exposed to air pollutants. Lichens, which derive their nutrients directly from the atmosphere, are especially sensitive to changes in air quality due to their unique anatomical, morphological, and physiological characteristics. This sensitivity allows them to act as natural indicators of air pollution (Conti & Cecchetti, 2001;

Will-Wolf et al., 2017). Additionally, lichens lack a cuticle, which enables them to accumulate pollutants in their intracellular sites, making them highly responsive to harmful air contaminants (Correa-Ochoa et al., 2021). Lichens are particularly sensitive to nitrogen and sulfur oxides in the air. Traditional biomonitoring systems are often costly, requiring significant maintenance and facing logistical challenges when deployed in difficult terrains. In contrast, lichens offer a significant advantage, making them an increasingly popular choice for air pollution monitoring. Unlike conventional systems, lichens are not limited to specific chemical compounds or pollutants, allowing them to assess overall atmospheric air quality (Will-Wolf et al., 2015). The Lichen Purity Index is a reliable method for evaluating air quality, based on the diversity and health of lichen species present in a particular area. The greater the diversity and abundance of lichen species, the purer the air in that area (Will-Wolf et al., 2015). The presence of a high number of healthy lichen species typically indicates lower levels of air pollution, while a lack of lichen diversity or abundance may reflect higher pollution levels. (Will-Wolf et al., 2015).

8.4.2 Physiological response methods of plants to air pollution

Physiological parameters such as ascorbic acid content, total chlorophyll index, relative water content, and pH of leaf extract are commonly used to assess a plant's vulnerability to air pollutants. These four physiological variables are combined to calculate the plant's APTI, which provides insight into its tolerance to air pollution. Plants with higher APTI values are more resistant to air pollution, while those with lower APTI values are more sensitive and delicate, making them effective bioindicators of pollution. Based on APTI values, plants can be categorized into classes, such as tolerant, moderately tolerant, transitional, and sensitive (Joshi & Swami, 2009). Ascorbic acid, an antioxidant, plays a crucial role in protecting plants from oxidative stress caused by pollutants and environmental factors such as fluctuating temperatures, drought, and metabolic processes. Exposure to environmental stress triggers the production of reactive oxygen species (ROS) in plants, which can damage tissues. In response, plants reduce water loss from their leaves, and the water content in plant tissues helps maintain physiological balance during pollution stress. Therefore, the water content in plants can indicate the level of air pollution they are exposed to. The pH of plant tissues also reflects the degree of air pollution, as pollutants in the air combine with rainwater to form mixtures with varying pH levels depending on the types of pollutants. The total chlorophyll index is another critical parameter in assessing plant productivity. Chlorophyll plays a key role in photosynthesis, and any reduction in chlorophyll content due to pollution directly impacts plant health and growth. As a result, measuring chlorophyll content is a valuable tool for evaluating the effects of air pollutants on plants (Klumpp et al., 2000).

8.4.3 Air Pollution Tolerance Index in assessing urban air pollution

The Air Pollution Tolerance Index (APTI) is used to assess a plant's ability to withstand air pollution (Elhadad et al., 2026). While there are various methods for monitoring air pollution, plants provide a natural and reliable means of assessing air quality. Due to their significant role in maintaining ecological balance, plants are particularly vulnerable to pollution. The effects of pollutants are most evident in the leaves, which are the plant's primary interface with the air and the most abundant part of the plant. However, air pollution can also affect other aspects of plant development, such as seed germination, stem length, and flower production. Depending on the plant's ability to tolerate pollutants, it can serve as a biomonitor for air pollution. The key parameters for evaluating air pollution tolerance include ascorbic acid content, chlorophyll content, leaf extract pH, and relative water content. To determine the APTI of a plant, the following formula is used: (Singh & Verma, 2007) Where AA= Ascorbic acid in mg/g T= Total Chlorophyll in mg/g P= pH of leaf extract R= Relative water content in mg/g.

8.4.4 Stomatal conductance in response to urban air pollution

Stomata are small openings on the epidermal surface of leaves that control the exchange of gases between the plant and the atmosphere (Alushi & Veizi, 2020). They regulate the uptake of CO_2, the loss of water through transpiration, and the cooling effect produced by evaporation. Stomatal apertures continuously adjust in response to external factors such as light, water content, hormonal signals, and temperature, balancing the assimilation of CO_2 with the transpiration of water (Franks & Farquhar, 2001; Faralli et al., 2019). Stomatal conductance is influenced by both behavioral and anatomical factors, but significant variation exists among different leaves and species, leading to differences in stomatal function (Lin et al., 2015). Guard cells, which surround each stoma, regulate its opening and closing (Franks & Beerling, 2009). The two main anatomical features that determine maximum stomatal conductance are stomatal density and the maximum pore area. However, the actual or operational stomatal conductance is usually a fraction of this maximum, as the factors controlling stoma opening and closure often limit the observed stomatal conductance (Franks & Farquhar, 2001; Franks & Beerling, 2009). Stomatal conductance directly impacts the rate of photosynthesis and the water use efficiency of the plant. Research has shown a strong correlation between stomatal conductance and photosynthetic capacity. Several studies have highlighted how increases in stomatal conductance can enhance photosynthesis and overall plant yield. In urban environments, where plants are exposed to air pollutants, poor air quality can reduce stomatal conductance. Even low concentrations of atmospheric pollutants can interfere with stomatal function, disrupting the water balance of affected leaves and plants (Robinson et al., 1998).

8.4.5 Annual ring analysis in assessing urban air quality

Annual rings are formed by the radial growth of trees, with each ring representing one year of growth. Tree ring analysis, or dendrochronology, is a valuable tool for studying tree growth rates, age, and the effects of environmental factors, including climate, on forest dynamics (UNEP, 2001). The growth rings are typically determined by counting colored layers in the wood, especially in plantation trees or those in botanical gardens where the planting dates are known. When the planting dates are unknown, techniques such as cambial marking are used, where a mechanical injury to the cambium triggers a wound response, creating a scar that serves as a time marker for growth analysis (Worbes, 1995; Verheyden et al., 2004). The width of annual tree rings can be influenced by various ecological factors, including climate, soil, and air pollution. In areas with high levels of air pollution, reductions in ring width have been observed. Methods that measure changes in the diameter of annual rings are commonly used to assess the impact of air pollution on tree growth (Tolunay, 2003). Pollutants such as NOx, O_3, and heavy metals can significantly affect tree growth and physiological processes, leading to narrower rings due to reduced growth. Physiological stress caused by pollutants can impair nutrient uptake, photosynthesis, and overall tree health. Additionally, tree rings can accumulate heavy metals like lead, cadmium, and mercury, as well as sulfur and nitrogen compounds. Analyzing the chemical composition of tree rings provides valuable insights into pollutant concentrations over time in a specific area (Tolunay, 2003).

8.5 Effects of air pollutants on plants

The urban environment is characterized by a complex mixture of air pollutants, with the composition varying over time and across different cities due to differences in emission sources and patterns. The nature of urban air pollution has significantly changed in recent decades (Honour et al., 2009). During the industrial revolution, major pollutants such as sooty smoke and SO_2 dominated the air. However, with the advent of fossil fuels, which power many automobiles, the release of numerous toxic air pollutants has become more prevalent (Perera, 2017). These pollutants have both direct and indirect effects on living organisms, especially higher plants, due to their immobility. Numerous studies have highlighted the impacts of traffic-related emissions on vegetation at various distances from roadways, with varying traffic densities. Research by Angold (1997) in England, Bernhardt-Romermann et al. (2006) in Germany, Gratani et al. (2000) in Italy, Viskari et al. (2000) in Finland, and Oyedeji et al. (2016) in Nigeria all indicate that plants act as sinks for air pollutants, showing a direct correlation between traffic density and changes in stomatal conductance, photosynthetic activity, chlorophyll content, leaf senescence, and pollutant concentrations. The direct effects of air pollution on plants involve the uptake of harmful pollutants such as carbon monoxide (Wang & Liao, 2016), ozone, sulfur dioxide, nitrogen oxides (Elkiey & Ormrod, 1981; Winner & Atkinson, 1986), volatile organic compounds (Abeles & Heggestad, 1973; Cape, 2003), and particulate matter (Rai, 2016). These pollutants disrupt key plant functions, including stomatal conductance, chlorophyll formation, photosynthesis, respiration, and other cellular activities, while also causing damage to both internal and exposed tissues (Winner & Atkinson, 1986; Gratani et al., 2000). Indirect effects arise from the impact of air pollution on the climate, which leads to changes in radiation, temperature, and precipitation. These climatic changes ultimately affect plant growth and productivity (Burney & Ramanathan, 2014). Air pollutants exert varied effects on plants, potentially altering their physiology, morphology, and internal tissue structure (anatomy), which disrupts biochemical processes and causes injury to tissues and organs. The extent of injury depends not only on the type and concentration of 24 pollutants but also on other factors such as exposure duration, the nature of exposure (e.g., gaseous vs particulate, dry vs wet deposition), the plant species, its stage of growth, and the climatic conditions that influence pollutant accumulation. These factors also determine whether a plant will be more susceptible or resistant to pollutant-induced damage (OMAFRA, 2022).

8.5.1 Physiological effects of air pollutants

There is growing concern that air pollutants, especially sulfur dioxide, ozone, and nitrogen oxides, can alter the physiological functions of plants and disrupt their growth patterns. These airborne contaminants can negatively affect a range of biochemical and physiological processes, including cellular enzymatic activities. Certain plant species are highly sensitive to these pollutants, exhibiting clear and measurable symptoms (Gupta, 2016). Air pollution influences plant processes such as pigment synthesis, sugar content, proline concentration, and the activities of enzymes and antioxidants (Aier et al., 2024), all of which directly impact stomatal conductance. This, in turn, affects photosynthesis, carbon allocation, respiration, and plant-water relations (Darrall, 1989; Gupta, 2016). Fig. 8.3 illustrates the physiological processes in plants that are impacted by air pollution Fig. 8.3.

8.5.1.1 Effects of air pollution on stomatal conductance

Stomatal conductance measures the rate of gas exchange (such as CO_2, oxygen, and water vapor) through the stomata of plants. It reflects the extent of stomatal opening and is a key determinant of plant-water relations

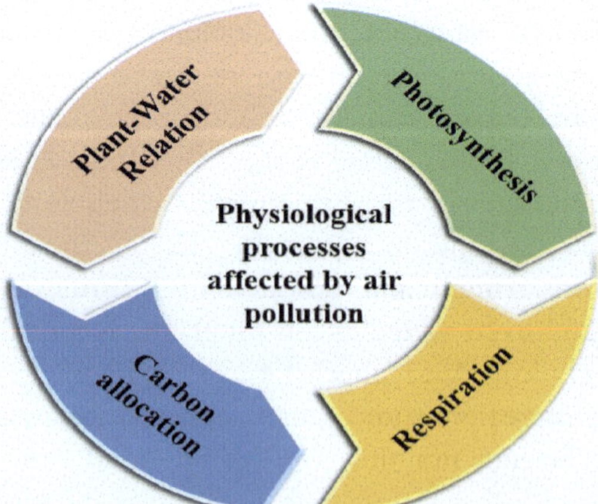

Figure 8.3 Major physiological processes influenced by stomatal conductance and air pollution effects. Source: Authors.

(Giménez et al., 2013). The rate of gaseous exchange between stomata and the atmosphere is influenced by the water status of guard cells and environmental conditions. The opening and closing of stomata regulate turgor pressure in guard cells, a process governed by active potassium ion transport. This fluctuation in turgor can be impacted by air pollutants (Giménez et al., 2013). Since stomata are the primary entry point for 26 airborne pollutants into plant tissues, their response—through changes in stomatal conductance and resistance—plays a crucial role in controlling pollutant concentrations and mitigating their toxic effects on plants (Kropff, 1987). Robinson et al. (1998) noted that even low concentrations of airborne pollutants can disrupt stomatal regulation, affecting the water balance of plants. They found that elevated concentrations of SO_2 and O_3 led to a decrease in stomatal conductance by inducing stomatal closure. Early studies, such as those by Rosen et al. (1978), also observed increased stomatal resistance (reduced stomatal conductance) in grapevines exposed to SO_2 and O_3. Additionally, Rao (1983) reported a significant decrease in stomatal conductance in pea plants after exposure to 0.5 ppm SO_2, with a further reduction after extended exposure. Trees, particularly in urban areas, serve as important air pollutant filters, absorbing significant quantities of gaseous pollutants through their stomata (Kwak et al., 2020). These pollutants, once absorbed into the mesophyll, can react with plant cells to produce ROS, which damage DNA, RNA, amino acids, proteins, and enzymes (Kwak et al., 2020). The nature of leaf surfaces, such as their wetness, can also influence stomatal conductance and the plant's response to air pollutants. According to Buckley (2017), wet leaf surfaces may lead to partial or complete stomatal closure, thereby reducing stomatal conductance. SO_2 and O_3 are known to have varying effects on stomatal conductance, with higher concentrations typically reducing conductance, while lower concentrations may have the opposite effect (Robinson et al., 1998). The impact of air pollution on stomatal conductance is complex, influenced by factors such as pollutant type and concentration, plant species, and environmental conditions (Xu et al., 2016).

8.5.1.2 Effects of air pollution on photosynthesis

Photosynthesis is a critical process for plant growth, survival, and biomass production (Rabinowitch & Govindjee, 1965). Air pollution can affect both stomatal conductance and photosynthesis in plants. Pollutants such as ozone, SO_2, NOx, and VOCs (e.g., peroxyacetylnitrate and aldehyde compounds) can trigger the production of ROS, which damage the photosynthetic machinery and induce oxidative stress in plants (Dhir, 2016). These pollutants also interfere with the synthesis of photosynthetic pigments (chlorophyll and carotenoids) and enzymes involved in photosynthesis (Taylor et al., 1965). Chlorophyll, essential for light absorption and energy transfer in photosynthesis, plays a crucial role in driving adenosine triphosphate (ATP) and nicotinamide adenine dinucleotide phosphate (NADPH) production, oxygen generation, and carbon fixation (Johnson, 2016; Mandal & Dutta, 2020). Excessive exposure to pollutants such as ozone and SO_2 can lead to chlorophyll degradation, particularly in areas with high pollution levels, such as roadsides and industrial zones (Chaurasia et al., 2013; Rai & Panda, 2014; Hariram et al., 2018). Chlorophyll may degrade to phaeophytin, further impairing photosynthesis (Talebzadeh & Valeo, 2022). Studies by Ziegler (1975) concluded that air pollutants such as SO_2, O_3, and peroxylacetyl nitrate compete with CO_2 for binding sites on the carboxylating enzyme RuBisCO, thus inhibiting net photosynthesis. Even at low concentrations, pollutants can suppress plant growth by affecting photosynthesis and respiration (Taylor et al., 1965; Gonzalez & Pignata, 1994). However, some species exhibit enhanced growth in response to certain pollutants, as demonstrated by Jiechen et al. (2021), who observed improved growth in tobacco exposed to 8 $\mu L\ L^{-1}$ NO_2.

8.5.1.3 Effects of air pollution on carbon allocation and plant-water relations

Carbon allocation is the process by which the products of photosynthesis are distributed to various plant functions (Darrall, 1989). This distribution —among aboveground parts (leaves, stems, wood) and belowground parts—affects plant growth and plays a key role in the terrestrial carbon cycle. Understanding how air pollution influences carbon allocation is essential for predicting plant responses to environmental stressors and their effects on ecosystem functioning (Brüggemann et al., 2011; Dietze et al., 2014; Hartmann et al., 2020). Although research has shown that both climate change and air pollution affect carbon allocation in plants, findings on the extent and mechanisms of these effects remain inconsistent. Winner and Atkinson (1986) reported that air pollution-induced damage to leaf tissues alters carbon allocation patterns. Some models, such as the CO_2 fertilization hypothesis, suggest that rising CO_2 levels can increase photosynthesis and water-use efficiency across plant species globally (Schimel et al., 2015; Piao et al., 2020). However, tree-ring studies and isotope analysis offer less support for the CO_2 fertilization hypothesis, instead attributing increases in carbon allocation to reduced disturbances in vegetated areas (Peñuelas et al., 2017). Darrall (1989) found variability in the effects of air pollutants (O_3, SO_2, NOx, hydrogen fluoride) on carbon allocation, with certain pollutants favoring aboveground carbon allocation at the expense of root growth. This imbalance can limit water uptake and negatively affect plant-water relations, ultimately impacting photosynthesis and carbon allocation to sinks (Hartmann et al., 2020). Air pollution also affects plant-water relations by influencing stomatal conductance, root hydraulic functions, and transpiration. Changes in stomatal behavior, such as the response to ambient CO_2 levels, can either benefit or harm the plant, depending on air quality. Studies have shown that NO_2 and SO_2 can temporarily increase transpiration by stimulating stomatal opening (Ashenden, 1979; Neighbour et al., 1988). Saxe (1986) observed that NO, among nitrogen oxides, significantly reduced photosynthesis and transpiration under elevated CO_2 conditions. In contrast, O_3 has been linked to reduced stomatal conductance and root growth, leading to increased water stress in plants (McLaughlin & Downing, 1995; Showman, 1991; Runeckles & Chevone, 1992). Hoshika et al. (2012) found that trees exposed to higher O_3 levels exhibited decreased photosynthesis, carbon assimilation, and water loss, highlighting the detrimental effects of ozone on plant-water relations.

8.5.1.4 Effects of air pollution on cellular and dark respiration

Cellular respiration, also known as mitochondrial respiration, is a metabolic process that produces ATP, meeting the energy demands for plant growth and development (Wang et al., 2001). This process involves the enzymatic conversion of glucose and oxygen into ATP and CO_2. Most respiration in trees occurs via the cytochrome-mediated pathway, though an alternative, nonphosphorylating pathway can also occur (Sprugel et al., 1995), producing less ATP. Air pollutants can directly or indirectly affect respiration in plants by influencing stomatal opening, enzyme activities, and stress responses (Muneer et al., 2014). Pollutants such as SO_2 and NO_2 can acidify leaf cells, disrupting metabolic activities, including mitochondrial function (Pfanz et al., 1987; Santo-Domingo & Demaurex, 2012). The presence of ROS due to environmental stressors can damage plant cells, leading to oxidative stress. Consequently, plants may increase antioxidant production to counteract the negative effects of pollutants (Mehla et al., 2017). While respiration is essential for plant growth, excessive respiration can lead to poor growth and carbon allocation (Gonzalez-Meler et al., 2004). SO_2 has been shown to increase dark respiration rates in several plant species, including lichens (Türk et al., 1974), broad beans (Black & Unsworth, 1979), and Scots pine (Katainen et al., 1987). Ozone exposure has also been linked to elevated dark respiration rates (Yang et al., 1983). 30 Conversely, NO_2 has been shown to reduce dark respiration in soybean (Carlson, 1983) and Douglas fir (Van Hove et al., 1992).

8.5.2 Morpho-anatomical effects of air pollutants

Air pollution can significantly damage the morphological and anatomical structures of plants. The extent of this damage is influenced by various factors, including the plant species, type and concentration of the pollutant, and duration of exposure. Pollutants such as O_3, peroxyacetyl nitrate (PAN), NOx, and SO_2 affect plant morphology through different mechanisms, including direct injury to plant organs and interference with physiological and biochemical processes, which in turn disrupt photosynthesis and growth. Heavy metals, a class of pollutants, are often deposited onto the soil and first affect the root system by obstructing the plant's ability to absorb essential nutrients and water (Umar et al., 2025). These metals and other particulate matter may also settle on leaf surfaces, blocking stomata, which impedes CO_2 uptake, carbon assimilation, and overall biomass production (Hubai et al., 2021).

Pollutants not only clog stomata, hindering photosynthesis but also cause damage to photosynthetic tissues, leading to symptoms such as leaf discoloration, reduced leaf area, and stunted growth. Leaf tissues, due to their direct exposure to pollutants and the presence of stomata through which these contaminants enter, are often the most affected. In polluted environments, plants may show morphological deformities, such as stomatal clogging, degradation of wax and lipid layers on leaves, necrosis, chlorosis, leaf burn, leaf folding or rolling, and premature leaf drop. Additionally, Chrabąszcz and Mróz (2017) reported the loss of bark in species like maple (Acer pseudoplatanus), beech (Fagus sylvatica), and elm (Ulmus sp.) as a result of air pollution. Fig. 8.4 illustrates the various detrimental effects of air pollutants on different plant structures Fig. 8.4.

Air toxicants such as O_3, SO_2, and NOx can cause significant damage to plants due to their ability to dissolve in the cellular fluids of leaves, triggering oxidative reactions that result in visible injuries. Once these pollutants dissolve in the apoplastic fluid of plant cells, they generate ROS,

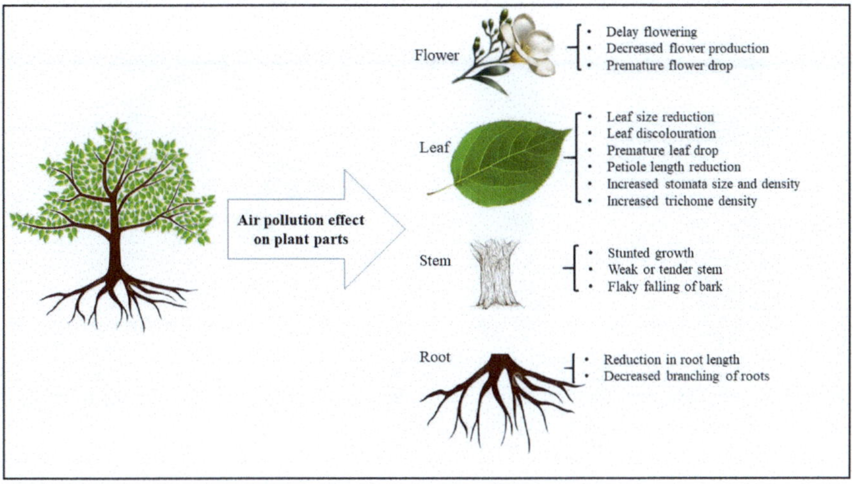

Figure 8.4 Effects of air pollution on various plant parts. Source: Authors.

which are highly reactive molecules capable of causing cellular damage (Fiscus et al., 2005; Sharps et al., 2021). While plants possess natural defense mechanisms, such as antioxidants, such as ascorbic acid and carotenoids (e.g., β-carotene, lycopene, xanthophyll, lutein, and zeaxanthin), these systems can become overwhelmed with acute or prolonged exposure to air pollutants. When ROS levels exceed the plant's detoxification capacity, oxidative damage occurs, leading to cell death. This process manifests as discoloration in the affected leaf areas, such as chlorosis, necrosis, and leaf burning in more severe cases (Gupta, 2016).

Among these air toxicants, ozone is particularly notorious for causing significant morphological damage to plants. The initial signs of ozone pollution often appear as stippling, primarily affecting the upper parts of older leaves. As exposure continues, this stippling may progress to leaf yellowing (chlorosis) and necrotic lesions (Fig. 8.5A). Felzer et al. (2007) confirmed that prolonged exposure to even low concentrations of ozone results in changes such as bronzing, chlorosis, and premature leaf senescence. High concentrations of ozone can cause the flaking of plant tissues. The extent of ozone-induced injury is well documented, and an injury index can be used to assess pollutant concentration and exposure duration. Fig. 8.5B shows that ozone injury typically occurs between leaf veins, sparing the veins themselves at all levels of damage.

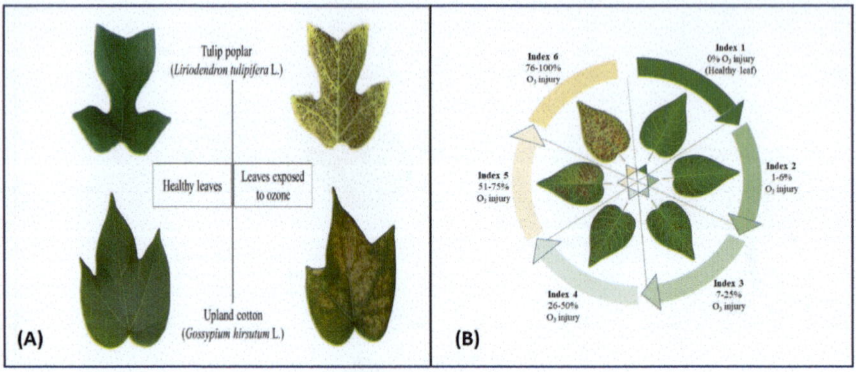

Figure 8.5 (A). Discoloration of leaf tissues exposed to ozone (B). Degrees of leaf injury due to ozone exposure. Source: Authors.

Air pollutants can interact in complex ways (Reinert et al., 1975; Schubert, 1984), leading to symptoms that may be difficult to attribute to a single contaminant. For example, pollutants such as chlorine gas, hydrochloric acid fumes, NO_2, and PAN can produce damage that resembles that caused by SO_2 when acting alone (Barrett & Benedict, 1970). NO_2, for instance, also causes interveinal chlorosis, starting with yellowing at the leaf margins and progressing to necrotic brown or black patches (Taylor & Eaton, 1966; Jiechen et al., 2021), much like SO_2. However, NO_2 damage affects both young and mature leaves, whereas SO_2 primarily impacts older leaves first. A study by Furukawa and Totsuka (1983) found that younger sunflower leaves were more tolerant to SO_2 than older leaves. Therefore, it may be necessary to sample various nearby plant species to accurately identify the specific air contaminants responsible for observed damage in plants.

8.6 Conclusion

Human exposure to air pollution has significantly increased due to urbanization. As cities expand and populations concentrate in urban areas, the demand for transportation, industry, construction, and energy rises, leading to a corresponding increase in the generation of air pollutants. Air pollution has a detrimental impact on the environment—affecting air, soil, and water—and poses various risks to all living organisms, including both plants and animals. In humans and animals, exposure to pollutants such as fine $PM_{2.5}$, NO_2, SO_2, carbon monoxide, VOCs, and O_3 can trigger respiratory and cardiovascular diseases, including asthma, chronic obstructive pulmonary disease, and cancer. Plants play a vital role in monitoring pollution and serve as effective indicators of air quality. Lichens, for example, are particularly sensitive to SO_2, mosses to heavy metals and particulate matter, and trees and shrubs reveal the presence of a variety of pollutants, often exhibiting distinct morphological changes. Monitoring air pollution is crucial for preventing environmental and health hazards. In urban centers, plants serve an important function in indicating air quality and mitigating the effects of air pollution. They act as natural sinks, absorbing and reducing the concentrations and toxicity of various pollutants, thereby helping to improve the environment.

References

Abeles, F. B., & Heggestad, H. E. (1973). Ethylene: An urban air pollutant. *Journal of the Air Pollution Control Association, 23*(6), 517–521.

Abhijith, K. V., & Kumar, P. (2019). Field investigations for evaluating green infrastructure effects on air quality in open-road conditions. *Atmospheric Environment, 201*, 132–147.

Agbaire, P. O., & Esiefarienrhe, E. (2009). Air Pollution Tolerance Indices (APTI) of some plants around Otorogun gas plant in Delta State, Nigeria. *Journal of Applied Science and Environmental Management, 13*, 11–14.

Aier, J., Panda, K. K., & Kumar, S. (2024). Impact of air pollution on plant physiology and ecosystem health. *Journal of Chemical Health Risks, 14*(3), 3167–3173.

Alushi, I., & Veizi, X. (2020). Effects of air pollution on stomatal responses, including paleoatmospheric CO2 concentration, in leaves of *Hedera helix*. *Albanian Journal of Agricultural Sciences, 19*(1), 21–28.

Anenberg, S. C., Mohegh, A., Goldberg, D. L., Kerr, G. H., Brauer, M., Burkart, K., Hystad, P., Larkin, A., Wozniak, S., & Lamsal, L. (2022). Long-term trends in urban NO2 concentrations and associated paediatric asthma incidence: Estimates from global datasets. *The Lancet Planetary Health, 6*(1), e49–e58.

Angold, P. G. (1997). The impact of a road upon adjacent heathland vegetation: Effects on plant species composition. *Journal of Applied Ecology, 34*, 409–417.

Anjaneyulu, M., Harikrishna, M., & Chenchuobulu, S. (2006). Modeling ambient carbon monoxide pollutant due to road traffic. *World Academy of Science Engineering and Technology, 17*, 103–106.

Ashenden, T. W., & Bell, S. A. (1987). The effect of sulfur dioxide on photosynthesis and stomatal behaviour of five species of urban trees. *Environmental Pollution, 43*(4), 293–304.

Ashenden, T. W. (1979). Effects of SO2 and NO2 pollution on transpiration in Phaseolus vulgaris L. *Environmental Pollution, 18*, 45–50.

Astel, A., Astel, K., & Biziuk, M. (2008). PCA and multidimensional visualization techniques united to aid in the bioindication of elements from transplanted Sphagnum palustre moss exposed in the Gdansk City Area. *Environmental Science and Pollution Research, 15*(1), 41–50. https://doi.org/10.1065/espr2007.05.42.

Badami, M. G. (2005). Transport and urban air pollution in India. *Environmental Management, 36*(2), 195–204.

Bailey-Serres, J., Parker, J. E., Ainsworth, E. A., Oldroyd, G. E., & Schroeder, J. I. (2019). Genetic strategies for improving crop yields. *Nature, 575*(7781), 109–118. https://doi.org/10.1038/s41586-019-1679-0.

Barrett, T. W., & Bendict, H. M. (1970). Sulfur dioxide. In Jacobson J. S., & Hill A. C. (Eds.). *Recognition of air pollution injury to vegetation: a pictorial atlas*. Pittsburg and Pennsylvania: C1-C17, Air Pollution Control Association.

Beatley, T. (2016). *Handbook of biophilic city planning and design*. Washington, DC: Island Press.

Berardi, U., GhaffarianHoseini, A., & GhaffarianHoseini, A. (2014). State-of-the-art analysis of the environmental benefits of green roofs. *Applied Energy, 115*, 411–428.

Bernhardt-Romermann, M., Kirchner, M., Kudernatch, T., Jakobi, G., & Fischer, A. (2006). Changed vegetation composition in coniferous forest near motorways in southern Germany: The effects of traffic-born pollution. *Environmental Pollution, 143*, 572–581.

Bikis, A. (2023). Urban air pollution and greenness in relation to public health. *Journal of Environmental and Public Health*. https://doi.org/10.1155/2023/8516622.

Black, V. J., & Unsworth, M. H. (1979). Effects of low concentrations of sulphur dioxide on net photosynthesis and dark respiration of *Vicia faba*. *Journal of Experimental Botany, 30*, 473–483.

Boquette, M. T., Fernandez, J. A., Carballeira, A., & Aboal, J.R. (2013). Assessing the tolerance of the terrestrial moss *Psuedoscleropodium purum* to high levels of atmospheric heavy metals: A reciprocal transplant study.

Brüggemann, N., Gessler, A., Kayler, Z., Keel, S., Badeck, F., Barthel, M., Boeckx, P., Buchmann, N., Brugnoli, E., & Esperschütz, J. (2011). Carbon allocation and carbon isotope fluxes in the plant–soil–atmosphere continuum: A review. *Biogeosciences, 8*, 3457–3489.

Buckley, T. N. (2017). Modeling stomatal conductance. *Plant Physiology, 174*(2), 572–582. https://doi.org/10.1104/pp.16.01772.

Burney, J., & Ramanathan, V. (2014). Recent climate and air pollution impacts on Indian agriculture. *PNAS, 111*(46), 16319–16324.

Calfapietra, C., Fares, S., & Loreto, F. (2013). Volatile organic compounds from Italian vegetation and their interaction with ozone. *Environmental Pollution, 181*, 25–35. https://doi.org/10.1016/j.envpol.2013.05.016.

Cape, J. N. (2003). Effects of airborne volatile organic compounds on plants. *Environmental Pollution (Barking, Essex 1987), 122*(1), 145–157.

Carlson, R. W. (1983). Interaction between SO2 and NO2 and their effects on photosynthetic properties of soybean Glycine max. *Environmental Pollution Series A, Ecological and Biological, 32*(1), 11–38.

Carneiro, M. F. H., Ribeiro, F. Q., Fernandes-Filho, F. N., Lobo, D. J. A., Barbosa, F., Rhoden, C. R., et al. (2011). Pollen abortion rates, nitrogen dioxide by passive diffusive tubes and bioaccumulation in tree barks are effective in the characterization of air pollution. *Environmental and Experimental Botany, 722*, 272–277. https://doi.org/10.1016/j.envexpbot.2011.04.001.

Cen, S. (2015). Biological monitoring of air pollutants and its influence on human beings. *Open Biomedical Engineering Journal, 9*, 219–223.

Chakrabortty, S., & Paratkar, G. T. (2006). Biomonitoring of trace element air pollution using mosses. *Aerosol and Air Quality Research, 6*(3), 247–258.

Chaurasia, S., Karwariya, A., & Gupta, A. (2013). Effect of cement industry pollution on chlorophyll content of some crops at Kodinar, Gujarat, India. *Proceedings of the International Academy of Ecology and Environmental Sciences, 3*(4), 288–295.

Chen, L., Liu, C., Zhang, L., Zou, R., & Zhang, Z. (2017). Variation in tree species ability to capture and retain airborne fine particulatematter (PM2.5). *Scientific Reports, 7*(1), 3206–3211.

Choochuay, C., Pongpiachan, S., Tipmanee, D., et al. (2020). Impacts of PM2.5 sources on variations in particulate chemical compounds in ambient air of Bangkok, Thailand. *Atmospheric Pollution Research, 11*(9), 1657–1667.

Chrabąszcz, M., & Mróz, L. (2017). Tree bark, a valuable source of information on air quality. *Polish Journal of Environmental Studies, 26*(2), 453–466. https://doi.org/10.15244/pjoes/65908.

Cohen, A. J., Brauer, M., Burnett, R., Anderson, H. R., Frostad, J., Estep, K., Balakrishnan, K., Brunekreef, B., Dandona, L., Dandona, R., Feigin, V., Freedman, G., Hubbell, B., Jobling, A., Kan, H., Knibbs, L., Liu, Y., Martin, R., Morawska, L., ... Forouzanfar, M. H. (2017). Estimates and 25-year trends of the global burden of disease attributable to ambient air pollution: An analysis of data from the Global Burden of Diseases Study 2015. *The Lancet, 389*(10082), 1907–1918.

Conti, M. E., & Cecchetti, G. (2001). Biological monitoring: Lichens as bioindicators of air pollution assessment—A review. *Environmental Pollution, 114*(3), 471–492.

Correa-Ochoa, M., Mejia-Sepulveda, J., Saldarriaga-Molina, J., Castro-Jimenez, C., & A-Gil, D. (2021). Evaluation of air pollution tolerance index and anticipated performance index of six plant species, in an urban tropical valley: Medellin, Colombia. *Environmental Science and Pollution Research, 5*, 7952–7971.

Darrall, N. M. (1989). The effect of air pollutants on physiological processes in plants. *Plant, Cell & Environment, 12*, 1–30.

Dhir, B. (2016). Air pollutants and photosynthetic efficiency of plants. In U. Kulshrestha, & P. Saxena (Eds.). *Plant Responses to Air Pollution* Singapore: Springer. https://doi.org/10.1007/978-981-10-1201-3_7.

Dietze, M. C., Sala, A., Carbone, M. S., Czimczik, C. I., Mantooth, J. A., Richardson, A. D., & Vargas, R. (2014). Nonstructural carbon in woody plants. *Annual Review of Plant Biology, 65*, 667–687.

Dockery, D. W., & Pope, C. A. (2006). Health effects of fine particulate air pollution: Lines that connect. *Journal of the Air and Waste Management Association, 56*(6), 709–742.

Elhadad, S. M., Ea, S., Saleh, I. H., & Omar, M. Y. (2026). Integrating multivariate analysis and Air Pollution Tolerance Index (APTI) to evaluate four ornamental plants for sustainable indoor air phytoremediation. *Scientific Reports, 16*(1), 16013. https://doi.org/10.1038/s41598-026-50763-0.

Elkiey, T., & Ormrod, D. P. (1981). Absorption of ozone, sulphur dioxide, and nitrogen dioxide by petunia plants. *Environmental and Experimental Botany, 21*(1), 63–70.

Escobedo, F. J., Wagner, J. E., Nowak, D. J., De la Maza, C. L., Rodriguez, M., & Crane, D. E. (2008). Analyzing the cost effectiveness of Santiago, Chile's policy of using urban forests to improve air quality. *Journal of Environmental Management, 86*(1), 148–157.

Faralli, M., Matthews, J., & Lawson, T. (2019). Exploiting natural variation and genetic manipulation of stomatal conductance for crop improvement. *Current Opinion in Plant Biology, 49*, 1–7.

Fatoba, P. O., Ogunkunle, C. O., Oyedeji, S., & Salawudeen, M. B. (2013). Heavy metal depositions around some petroleum product depots in nigeria, using mosses as biomonitors. *The Bioscientist Journal, 1*(1), 99–105.

Felzer, B. S., Cronin, T., Reilly, J. M., Melillo, J. M., & Wang, X. (2007). Impacts of ozone on trees and crops. *Comptes Rendus Geoscience, 339*(11–12), 784–798.

Fernández, J. A., Aboal, J. R., & Carballeira, A. (2000). Use of native and transplanted mosses as complementary techniques for biomonitoring mercury around an industrial facility. *Science of the Total Environment, 256*, 151–161.

Figueira, R., Sergio, C., & Sousa, A. J. (2002). Distribution of trace metals in moss biomonitors and assessment of contamination sources in Portugal. *Environmental Pollution, 118*(1), 153–163.

Fiscus, E. L., Booker, F. L., & Burkey, K. O. (2005). Crop responses to ozone: Uptake, modes of action, carbon assimilation and partitioning. *Plant, Cell and Environment, 28*, 997–1011.

Franks, P. J., & Beerling, D. J. (2009). Maximum leaf conductance driven by CO_2 effects on stomatal size and density over geologic time. *Proceedings of the National Academy of Sciences of the United States of America, 106*, 10343–10347.

Franks, P. J., & Farquhar, G. D. (2001). The effect of exogenous abscisic acid on stomatal development, stomatal mechanics, and leaf gas exchange in *Tradescantia virginiana. Plant Physiology, 125*, 935–942.

Furukawa, A., & Totsuka, T. (1983). Effect of SO2 on photosynthesis in sunflower leaves: Age- dependent inhibition. *Environment Control in Biology, 21*(2), 43–49. https://doi.org/10.2525/ecb1963.21.43.

Gómez-Arroyo, S., Zavala-Sánchez, M.Á., Alonso-Murillo, C. D., Cortés-Eslava, J., Amador- Muñoz, O., Jiménez-García, L. F., & Morton-Bermea, O. (2021). Moss (*Hypnum amabile*) as biomonitor of genotoxic damage and as bioaccumulator of atmospheric pollutants at five different sites of Mexico City and metropolitan area. *Environmental Science and Pollution Research International, 28*(8), 9849–9863. https://doi.org/10.1007/s11356-020-11441-4.

Galsomiès, L., Letrouit, M. A., Deschamps, C., Savanne, D., & Avnaim, M. (1999). Atmospheric metal deposition in France: Initial results on moss calibration from the 1996 biomonitoring. *Science of The Total Environment, 232*(1-2), 39–47.

Gerdol, R., Bragazza, L., & Marchesini, R. (2002). Element concentrations in the forest moss *Hylocomium splendens*: Variation associated with altitude, net primary production and soil chemistry. *Environmental Pollution, 116*(1), 129–135.

Gilbert, O. L. (1971). The effect of airborne fluorides on lichens. *The Lichenologist, 5*(1-2), 26–32.

Gillner, S., Vogt, J., & Roloff, A. (2015). Climate change and urban trees: Effects on growth, pests, and diseases. *Urban Forestry and Urban Greening, 14*(2), 234–243.

Giménez, C., Gallardo, M., & Thompson, R. B. (2013). Plant–Water Relations. In *Reference module in earth systems and environmental sciences*. Elsevier. https://doi.org/10.1016/B978-0-12-409548-9.05257-X.

Giordano, S., Spagnuolo, V., & Capozzi, F. (2021). Biomonitoring of air pollution. *Atmosphere, 12*(4), 433. https://doi.org/10.3390/atmos12040433.

Gonzalez, C. M., & Pignata, M. L. (1994). The Influence of air pollution on soluble proteins, chlorophyll degradation, MDA, sulphur and heavy metals in a transplanted lichen. *Chemistry and Ecology, 9*(2), 105–113.

Gonzalez-Meler, M. A., Taneva, L., & Trueman, R. J. (2004). Plant respiration and elevated atmospheric CO_2 concentration: Cellular responses and global significance. *Annals of Botany, 94*(5), 647–656.

Gratani, L., Crescente, M. F., & Petruzzi, C. (2000). Relationship between leaf life-span and photosynthetic activity of Quercus ilex in polluted urban areas (Rome). *Environmental Pollution, 110*, 19–28.

Grodzińska, K., & Szarek-Łukaszewska, G. (2001). Response of mosses to the heavy metal deposition in Poland - An overview. *Environmental Pollution, 114*(3), 443–451.

Gupta, A. (2016). Effect of air pollutants on plant gaseous exchange process: Effect on stomata and respiration. In U. Kulshrestha, & P. Saxena (Eds.). *Plant Responses to Air Pollution* (pp. 85–92). Singapore: Springer. https://doi.org/10.1007/978-981-10-1201-3_8.

Halliwell, B., & Gutteridge, J. M. (2015). *Free radicals in biology and medicine*. Oxford University Press.
Hariram, M., Sahu, R., & Elumalai, S. P. (2018). Impact assessment of atmospheric dust on foliage pigments and pollution resistances of plants grown nearby coal based thermal power plants. *Archives of Environmental Contamination and Toxicology, 74*, 56–70.
Hartmann, H., Bahn, M., Carbone, M., & Richardson, A. D. (2020). Plant carbon allocation in a changing world – challenges and progress: Introduction to a virtual issue on carbon allocation. *New Phytologist, 227*(4), 981–988.
Honour, S. L., Bell, J. N. B., Ashenden, T. W., Cape, J. N., & Power, S. A. (2009). Responses of herbaceous plants to urban air pollution: Effects on growth, phenology and leaf surface characteristics. *Environmental Pollution, 157*(4), 1279–1286. https://doi.org/10.1016/j.envpol.2008.11.049.
Hoshika, Y., Omasa, K., & Paoletti, E. (2012). Whole-tree water use efficiency is decreased by ambient ozone and not affected by O3-induced stomatal sluggishness. *PLoS One, 7*(6), e39270. https://doi.org/10.1371/journal.pone.0039270.
Hubai, K., Kováts, N., & Teke, G. (2021). Effects of urban atmospheric particulate matter on higher plants using *Lycopersicon esculentum* as model species. *SN Applied Sciences, 3*, 770. https://doi.org/10.1007/s42452-021-04745-8.
Jafari, A. J., Charkhloo, E., & Pasalari, H. (2021). Urban air pollution control policies and strategies: A systematic review. *Journal of Environmental Health Science and Engineering, 19*, 1911–1940. https://doi.org/10.1007/s40201-021-00744-4.
Jiang, Y., Fan, M., Hu, R., Zhao, J., & Wu, Y. (2018). Mosses are better than leaves of vascular plants in monitoring atmospheric heavy metal pollution in urban areas. *International Journal of Environmental Research and Public Health, 15*(6), 1105. https://doi.org/10.3390/ijerph15061105.
Jiechen, W., Yue, W., Huihui, Z., Dandan, G., & Guangyu, S. (2021). Atmospheric nitrogen dioxide at different concentrations levels regulates growth and photosynthesis of tobacco plants. *Journal of Plant Interactions, 16*(1), 422–431. https://doi.org/10.1080/17429145.2021.1967474.
Johnson, M. P. (2016). Photosynthesis. *Essays in Biochemistry, 60*(3), 255–273. https://doi.org/10.1042/EBC20160016.
Jones, K. C., & Duarte-Davidson, R. (1997). Transfers of airborne PCDD/Fs to bulk deposition collectors and herbage. *Environmental Science & Technology, 31*(10), 2937–2943.
Joshi, P. C., & Swami, A. (2009). Air pollution induced changes in the photosynthetic pigments of selected plant species. *Journal of Environmental Biology, 30*, 295–298.
Jovan, S. E., Monleon, V. C., Donovan, G. H., Gatziolis, D., & Amacher, M. C. (2021). Small- scale distributions of polycyclic aromatic hydrocarbons in urban areas using geospatial modeling: A case study using the moss Orthotrichum lyellii in Portland, Oregon, U.S.A. *Atmospheric Environmental, 256*, 118433. https://doi.org/10.1016/j.atmosenv.2021.118433.
Jovan, S. E., Zuidema, C., Derrien, M. M., Bidwell, A. L., Brinkley, W., Barhil, R., Smith, R. J., Barnhill, R., Gould, L., Rodríguez, A. J., Amacher, M. C., Abel, T. D., & Lopez, P. (2022). Heavy metals in moss guide environmental justice investigation: A case study using community science in Seattle, WA, USA. *Ecosphere, 13*(6), e4109. https://doi.org/10.1002/ecs2.4109.
Jovan, S. (2008). Lichen bioindication of biodiversity, air quality, and climate: Baseline results from monitoring in Washington, Oregon, and California. General Technical Report PNW-GTR-737. U.S. Department of Agriculture, Forest Service, Pacific Northwest Research Station, Portland, Oregon, USA, 115 p.

Katainen, H.-S., Maikinen, E., Jokinen, J., Karjalainen, R., & Kellomaki, S. (1987). Effects of SO_2 on the photosynthetic and respiration rates in scots pine seedlings. *Environmental Pollution, 46*, 241–251.

Klumpp, G., Furlan, C. M., Domingos, M., & Klumpp, A. (2000). Response of stress indicators and growth parameters of *Tibouchina pulchra* exposed to air and soil pollution near the industrial complex of Cubatao, Brazil. *Science of The Total Environment, 246*(1), 79–91. https://doi.org/10.1016/S0048-9697(99)00453-2.

Koul, M., & Thakur, P. (2016). Plantation of shrubs in urban areas to contest air pollution. *Proceedings of the National Seminar on Water and Air Quality in Urban Ecosystem, 22*, 62–67.

Kropff, M. J. (1987). Physiological effects of sulphur dioxide. 1. The effect of SO_2 on photosynthesis and stomatal regulation of *Vicia faba* L. *Plant, Cell and Environment, 10*, 753–760.

Kumie, A., Worku, A., Tazu, Z., et al. (2021). Fine particulate pollution concentration in Addis Ababa exceeds the WHO guideline value: Results of 3 years of continuous monitoring and health impact assessment. *Environmental Epidemiology, 5*(3), e155.

Kwak, M. J., Lee, J. K., Park, S., Lim, Y. J., Kim, H., Kim, K. N., Je, S. M., Park, C. R., & Woo, S. Y. (2020). Evaluation of the Importance of some East Asian tree species for refinement of air quality by estimating air pollution tolerance index, anticipated performance index, and air pollutant uptake. *Sustainability, 12*(7), 3067. https://doi.org/10.3390/su12073067.

Li, X., Stringer, L. C., & Dallimer, M. (2022). The impacts of urbanisation and climate change on the urban thermal environment in Africa. *Climate, 10*, 164. https://doi.org/10.3390/cli10110164.

Li, F., Yigitcanlar, T., Li, W., Nepal, M., Nguyen, K., & Dur, F. (2024). Understanding urban heat vulnerability: Scientometric analysis of five decades of research. *Urban Climate, 56*, 102035. https://doi.org/10.1016/j.uclim.2024.102035.

Lin, Y. S., Medlyn, B. E., Duursma, R. A., Prentice, I. C., Wang, H., & Baig, S. (2015). Optimal stomatal behaviour around the world. *Nature Climate Change, 5*, 459.

Liu, J., & Chen, F. (2016). Biology and industrial applications of Chlorella: Advances and prospects. *Advances in Biochemical Engineering/Biotechnology, 153*, 1–35.

Lounamaa, J. (1956). Trace elements in plants growing wild on different rocks in Finland. A semi-quantitative spectrographic survey. *Annales Botanici Societatis Zoologicæ Fennicæ "Vanamo" 29*, 1–196.

Mandal, R., & Dutta, G. (2020). From photosynthesis to biosensing: Chlorophyll proves to be a versatile molecule. *Sensors International, 1*, 100058. https://doi.org/10.1016/j.sintl.2020.100058.

Mane, P. C., & Bhosle, A. B. (2012). Bioremoval of some metals by living algae *Spirogyra* sp. and *Spirullina* sp. from aqueous solution. *International Journal of Environmental Research, 6*(2), 571–576.

Manisalidis, I., Stavropoulou, E., Stavropoulos, A., & Bezirtzoglou, E. (2020). Environmental and health impacts of air pollution: A review. *Frontiers in Public Health, 8*, 14. https://doi.org/10.3389/fpubh.2020.00014.

Marć, M., Tobiszewski, M., Zabiegała, B., de la Guardia, M., & Namieśnik, J. (2015). Current air quality analytics and monitoring: A review. *Analytica Chimica Acta, 853*, 116–126.

McLaughlin, S. B., & Downing, D. J. (1995). Interactive effects of ambient ozone and climate measured on growth of mature forest trees. *Nature, 374*, 252–254.

Meeranayak, U. F. J., Nadaf, R. D., Toragall, M. M., Nadaf, U., & Shivasharana, C. T. (2020). The role of algae in sustainable environment: A review. *Journal of Algal Biomass Utilization, 11*(2), 28–34.

Mehla, N., Sindhi, V., Josula, D., Bisht, P., & Wani, S. H. (2017). An introduction to antioxidants and their roles in plant stress tolerance. In M. I. R. Khan, & N. A. Khan

(Eds.). *Reactive oxygen species and antioxidant systems in plants: Role and regulation under abiotic stress* (pp. 1–23). Singapore: Springer.
Mikhaylov, A. (2020). Lichens as indicators of atmospheric pollution in urban ecosystems. *Israel Journal of Ecology and Evolution, 67*(1-2), 60–68.
Moreira, T. C. L., Amato-Lourenço, L. F., da Silva, G. T., Saldiva, de A. C. D., de André, P. A., Barrozo, L. V., Singer, J. M., Saldiva, P. H. N., Saiki, M., & Locosselli, G. M. (2018). The use of tree barks to monitor traffic related air pollution: A case study in São Paulo–Brazil. *Frontiers in Environmental Science, 6*, 00072.
Muneer, S., Kim, T. H., Choi, B. C., Lee, B. S., & Lee, J. H. (2014). Effect of CO, NOx and SO2 on ROS production, photosynthesis and ascorbate-glutathione pathway to induce Fragaria×annasa as a hyperaccumulator. *Redox Biology, 2*, 91–98.
Murray, C. J., Aravkin, A. Y., Zheng, P., Abbafati, C., Abbas, K. M., Abbasi-Kangevari, M., ... Borzouei, S. (2020). Global burden of 87 risk factors in 204 countries and territories, 1990–2019: A systematic analysis for the Global Burden of Disease Study 2019. *The Lancet, 396*(10258), 1223–1249.
Ndlovu, N. B., Frontasyeva, M. V., Newman, & Maleka, R. T. (2021). Moss and lichen biomonitoring of atmospheric pollution in the Western Cape Province (South Africa). *American Journal of Analytical Chemistry, 10*, 86–102.
Neighbour, E. A., Cottam, D. A., & Mansfield, T. A. (1988). Effects of sulphur dioxide and nitrogen dioxide on the control of water loss by birch (*Betula* spp.). *New Phytologist, 108*, 149–157.
Ogunkunle, C. O., & Fatoba, P. O. (2012). Cellular compartmentalization and heavy metal load in the moss *Barbula lambarenensis* around a mega cement factory in southwest Nigeria. *Ife Journal of Science, 14*(2), 185–193.
Ogunkunle, C. O., Suleiman, L. B., Oyedeji, S., Awotoye, O. O., & Fatoba, P. O. (2015). Assessing the air pollution tolerance index and anticipated performance index of some tree species for biomonitoring environmental health. *Agroforestry Systems, 89*, 447–454.
Ogunkunle, C., Oyedeji, S., Adeniran, I. F., Olorunmaiye, K. S., & Fatoba, P. O. (2019). *Thuja occidentalis* and *Duranta repens* as indicators of urban air pollution in industrialized areas of southwest Nigeria. *Agriculturae Conspectus Scientificus, 84*(2), 193–202.
Oksanen, I. (2006). Ecological and biotechnological aspects of lichens. *Applied Microbiology and Biotechnology, 73*, 723–734.
OMAFRA. (2022). Effects of air pollution on agricultural crops. Ontario Ministry of Agriculture, Food and Rural Affairs, Guelph, Ontario N1G 4Y2, Canada. https://www.ontario.ca/page/effects-air-pollution-agricultural-crops.
Oyedeji, S., Eze, C. N., Ogunkunle, C. O., & Fatoba, P. O. (2016). Assessment of environmental quality of two industrial cities in North-Central Nigeria using metal pollution indices. *International Journal of Applied Bioengineering, 10*, 1–6. http://dx.doi.org/10.18000/ijabeg.10137.
Oyedeji, S., Agboola, O. O., Oyekunle, J. K., Animasaun, D. A., & Fatoba, P. O. (2019). Assessment of tree species resistance to air pollution around a metal-scrap recycling factory using air pollution tolerance index and anticipated performance index. *Ruhuna Journal of Science, 10*(1), 65–76.
Parus, A., & Karbowska, B. (2020). Marine algae as natural indicator of environmental cleanliness. *Water, Air, and Soil Pollution, 231*, 1–8.
Peñuelas, J., Ciais, P., Canadell, J. G., Janssens, I. A., Fernández Martínez, M., Carnicer, J., Obersteiner, M., Piao, S., Vautard, R., & Sardans, J. (2017). Shifting from a fertilization dominated to a warming dominated period. *Nature Ecology and Evolution, 1*, 1438–1445.
Pearson, L., & Skye, E. (1965). Air pollution affects pattern of photosynthesis *in Parmelia sulcata*, a corticolous lichen. *Science (New York, N.Y.), 148*(3677), 1600–1602.

Perera, F. (2017). Pollution from fossil-fuel combustion is the leading environmental threat to global pediatric health and equity: Solutions exist. *International Journal of Environmental Research and Public Health, 15*(1), 16. https://doi.org/10.3390/ijerph15010016.

Pfanz, H., Martinoia, E., Lange, O.-L., & Heber, U. (1987). Flux of SO_2 into leaf cells and cellular acidification by SO_2. *Plant Physiology, 85*(4), 928–933.

Piao, S., Wang, X., Park, T., et al. (2019). Characteristics, drivers and feedbacks of global greening. *Nature Reviews Earth & Environment, 1*(1), 14–27. https://doi.org/10.1038/s43017-019-0001-x.

Piracha, A., & Chaudhary, M. T. (2022). Urban air pollution, urban heat island and human health: A review of the literature. *Sustainability, 14*, 9234. https://doi.org/10.3390/su14159234.

Poikolainen, J., Kuusinen, M., & Mikkola, K. (2000). Epiphytic lichens as indicators of air quality. In E. Mälkönen (Vol. Ed.), Forest condition in a changing environment. *Forestry Sciences: Vol. 65*. Dordrecht: Springer. https://doi.org/10.1007/978-94-015-9373-1_18.

Pongpiachan, S., & Iijima, A. (2016). Assessment of selected metals in the ambient air PM 10 in urban sites of Bangkok Thailand). *Environmental Science and Pollution Research, 23*(3), 2948–2961.

Rabinowitch, E., & Govindjee (1965). The role of chlorophyll in photosynthesis. *Scientific American, 213*, 74–83. https://doi.org/10.1038/scientificamerican0765-74.

Rahman, M. M., Adzkia, U., Nandika, D., Siregar, I. Z., & Karlinasari, L. (2022). Urban tree bark analysis for monitoring of air pollution level in Jakarta business district. *IOP Conference Series: Earth and Environmental Science, 1109*, 012052. 10.1088/1755-1315/1109/1/012052.

Rai, P. K., & Panda, L. L. (2014). Dust capturing potential and air pollution tolerance index (APTI) of some roadside tree vegetation in Aizawl, Mizoram, India: An Indo-Burma hot spot region. *Air Quality, Atmosphere and Health,* 7, 93–101.

Rai, P. K. (2016). Impacts of particulate matter pollution on plants: Implications for environmental biomonitoring. *Ecotoxicology and Environmental Safety, 129*, 120–136. https://doi.org/10.1016/j.ecoenv.2016.03.012.

Rao, D. N. (1983). Sulphur dioxide pollution versus plant injury with special reference to fumigation and precipitation. In *Proceedings Symposium on Air Pollution Control, vol. 1*. New Delhi: Indian Association for Air pollution Control, 91–96.

Reinert, R. A., Heagle, A. S., & Heck, W. W. (1975). Plant responses to pollutant combinations. In J. B. Mudd, & T. T. Kozlowski (Eds.). *Responses of plants to air pollution* (pp. 159–177). New York: Academic Press.

Ristić, S., Šajn, R., & Stamenković, S. (2021). Lichens as the main indicator in biological monitoring of air quality. In B. Balabanova, & T. Stafilov (Eds.). *Contaminant levels and ecological effects. Emerging contaminants and associated treatment technologies* (pp. 101–129). Cham: Springer. https://doi.org/10.1007/978-3-030-66135-9_4.

Robinson, M. F., Heath, J., & Mansfield, T. A. (1998). Disturbances in stomata behavior caused by air pollutants. *Journal of Experimental Botany, 49*, 461–469.

Rosen, P. M., Musselman, R. C., & Kender, W. J. (1978). Relationship of stomatal resistance to sulfur dioxide and ozone injury in grapevines. *Scientia Horticulturae, 8*(2), 137–142.

Runeckles, V. C., & Chevone, B. I. (1992). Tree responses to ozone. In A.S. Lefohn (ed.), Surface-level ozone exposures and their effects on vegetation, Lewis Publishers Inc, Chelsea, MI, USA (1992), pp. 189-270.

Salo, H., Paturi, P., & Mäkinen, J. (2016). Moss bag (*Sphagnum papillosum*) magnetic and elemental properties for characterising seasonal and spatial variation in urban pollution. *International Journal of Environmental Science and Technology, 13*, 1515–1524. https://doi.org/10.1007/s13762-016-0998-z.

Santo-Domingo, J., & Demaurex, N. (2012). Perspectives on: SGP symposium on mitochondrial physiology and medicine: The renaissance of mitochondrial pH. *The Journal of General Physiology, 139*(6), 415–423.
Saxe, H. (1986). Effects of NO, NO_2 and CO_2 on net photosynthesis, dark respiration and transpiration of pot plants. *New Phytologist, 103*, 185–197.
Schimel, D., Stephens, B. B., & Fisher, J. B. (2015). Effect of increasing CO_2 on the terrestrial carbon cycle. *Proceedings of the National Academy of Sciences of United States of America, 112*, 436–441.
Schubert, T. S. (1984). *Sulfur dioxide injury to plants. Plant Pathology Circular No. 257*. Florida, USA: Bureau of Plant Pathology, Florida Dept. Agric. & Consumer Services. https://ccmedia.fdacs.gov/content/download/11263/file/pp257.pdf.
Sharps, K., Hayes, F., Harmens, H., & Mills, G. (2021). Ozone-induced effects on leaves in African crop species. *Environmental Pollution, 268*, 115789. https://doi.org/10.1016/j.envpol.2020.115789.
Shing, W. L., Hwang, T. Y., Yi, K. W., Han, L. J., & Hock, O. G. (2018). Using the responses of green algae Spirogyra as bioindicator for metals and pesticides pollution. *Journal of Environmental Science and Management, 21*(2), 1–6.
Shiva-Nagendra, S. M., Khare, M., Schlink, U., & Peter, A. E. (2021). Introduction to Urban Air Pollution. In S. M. Shiva-Nagendra, U. Schlink, A. Müller, & M. Khare (Eds.). Urban air quality monitoring, modelling and human exposure assessment. *Springer transactions in civil and environmental engineering*Singapore: Springerhttps://doi.org/10.1007/978-981-15-5511-4_1.
Showman, R. E. (1991). A comparison of ozone injury in vegetation during moist and drought years. *Journal of the Air & Waste Management Association, 41*, 63–64.
Sicard, P., Agathokleous, E., Anenberg, S. C., De-Marco, A., Paoletti, E., & Calatayud, V. (2023). Trends in urban air pollution over the last two decades: A global perspective. *Science of The Total Environment, 858*, 160064.
Singh, S. N., & Agrawal, M. (2007). Air pollution and its impact on plants. In S. N. Singh, & R. D. Tripathi (Eds.). *Environmental bioremediation technologies* (pp. 353–373). Berlin, Heidelberg: Springer. https://doi.org/10.1007/978-3-540-34793-4_16.
Singh, S. N., & Verma, A. (2007). Phytoremediation of air pollutants: a review. In *Environmental bioremediation technologies*. Berlin, Heidelberg: Springer, 293–314. https://doi.org/10.1007/978-3-540-34793-4_16.
Singh, A., Gupta, H., Gupta, K., Singh, P., Gupta, V., & Sharma, R. (2007). A comparative study of air pollution in Indian cities. *Bulletin of Environmental Contamination and Toxicology, 78*(5), 411–416.
Sprugel, D. G., Ryan, M. G., Brooks, J. R., Vogt, K. A., & Martin, T. A. (1995). Respiration from the organ level to the stand. In W. K. Smith, & T. M. Hinckley (Eds.). *Physiological ecology: Resource physiology of conifers* (pp. 255–299). Academic Press. https://doi.org/10.1016/B978-0-08-092591-2.50013-3.
Srivastava, S., & Gupta, S. (2020). Application of algae in air pollution control technique. *International Research Journal of Engineering and Technology, 7*(3), 3613–3618.
Srivastava, A., & Rao, B. P. S. (2011). Urban air pollution modeling. In D. Popovic (Ed.). *Air quality-models and applications*Morn Hill: InTechhttps://doi.org/10.5772/16776.
Świsłowski, P., & Rajfur, M. (2024). Assessment of the air quality in an industrial zone using active moss biomonitoring. *International Journal of Environmental Science and Technology, 21*, 4213–4222. https://doi.org/10.1007/s13762-023-05276-y.
Türk, R., Wirth, V., & Lange, O. L. (1974). CO2-Gaswechsel Untersuchungen zur SO_2-Resistenz von Flechten (CO exchange measurements for determination of SO resistance of lichens). *Oecologia, 15*, 33–64.

Talebzadeh, F., & Valeo, C. (2022). Evaluating the effects of environmental stress on leaf chlorophyll content as an index for tree health. *IOP Conference Series: Earth and Environmental Science, 1006*, 012007.

Taylor, O. C., & Eaton, F. M. (1966). Suppression of plant growth by nitrogen dioxide. *Plant Physiology, 41*(1), 132–135.

Taylor, O. C., Cardiff, E. A., & Mersereau, J. D. (1965). Apparent photosynthesis as a measure of air pollution damage. *Journal of the Air Pollution Control Association, 15*(4), 171–173.

Terekhina, N. V., & Ufimtseva, M. D. (2020). Leaves of trees and shrubs as bioindicators of air pollution by particulate matter in Saint Petersburg. *Geography, Environment, Sustainability, 13*(1), 224–232. https://doi.org/10.24057/2071-9388-2019-65.

Tiwari, S., & Tiwari, M. (2006). Air pollution tolerance indices of few plants growing near Raigarh (India). *Journal of Environmental Research and Development, 1*(2), 129–135.

Tiwary, A., & Kumar, P. (2020). Integrated dispersion-deposition modelling for air pollutant reduction via green infrastructure at an urban scale. *Science of the Total Environment, 723*, 138078.

Tolunay, D. (2003). Air pollution effects on annual ring widths of forest trees in mountainous land of Izmir (Turkey). *Water, Air, & Soil Pollution, 3*, 227–242.

Tripathi, D. P., & Nema, A. K. (2023). Seasonal variation of biochemical parameters and air pollution tolerance index (APTI) of selected plant species in Delhi city, and detailed meta-analysis from Indian metropolitan cities. *Atmospheric Environment, 309*, 119862. https://doi.org/10.1016/j.atmosenv.2023.119862.

Umar, A. W., Naeem, M., Hussain, H., et al. (2025). Starvation from within: How heavy metals compete with essential nutrients, disrupt metabolism, and impair plant growth. *Plant Science, 353*, 112412. https://doi.org/10.1016/j.plantsci.2025.112412.

UNECE - ICP Vegetation. (2020). Heavy metals, nitrogen and POPs in European mosses: 2020 survey 27. https://unece.org/fileadmin/DAM/env/documents/2020/AIR/EMEP_WGE_Joint_Session/ECE_EB.AIR_GE.1_2020_14_E.pdf.

UNEP. (2001). The Eastern African coastal resources atlas: Kenya. Eastern African Database and Atlas Project (EAF/14), United Nations Environment Programme, Nairobi, Kenya.

United Nations Environment Programme. (2021). One in three countries in the world lack any legally mandated standards for outdoor air quality. https://www.unep.org/news-and-stories/press-release/one-three-countries-world-lack-any-legally-mandated-standards.

United States Environmental Protection Agency (USEPA). (2024). Urbanisation-overview. Retrieved September 26, 2024 from https://www.epa.gov/caddis/Urbanisation-overview.

Van Hove, L. W. A., Bossen, M. E., Mensink, M. G. J., & van Kooten, O. (1992). Physiological effects of a long term exposure to low concentrations of NH3, NO2 and SO2 on Douglas fir (*Pseudotsuga menziesii*). *Physiologia Plantarum, 86*, 559–567.

Vardoulakis, S., Fisher, B. E., Pericleous, K., & Gonzalez-Flesca, N. (2011). Modelling air quality in street canyons: A review. *Atmospheric Environment, 37*(2), 155–182. https://doi.org/10.1016/S1352-2310(02)00857-9.

Varela, Z., Aboal, J. R., Carballeira, A., Real, C., & Fernández, J. A. (2014). Use of a moss biomonitoring method to compile emission inventories for small-scale industries. *Journal of Hazardous Materials, 275*, 72–78.

Vats, S., Singh, A., Koul, M., & Uniyal, P. (2010). Study on the metal absorption by two mosses in Delhi Region (India). *Marsland Press Journal of American Science, 6*.

Vergel, K., Zinicovscaia, I., Yushin, N., Chaligava, O., Nekhoroshkov, P., & Grozdov, D. (2022). Moss biomonitoring of atmospheric pollution with trace elements in the Moscow Region, Russia. *Toxics, 10*(2), 66. https://doi.org/10.3390/toxics10020066.

Verheyden, A., Kairo, J. G., Beeckman, H., & Koedam, N. (2004). Grpwth rings, growth ring formation and age determination in the mangrove *Rhizophora mucronata*. *Annals of Botany, 94*, 59–66. https://doi.org/10.1093/aob/mch115.

Viskari, E. L., Kossi, S., & Holopainen, J. K. (2000). Norway spruce and spruce shoot aphid as indicators of traffic pollution. *Environmental Pollution, 107*, 305–314.

Vuković, G., Urošević, M. A., Goryainova, Z., Pergal, M., Škrivanj, S., Samson, R., & Popović, A. (2015). Active moss biomonitoring for extensive screening of urban air pollution: Magnetic and chemical analyses. *Science of The Total Environment, 521-522*, 200–210. https://doi.org/10.1016/j.scitotenv.2015.03.085.

Vuković, G., Urošević, M. A., Razumenic, I., Antanasijević, D. Z., & Popović, A. R. (2015). Air quality monitoring using moss biomonitors: A case study in urban areas of Belgrade & the surrounding region. *Environmental Science and Pollution Research, 22*(18), 14116–14133. https://doi.org/10.1007/s11356-015-4620-0.

Wang, M., & Liao, W. (2016). Carbon monoxide as a signaling molecule in plants. *Frontiers in plant science*, 7, 572. https://doi.org/10.3389/fpls.2016.00572.

Wang, X., Lewis, D. J., Tissue, D. T., Seemann, J. R., & Griffin, K. L. (2001). Effects of elevated atmospheric CO_2 concentration on leaf dark respiration of *Xanthium strumarium* in light and in darkness. *Proceedings of the National Academy of Sciences of the United States of America, 98*(5), 2479–2484. https://doi.org/10.1073/pnas.051622998.

Wang, H., Fu, L., Zhou, Y., Du, X., & Ge, W. (2010). Trends in vehicular emissions in China's mega cities from 1995 to 2005. *Environmental Pollution, 158*(2), 394–400.

Will-Wolf, S., Jovan, S., Neitlich, P., Peck, J. E., & Rosentreter, R. (2015). Lichen-based indices to quantify responses to climate and air pollution across northeastern U.S.A. *The Bryologist, 118*(1), 59–82.

Will-Wolf, S., Jovan, S., & Amacher, M. C. (2017). Lichen elemental content bioindicators for air quality in upper Midwest, USA: A model for large-scale monitoring. *Ecological Indicators, 78*, 253–263.

Winner, W. E., & Atkinson, C. J. (1986). Absorption of air pollution by plants, and consequences for growth. *Trends in Ecology & Evolution (Personal Edition), 1*(1), 15–18.

Wolff, C. W., & Perry, R. L. (2010). The effects of carbon monoxide exposure on plants: A review of research and applications. *Journal of Environmental Management, 91*(8), 1829–1842. https://doi.org/10.1016/j.jenvman.2010.03.021.

Worbes, M. (1995). How to measure growth dynamics in tropical trees - A review. *International Association of Wood Anatomists Journal, 16*, 337–351.

World Health Organization (WHO). (2019). Air pollution in the South-East Asia Region. WHO Regional Office for South-East Asia. Retrieved September 26, 2024 from https://www.who.int/southeastasia/health-topics/air-pollution.

World Health Organization (WHO). (2022). *Air quality guidelines: Global up*date 2021. https://www.who.int/publications/i/item/9789240034228.

World Heart Foundation. (2024). World Heart Report 2024: Air pollution and CVD. Retrieved September 26, 2024 from https://world-heart-federation.org/what-we-do/air-pollution/.

Xu, Z., Jiang, Y., Jia, B., & Zhou, G. (2016). Elevated-CO_2 response of stomata and its dependence on environmental factors. *Frontiers Plant Science, 13*(7), 657. https://doi.org/10.3389/fpls.2016.00657.

Yang, Y. S., Skelly, B. J. M., Chevone, I., & Birch, J. B. (1983). Effects of short-term ozone exposure on net photosynthesis, dark respiration, and transpiration of three eastern white pine clones. *Environment International, 9*(4), 265–269. https://doi.org/10.1016/0160-4120(83)90084-3.

Yang, J., Oh, S. O., & Hur, J. S. (2023). Lichen as bioindicators: Assessing their response to heavy metal pollution in their native ecosystem. *Mycobiology, 51*(5), 343–353.

Zhang, J., Wei, Y., & Fang, Z. (2019). Ozone pollution: a major health hazard worldwide. *Frontiers in Immunology, 10*, 2518. https://doi.org/10.3389/fimmu.2019.02518.

Zhou, T., Sun, J., & Yu, H. (2017). Temporal and spatial patterns of China's main air pollutants: Years 2014 and 2015. *Atmosphere, 8*(8), 137.

Ziegler, I. (1975). The effect of SO_2 pollution on plant metabolism. In F. A. Gunther (Ed.). *Residue reviews. Residue reviews, vol 56* (pp. 79–105). New York: Springer-Verlag. https://doi.org/10.1007/978-1-4613-9388-7_2.

CHAPTER NINE

Urban sustainability: plants help clean up the mess, what helps the plants?

Shivani Gupta[1,2], Supriya Tiwari[1], and Pradeep Kumar Mishra[2]
[1]Department of Botany, Laboratory of Ecotoxicology, Institute of Science, Banaras Hindu University, Varanasi, Uttar Pradesh, India,
[2]Department of Chemical Engineering & Technology, Green Processing & Synthesis Lab, Indian Institute of Technology, Banaras Hindu University, Varanasi, Uttar Pradesh, India

9.1 Introduction

Rapid urbanization has transformed cities into epicenters of air pollution, with elevated concentrations of tropospheric ozone (O_3), particulate matter ($PM_{2.5}$ and PM_{10}), nitrogen oxides (NO_x), sulfur dioxide (SO_2), and volatile organic compounds posing serious risks to human health and ecosystem functioning (Liang et al., 2019). Urban air pollution is now recognized as one of the most critical global environmental and public health challenges of the 21st century (Sicard et al., 2023). According to the Global Burden of Disease assessment, exposure to ambient air pollution was responsible for approximately 4.14 million nonaccidental premature deaths globally in 2019 (Vos et al., 2020), and projections suggest that pollution-related mortality could rise substantially by mid-century if effective mitigation strategies are not implemented (Lelieveld et al., 2015). With more than half of the world's population currently residing in urban areas, and nearly 70% projected to live in cities by 2050, the spatial scale and intensity of exposure to polluted air are expected to increase further (United Nations, 2018).

Among urban air pollutants, ground-level ozone (O_3), fine particulate matter ($PM_{2.5}$), and nitrogen dioxide (NO_2) are considered particularly harmful (Adjei & Afriyie, 2025). Fine and ultrafine particulate matter (PM) are of special concern because of their size-dependent toxicity; $PM_{2.5}$ and $PM_{0.1}$ can penetrate deep into the respiratory system and even enter the bloodstream (Schraufnagel, 2020). In rapidly developing regions such as East Asia and India, pollutant emissions have increased substantially over

Urban Air Pollution and Plant Responses
DOI: https://doi.org/10.1016/B978-0-443-40320-0.00004-8

recent decades (Lin et al., 2017), and billions of people remain exposed to concentrations exceeding World Health Organization guidelines (Southerland et al., 2022). Beyond human health, air pollution exerts profound impacts on biodiversity, nutrient cycling, and primary productivity, thereby influencing the sustainability of urban ecosystems (Ogwu et al., 2024).

In response to this escalating crisis, urban green infrastructure has emerged as a widely advocated nature-based solution. Plants function as natural air filters by absorbing gaseous pollutants such as O_3 and NO_2 through stomata and capturing PM on leaf surfaces (Joshi et al., 2023). Urban trees, roadside plantations, green belts, and green roofs also contribute to carbon sequestration, microclimate regulation, and enhancement of physical and psychological well-being, positioning vegetation as a critical component of sustainable city planning (Kumar et al., 2023).

However, an important paradox underlies the use of vegetation as a tool for air pollution mitigation. While plants act as sinks for airborne pollutants, they are simultaneously exposed to chronic oxidative stress induced by O_3, NO_2, and PM (Antenozio et al., 2024). Tropospheric ozone, in particular, is highly phytotoxic; it enters leaves through stomata and initiates the generation of reactive oxygen species, leading to membrane lipid peroxidation, chlorophyll degradation, impaired photosynthesis, accelerated senescence, and reduced growth (Tiwari & Agrawal, 2018). These impacts are often exacerbated in urban environments by cooccurring stressors such as elevated temperature, soil compaction, nutrient imbalance, and water limitation. Moreover, pollutant removal efficiency is strongly influenced by plant functional traits, including leaf morphology, surface roughness, wax composition, stomatal density, and canopy architecture, indicating that not all vegetation contributes equally to pollution mitigation (Singh et al., 2022).

This dual role of urban vegetation as both mitigator and victim of air pollution (Conceptually illustrated in Fig. 9.1) raises a critical yet often overlooked question in urban sustainability discourse: *if plants help clean up the urban air pollution mess, what helps the plants survive and function under polluted atmospheres?* Addressing this question is essential to ensure that green infrastructure strategies remain effective over the long term. Plant tolerance to air pollution is governed by intricate biochemical and physiological mechanisms, particularly redox regulation and antioxidant defense systems that detoxify reactive oxygen species. Additionally, external factors such as soil quality, nutrient availability, microbial

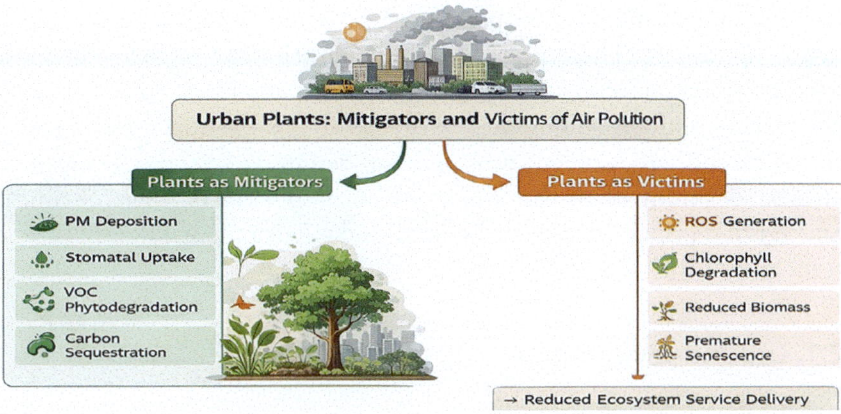

Figure 9.1 Urban plant areas are both mitigators and victims of air pollution.

associations, and adaptive plant traits significantly influence resilience under polluted conditions (Rao et al., 2025).

This chapter examines urban air pollution through the dual lens of plant-mediated pollution mitigation and plant vulnerability to pollution stress. It explores how plants contribute to urban air quality improvement, how air pollutants impair plant health, and, crucially, what biological and management strategies can be employed to support plants in polluted urban ecosystems. By reframing plants not merely as passive tools but as living infrastructure requiring resilience-building support, this chapter aims to contribute to more effective and sustainable approaches to urban sustainability. Understanding the scale and complexity of urban air pollution requires a closer examination of the specific pollutants that dominate metropolitan atmospheres and the anthropogenic processes that sustain them. Identifying the primary chemical constituents and their emission pathways provides the necessary foundation for evaluating both ecological impacts and mitigation strategies.

9.2 Major urban air pollutants and their sources

Urban air pollution is characterized by a complex mixture of gaseous and particulate contaminants originating primarily from anthropogenic activities. Among these, ground-level ozone (O_3), particulate matter ($PM_{2.5}$ and PM_{10}), nitrogen oxides (NO_x), and sulfur dioxide (SO_2) are of particular concern due to their prevalence, toxicity, and ecological impacts (Karri et al., 2024). These pollutants differ substantially in their chemical nature, emission sources, atmospheric formation processes, and ecological consequences. Understanding these characteristics is essential for evaluating their effects on urban vegetation and for designing effective mitigation strategies. A summary of the major urban air pollutants, their chemical characteristics, principal emission sources, and environmental significance is presented in Table 9.1.

9.2.1 Ground-level ozone

Tropospheric ozone (O_3) is the third most potent greenhouse gas in the atmosphere and a major surface air pollutant with significant implications for climate, atmospheric chemistry, human health, and ecosystem functioning (IPCC, 2007; Vongelis et al., 2025). Unlike stratospheric ozone, which protects the Earth from ultraviolet radiation, tropospheric ozone is a secondary pollutant formed through complex photochemical reactions in the lower atmosphere (Bernhard et al., 2023). It is also a major source of hydroxyl radicals (OH), which serve as key oxidants regulating atmospheric chemical processes. Ozone is not emitted directly into the atmosphere;

Table 9.1 Major urban air pollutants, their chemical nature, principal urban sources, key atmospheric formation characteristics, and environmental or ecological significance.

Pollutant	Chemical Nature	Major Urban Sources	Key Formation/Characteristics	Environmental and Ecological Significance
Ground-Level Ozone (O_3)	Secondary gaseous pollutant	Formed from reactions of NO_x and VOCs emitted by vehicular exhaust, industrial emissions, fuel combustion, and solvent use	Produced through photochemical reactions involving VOC oxidation, formation of peroxy radicals (RO_2/HO_2), and photolysis of NO_2 under sunlight	Strong oxidizing agent; damages plant tissues, reduces photosynthesis, contributes to climate forcing and atmospheric oxidative chemistry
Particulate Matter ($PM_{2.5}$ and PM_{10})	Mixture of solid and liquid particles suspended in air	Vehicular emissions, brake and tire wear, road dust resuspension, construction activities, biomass burning, coal power plants, and industrial processes	Classified by aerodynamic diameter: PM_{10} ≤10 μm, $PM_{2.5}$ ≤2.5 μm; often composed of heavy metals, black carbon, sulfates, nitrates, and organic compounds	Fine particles remain suspended for long periods, transport toxic compounds, impair plant surfaces, and affect atmospheric radiation balance
Nitrogen Oxides (NO_x)	Reactive gaseous nitrogen compounds (mainly NO and NO_2)	High-temperature combustion sources including vehicle engines, thermal power plants, diesel generators, and industrial boilers	Participate in atmospheric photochemical reactions; act as key precursors for tropospheric ozone and secondary particulate formation	Contribute to oxidative stress in plants, acid deposition, nutrient imbalance, and formation of photochemical smog
Sulfur Dioxide (SO_2)	Acidic gaseous pollutant	Combustion of sulfur-containing fossil fuels, coal-based power plants, metal smelting, petroleum refining, and industrial processes	Oxidizes in the atmosphere to form sulfate aerosols and sulfuric acid, contributing to acid deposition	Causes foliar injury, chlorosis, and reduced growth in plants; contributes to soil acidification and ecosystem degradation

rather, it is synthesized from its primary precursors-volatile organic compounds (VOCs) and nitrogen oxides (NO_x)-in the presence of sunlight (Gligorovski et al., 2015). VOCs include a broad range of compounds such as alkanes, alkenes, aromatic hydrocarbons, carbonyl compounds, alcohols, organic peroxides, and halogenated organics, while NO_x consists primarily of nitric oxide (NO) and nitrogen dioxide (NO_2) (Koppmann, 2020). Carbon monoxide (CO) and methane (CH_4) also contribute significantly to ozone formation by influencing radical chemistry (Isaksen et al., 2014).

The photochemical formation of ozone begins with the oxidation of VOCs or CO, leading to the production of organic peroxy (RO_2) and hydroperoxy (HO_2) radicals. These radicals react with NO to form NO_2, which subsequently undergoes photolysis under solar radiation to produce ground-state oxygen atoms ($O(^3P)$). The reaction of these oxygen atoms with molecular oxygen (O_2) results in the formation of ozone (Illmann, 2025). The primary source of hydroxyl radicals in the atmosphere is the reaction of electronically excited oxygen atoms with water vapor, initiating radical-driven oxidation cycles that sustain ozone production (Saloglu et al., 2025). Ozone formation exhibits a complex and nonlinear relationship with precursor concentrations and is strongly influenced by meteorological and environmental factors, including solar radiation intensity, atmospheric mixing, temperature, and interactions with clouds and aerosols (Dewan & Lakhani, 2022). Two principal chemical regimes govern ozone formation dynamics. In low-NO_x regimes, typically observed in rural and suburban environments, ozone concentrations increase with rising NO_x levels and are relatively less sensitive to VOC concentrations. In contrast, high-NO_x or VOC-sensitive regimes, common in dense urban and industrial areas, display a different behavior: ozone levels decrease with increasing NO_x but increase with higher VOC concentrations (Wang et al., 2018). In such environments, excess NO_2 can scavenge hydroxyl radicals, limiting peroxy radical formation and thereby altering ozone production efficiency. Consequently, ozone formation in urban atmospheres is highly dependent on radical availability and sunlight intensity (Guo et al., 2024).

9.2.2 Particulate matter ($PM_{2.5}$ and PM_{10})

PM consists of solid and liquid particles suspended in the air, varying widely in size, composition, and origin. PM_{10} refers to particles with an aerodynamic diameter of ≤10 μm, while $PM_{2.5}$ represents fine particles ≤2.5 μm (Ngagine, 2022). Fine and ultrafine particles are particularly hazardous because of their ability to penetrate deep into the respiratory tract and persist in the atmosphere for extended periods (Sharma et al., 2024). In urban environments, major sources of PM include vehicular exhaust, tire and brake wear, construction activities, road dust resuspension, industrial emissions, biomass burning, and coal-based power plants. Construction dust and poorly regulated demolition activities significantly elevate coarse particle concentrations in rapidly expanding cities (Gupta, 2019). The composition of urban PM often includes heavy metals, polycyclic aromatic hydrocarbons, sulfates, nitrates, and black carbon, which enhance its toxicity (Ali et al., 2019).

9.2.3 Nitrogen oxides and sulfur dioxide

Nitrogen oxides (primarily nitric oxide, NO, and nitrogen dioxide, NO_2) are produced mainly from high-temperature combustion processes (Denisova et al., 2019). Vehicular engines, thermal power plants, industrial boilers, and diesel generators are major contributors to urban NO_x emissions. NO_x not only exerts direct toxic effects but also acts as a precursor for ozone and secondary particulate formation (Shahbazi et al., 2021). Sulfur dioxide (SO_2) is primarily emitted from the combustion of sulfur-containing fossil fuels, especially coal and heavy oils, as well as from metal smelting and certain industrial processes (Hung & Ashner, 2020). Although SO_2 emissions have declined in some developed regions due to regulatory controls, they remain significant in areas dependent on coal-based energy systems (Kuttippurath et al., 2022).

The chemical diversity and spatial variability of urban air pollutants necessitate equally diverse mitigation approaches. Among the available strategies, vegetation-based interventions have emerged as a biologically integrated solution capable of addressing both gaseous and particulate contaminants. To appreciate their functional relevance, it is essential to examine the mechanisms through which plants interact with polluted atmospheres.

9.3 Impact of urban air pollution on plant health

Air pollution effects on plants can broadly be categorized into acute and chronic impacts, along with secondary consequences that alter plant sensitivity to other environmental stresses as described in Fig. 9.2 (Rafie-Rad et al., 2024). Acute injury typically results from short-term exposure to high pollutant concentrations and manifests as visible foliar damage,

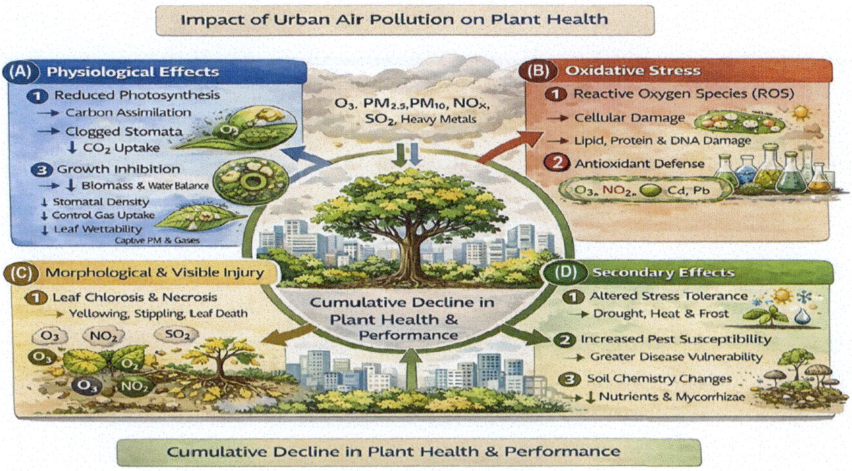

Figure 9.2 Impact of urban air pollution on plant health.

including lesions and necrotic spotting (Husen, 2022). In contrast, chronic injury arises from prolonged exposure to lower pollutant levels and is often expressed as reduced growth, suppressed yield, or impaired physiological functioning without conspicuous visible symptoms. Such chronic effects are more insidious and difficult to detect under field conditions (Patel et al., 2024). Secondary impacts include altered tolerance to abiotic stresses such as drought, heat, and frost, as well as increased susceptibility to pests and pathogens. Additionally, air pollution influences soil chemistry through acidification, nitrogen deposition, and reductions in microbial diversity, including mycorrhizal fungi, thereby indirectly affecting plant performance (Ahmad et al., 2019).

9.3.1 Physiological effects

9.3.1.1 Reduced photosynthesis and stomatal dysfunction

One of the earliest and most consistent physiological impacts of air pollutants is the reduction in photosynthetic activity (Dhir, 2016). Gaseous pollutants such as O_3, NO_2, and SO_2 enter the leaf through stomata, where they interfere with carbon assimilation and pigment stability (Husen, 2021). Ozone, in particular, causes a decline in net assimilation rate and disrupts chloroplast integrity. Stomatal closure under high pollutant stress reduces CO_2 availability, thereby suppressing carbon fixation (Nowroz et al., 2024).

PM deposition further compounds this stress by clogging stomatal pores and altering light absorption and reflection of photosynthetically active radiation (PAR), thereby disturbing normal metabolic processes (Singh & Verma, 2023). Elevated NO_2 concentrations may lead to nitrite accumulation and cellular acidification, resulting in inhibition of nitrogen assimilation and growth suppression (Feng et al., 2020).

9.3.1.2 *Altered transpiration and growth inhibition*

Air pollutants disrupt transpiration dynamics by degrading leaf surface waxes and increasing nonstomatal water loss (Singh & Singh, 2025). Structural changes such as altered epidermal cell size, increased stomatal density, and reduced leaf surface area have been reported in polluted environments. These alterations compromise water relations and thermoregulation (Rai et al., 2010). Chronic exposure leads to reduced biomass accumulation, shorter petiole length, smaller leaf area, delayed flowering, and diminished reproductive success (Jochner et al., 2015).

9.3.2 Biochemical and oxidative stress responses

9.3.2.1 *Reactive oxygen species generation under ozone and particulate matter exposure*

It is well documented that exposure to urban air pollutants triggers the overproduction of reactive oxygen species (ROS) in plant tissues (Rao et al., 2025). Ozone, being a powerful oxidizing agent, reacts with cellular components either directly or through radical-mediated pathways, leading to oxidative burst within leaf tissues (Tiwari & Agrawal, 2018). The interaction of O_3 with plant tissues occurs in three stages: diffusion through the boundary layer, stomatal uptake, and intracellular reactions involving detoxification and repair mechanisms (Wedow et al., 2021). Excess ROS accumulation results in lipid peroxidation of chloroplast membranes, protein oxidation, DNA damage, and enzyme inhibition (Grulke & Heath, 2020). PM and heavy metals such as Cd and Pb similarly induce oxidative stress, interfere with hormone balance, and disrupt nutrient uptake (Molina & Segura, 2021).

9.3.2.2 *Antioxidant defense system*

Plants possess a multitiered antioxidative system that counteracts ROS toxicity. Enzymatic antioxidants such as superoxide dismutase, catalase, ascorbate peroxidase, and glutathione reductase work in coordination with nonenzymatic antioxidants to restore cellular redox homeostasis

(Madheshiya et al., 2023). The balance between ROS production and antioxidant capacity determines plant tolerance; when ROS exceeds neutralizing capacity, oxidative injury becomes inevitable (Hasanuzzaman et al., 2020). Ozone exposure accelerates chlorophyll degradation, partly due to magnesium ion displacement and conversion of chlorophyll to pheophytin, altering light-harvesting efficiency. Decline in chlorophyll content and impairment of pigment stability are among the most consistent biochemical indicators of pollutant stress (Tiwari & Agrawal, 2018).

9.3.3 Morphological and visible injury symptoms

9.3.3.1 Leaf chlorosis, necrosis, and premature senescence

Visible foliar injury is often the most apparent manifestation of pollutant stress. Ozone-induced symptoms include stippling, pale yellow patches, interveinal chlorosis, bronzing, reddening, white patches, and necrotic lesions that eventually lead to premature senescence (Karmakar et al., 2022). Older leaves are typically more susceptible. SO_2 and NO_2 exposure can cause chlorosis, membrane damage, and chlorophyll degradation (Mulenga et al., 2020). In severe cases, whole-leaf necrosis and defoliation may occur. Heavy metal accumulation leads to leaf rolling, reduced stomatal density, increased cuticle thickness, and restricted water flow, resulting in water stress symptoms (Guo et al., 2023).

9.3.3.2 Reduced canopy density and biomass

Long-term exposure to polluted atmospheres results in reduced canopy density, smaller leaf area, and diminished biomass production. Early leaf fall and impaired reproductive development contribute to overall reduction in plant vigor and ecosystem service delivery (Emberson et al., 2018). Such structural decline compromises the ability of vegetation to capture pollutants effectively, thereby creating a feedback loop in which pollution weakens the very biological systems intended to mitigate it (Smith et al., 2013).

9.4 Role of plants in mitigating urban air pollution

Urban vegetation constitutes a critical biophysical interface between anthropogenic emission sources and the atmospheric environment. In rapidly urbanizing regions, where vehicular exhaust, industrial discharge, construction activities, and domestic combustion elevate ambient pollutant concentrations, plants function as natural biofilters capable of attenuating both particulate and gaseous contaminants (Ferrini et al., 2020). Unlike engineered air purification systems, vegetative mitigation operates continuously, requires minimal external energy input, and simultaneously delivers ecosystem cobenefits including carbon sequestration, microclimate regulation, biodiversity enhancement, and human health improvement (Singh et al., 2022). The capacity of plants to mitigate urban air pollution is mediated through a combination of physical interception, stomatal uptake, biochemical transformation, and rhizosphere-assisted degradation. These processes vary according to pollutant type, plant functional traits, and prevailing environmental conditions (Zayed et al., 2025).

9.4.1 Particulate matter interception and deposition

The removal of PM represents one of the most extensively documented ecosystem services provided by urban vegetation. Airborne particles are captured through impaction, gravitational settling, diffusion, and electrostatic attraction onto plant surfaces (Kim et al., 2023). Although all aerial plant organs contribute to atmospheric particle capture, leaves represent the primary interface for PM interception due to their extensive surface area and structural heterogeneity (del Carmen Redondo-Bermúdez et al., 2021). Leaf micromorphological traits play a decisive role in determining deposition efficiency. Among these, trichomes are frequently associated with enhanced PM accumulation, as they increase surface roughness, promote particle collision, and reduce resuspension (Roy et al., 2025). Similarly, surface features such as epicuticular waxes, grooves, crypts, ridges, furrows, and stomatal depressions facilitate the entrapment of both coarse and fine particles, including metal-associated fractions. In contrast, leaves with smooth surfaces generally exhibit lower retention capacity (Paull, 2020).

However, the relationship between individual traits and PM capture is not universally consistent (Li et al., 2019). While several studies report positive associations between trichome density and PM accumulation, others have observed neutral or even negative correlations in certain species, suggesting that particle retention depends on the interaction between surface structure, particle size, and aerodynamic conditions rather than a single morphological feature (Zhang et al., 2021).

Epicuticular wax characteristics further influence deposition dynamics. In many species, higher wax loads correspond with greater particle accumulation, yet wax chemistry and surface hydrophobicity can also reduce adhesion under certain conditions. Highly hydrophobic surfaces may limit initial particle attachment, whereas wax structures such as thin films, platelets, or tubules differ in their retention efficiency (Gawrońska & Bakera, 2015). Fine particles, particularly those $<10\ \mu m$, may become partially embedded within wax layers or incorporated into the cuticle, reducing the likelihood of wash-off. Because wax composition and ultrastructure vary with genotype, developmental stage, and environmental exposure, PM encapsulation represents a complex and dynamic process (Chen et al., 2024).

Leaf wettability also contributes to deposition patterns. Greater wettability on the adaxial (upper) surface often results in higher particle accumulation compared with the typically more hydrophobic abaxial surface (Li et al., 2021). In addition to micromorphology, leaf macrostructure influences aerodynamic interactions. Small, lobed, or intricately shaped leaves generate localized turbulence that enhances particle collision and retention (Barima et al., 2016; Wang et al., 2013). Needle-like leaves of conifers can similarly increase contact efficiency for fine particles, particularly during winter months when deciduous species are leafless, and pollution levels are elevated. Nonetheless, some conifer species exhibit sensitivity to pollutant-induced injury, limiting their suitability for highly contaminated environments (Cui & Adams, 2025).

Leaf shape has also been implicated in PM deposition efficiency, with certain forms, such as lanceolate leaves, potentially enhancing turbulence-mediated capture under windy conditions. Moreover, leaf developmental stage significantly affects retention capacity (Pace & Grote, 2020). Fully expanded leaves generally accumulate more PM than immature foliage, as surface properties-including wax structure, angle, and wettability-stabilize with maturation. Seasonal changes and chronic pollutant exposure may further modify wax thickness and surface characteristics, altering deposition efficiency over time. While surface deposition represents a critical initial barrier against airborne pollutants, the role of vegetation extends beyond passive interception (Ghosh et al., 2023). Internal physiological processes, including stomatal regulation and biochemical detoxification pathways, further determine the extent to which absorbed pollutants are neutralized or metabolically assimilated (Mulenga et al., 2020).

9.4.2 Gaseous pollutant uptake and biochemical assimilation

In addition to particulates, urban vegetation mitigates a range of gaseous pollutants, including sulfur dioxide (SO_2), nitrogen dioxide (NO_2), ozone

(O_3), and carbon dioxide (CO_2). These gases primarily enter leaf tissues through stomatal apertures during gas exchange. Once internalized, their fate depends on reactivity, solubility, and plant metabolic capacity (Meo et al., 2024). Carbon dioxide mitigation through photosynthetic fixation remains the most globally significant pathway, contributing to both short-term carbon assimilation and long-term sequestration in woody biomass and soil organic matter. In urban ecosystems, tree-based carbon storage plays an increasingly important role in offsetting localized emissions (Wasikoyo et al., 2026). NO_2 may be assimilated into nitrogen metabolism following conversion to nitrate and nitrite, subsequently incorporated into amino acids via nitrate and nitrite reductase pathways (Kumari et al., 2022). Sulfur dioxide can similarly enter sulfur assimilation pathways at subtoxic concentrations. However, assimilation capacity is species-dependent and influenced by physiological status and environmental context (Sharma et al., 2020).

9.4.3 Phytodegradation of volatile organic compounds

Vegetation contributes to VOC removal through stomatal absorption, cuticular sorption, and microbially mediated degradation in the rhizosphere. Following uptake, certain VOCs undergo enzymatic transformation through oxidative metabolic pathways often conceptualized under the "green liver" model (Kim et al., 2018). Phase I reactions, mediated by enzymes such as cytochrome P450 monooxygenases, introduce reactive functional groups, increasing compound solubility. Subsequent conjugation and compartmentalization reduce toxicity and facilitate sequestration or further metabolism (Molina & Segura, 2021). The rhizosphere substantially enhances VOC degradation efficiency. Root exudates stimulate diverse microbial consortia capable of mineralizing organic pollutants (Baloch et al., 2024). This plant–microbe synergy forms the basis of many biofiltration systems, including green walls and constructed botanical filters, underscoring the integrative role of vegetation in atmospheric remediation (Montaluisa-Mantilla et al., 2025).

9.5 What helps plants clean up urban air pollution?

The capacity of Plants to mitigate urban air pollution depends not merely on their presence, but on a suite of functional traits operating across multiple biological scales, from micromorphological leaf features to whole-canopy architecture and plant–microbe interactions (Buraerah et al., 2023). Pollutant removal efficiency is therefore governed by intrinsic plant characteristics, structural complexity, and ecological context. Understanding these enabling mechanisms is essential for optimizing urban green infrastructure performance (Weyens et al., 2015). The major factors influencing plant capacity for urban air pollution mitigation are summarized in Fig. 9.3.

9.5.1 Leaf-level traits enhancing pollutant capture and retention

Because most airborne PM initially deposits on leaf surfaces, micromorphological characteristics strongly influence removal efficiency (Roy et al., 2025). Species-specific variation in deposition velocity (Vd) has been shown to differ by orders of magnitude across taxa, driven largely by leaf surface properties and vegetation form. Surface roughness, trichome (leaf-hair) density, epicuticular wax content, and groove architecture collectively determine particle interception and retention capacity (Ossola & Farmer, 2024). Rough and hairy leaves create microscale aerodynamic resistance, slowing particle movement and enhancing impaction (Boontanom & Tafti, 2022). Experimental studies demonstrate that higher trichome density significantly increases both PM deposition velocity and total retention (Kim et al., 2020; Zhang et al., 2021). Grooved and ridged surfaces of

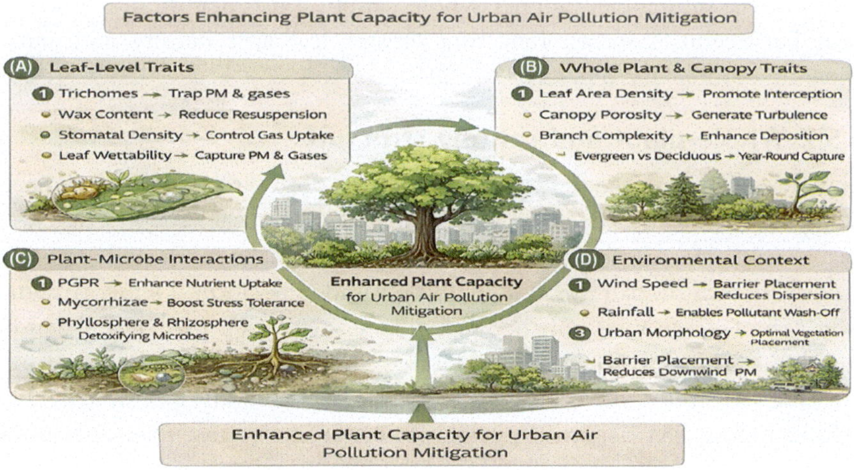

Figure 9.3 Factors enhancing plant capacity for urban air pollution mitigation.

leaves further trap particulate pollutants, particularly when groove dimensions exceed particle diameter (Shao et al., 2019).

Epicuticular waxes play a particularly critical role in long-term retention. Approximately half of the deposited PM may become embedded within the wax layer, depending on species and particle size. Wax content alone has been shown to explain up to 60%–90% of interspecific variation in encapsulated PM in controlled studies (Chen et al., 2024). Encapsulation reduces resuspension, thereby contributing to more permanent removal of PM from the atmosphere. Moreover, waxy surfaces exhibit lower wash-off rates during rainfall events, maintaining higher cumulative retention over time (Santos et al., 2021). Leaf architecture also influences resuspension dynamics. Shorter petiole length reduces leaf flutter under wind stress, minimizing re-entrainment of deposited particles (Zheng & Li, 2019). Complex foliage, particularly needle leaf conifers, enhances turbulence and particle capture through higher Stokes numbers associated with narrow needle geometry. Consequently, coniferous species often exhibit significantly higher deposition velocities compared to broadleaved deciduous trees (Zhang et al., 2020).

For gaseous pollutants, stomatal characteristics are central. Deposition is largely controlled by stomatal conductance (gs) and boundary layer resistance (Nargis & Singh, 2024). Stomatal density, size, and responsiveness to environmental drivers (radiation, humidity, soil moisture) regulate internal uptake (Prithviraj & Reshmi, 2025). Forest canopies typically demonstrate substantially higher whole-canopy NO_2 deposition velocities than herbaceous systems, reflecting greater leaf area density and aerodynamic complexity (Hicks et al., 2016).

9.5.2 Whole-plant and canopy structure

Beyond individual leaves, plant form and canopy architecture influence pollutant removal at the stand scale (Barwise & Kumar, 2020). High leaf area density increases the total interception surface while simultaneously modifying airflow patterns within and around vegetation. Dense, structurally complex canopies enhance turbulence, increasing particle impaction and diffusion (del Carmen Redondo-Bermúdez et al., 2021). Results from metaanalysis studies further reveal that evergreen woody species generally exhibit higher PM deposition velocities per unit leaf area than deciduous species, while shrubs often outperform trees on a leaf-area basis due to their dense foliage arrangement (Cai et al., 2017; Wang et al., 2024). Mixed vegetation systems, combining shrubs, trees, and climbers, provide vertical stratification and year-round surface availability, increasing cumulative pollutant capture (Aduse-Poku, 2025). Branch complexity further enhances deposition, particularly at moderate to high wind speeds. Needle leaf conifers, due to their fine-scale branching patterns, generate higher aerodynamic resistance and promote efficient capture of coarse and fine particles alike (Bonan, 2019). However, optimal canopy design must balance density and permeability; overly dense barriers may reduce airflow dispersion and inadvertently trap pollutants at the pedestrian level. Therefore plant structural traits must be evaluated in relation to local airflow dynamics and urban morphology (Ankaya, 2025).

9.5.3 Plant–microbe interactions and pollutant transformation

Vegetation does not act solely as a passive depositional surface. Phyllospheric (leaf-surface) and rhizospheric microbial communities associated with urban plants contribute to pollutant transformation and detoxification (Wei et al., 2017). Certain microbial taxa are capable of degrading organic pollutants, including components of PM and polycyclic aromatic hydrocarbons (Shahsavari et al., 2019). Root-associated microbes can immobilize heavy metals and enhance nutrient acquisition, indirectly strengthening plant stress tolerance (Upadhyay, 2025). Endophytic bacteria have been shown to improve tolerance to reactive nitrogen species by enhancing nitrogen assimilation pathways and moderating oxidative stress. These interactions suggest that pollutant mitigation is partly a biologically mediated process. The efficiency of urban vegetation may therefore depend not only on plant traits but also on the composition and functionality of associated microbial consortia (Monaco et al., 2024).

9.5.4 Influence of environmental context and green infrastructure design

The translation of beneficial plant traits into measurable air quality improvement depends strongly on spatial configuration and environmental context (Barwise & Kumar, 2020). Vegetation barriers positioned near traffic sources have been shown to reduce downwind $PM_{2.5}$ concentrations by more than 15%, particularly when dense shrub layers are established on the windward side. Long-term roadside studies demonstrate that PM reduction increases significantly after canopy maturation, with dense hedgerows achieving 35%–50% reductions in certain particulate fractions (Wang et al., 2019). Optimal barrier systems typically integrate permeable yet dense shrub layers with taller trees to generate turbulence without impeding vertical pollutant dispersion (Barwise & Kumar. 2021). Additionally, phenology plays a critical role: deciduous species exhibit seasonal variability in pollutant capture, whereas evergreen systems provide year-round functionality (Joshi et al., 2024). Rainfall and wind events influence the net removal of air pollutants by facilitating wash-off and resuspension. Species whose morphology supports strong initial capture while allowing periodic wash-off followed by re-accumulation may achieve higher long-term cumulative removal (Xu et al., 2017).

9.6 Urban green infrastructure as a strategic tool

Urban green infrastructure (GI) has emerged globally as a structured, nature-based response to deteriorating urban air quality (Idris et al., 2025). As an organized initiative to introduce ecological systems into built environments, GI integrates trees, shrubs, lawns, green roofs, and green walls into city planning frameworks to mitigate atmospheric pollutants and regulate greenhouse gas emissions (Malik & Singh, 2026). Foundational studies across North America, Europe, and Asia demonstrated that strategically placed vegetation significantly reduces urban pollutant loads while simultaneously delivering ecosystem cobenefits (Herath & Bai, 2024).

The air quality improvement achieved through GI operates via both direct and indirect mechanisms. Direct benefits arise from pollutant uptake and deposition (Hewitt et al., 2020). Gaseous contaminants such as ozone, nitrogen dioxide, and sulfur dioxide are absorbed through leaf stomata, where they may be detoxified or assimilated. PM is intercepted through dry deposition processes, adhering to leaf surfaces or being absorbed into wax layers, often aided by wind-driven impaction (Husen, 2021). The efficiency of this removal depends on canopy density, leaf morphology, surface roughness, and local meteorological conditions (Li et al., 2022). Indirect benefits stem from microclimatic regulation. Vegetation moderates urban heat islands through shading and evapotranspiration, reducing ambient temperatures during summer months. This cooling effect decreases electricity demand for air conditioning, which in turn lowers emissions from thermal power generation (Irfeey et al., 2023). Thus the cumulative impact of GI on air quality encompasses both pollutant removal and emission avoidance. Several studies emphasize that these combined effects can be quantified not only in physical pollutant flux units but also in economic terms, strengthening the policy relevance of GI investments (Akbari et al., 2001; Foster et al., 2014; Tallis et al., 2015).

To systematically quantify pollution mitigation capacity, researchers frequently employ dry deposition modeling approaches. Among the most widely applied tools is the urban forest effects (UFORE) model, which integrates field inventory data, meteorological parameters, and pollutant concentrations to estimate removal rates (Nowak et al., 2003; Ochoa-Alvarado et al., 2025). Building upon UFORE, the i-Tree Eco model was later developed by the United States Forest Service as a peer-reviewed, open-source platform capable of assessing pollutant removal, carbon sequestration, carbon storage, and associated monetary valuation (Alvarado, 2019; Jayasooriya et al., 2017). Initially implemented across multiple U.S. cities, i-Tree Eco has since been adapted for international contexts, including an Australian-compatible version integrating localized air quality and meteorological datasets (Jayasooriya et al., 2017; Lovell & Taylor, 2013).

While modeling frameworks provide valuable large-scale estimations, their outputs depend strongly on accurate species-specific parameters and local environmental inputs (Zappitelli et al., 2023). This highlights an important consideration: GI effectiveness is not uniform across urban landscapes. Spatial configuration, species composition, maintenance practices, and long-term plant health collectively determine performance (Ashinze et al., 2024). Poorly maintained or physiologically stressed vegetation may exhibit reduced pollutant uptake efficiency, underscoring the necessity of integrating ecological resilience into GI planning (Ogwu & Kosoe, 2025).

In essence, urban green infrastructure represents a strategic intersection of ecology, atmospheric science, and urban planning. When thoughtfully designed and supported by robust modeling and monitoring systems, GI functions as a cost-effective, multifunctional tool for improving urban air quality (Korkou et al., 2023). However, its sustained success depends not only on expanding green cover but also on ensuring that vegetation remains physiologically capable of delivering these ecosystem services under increasingly complex urban stress conditions (Coutts & Hahn, 2015).

9.7 Sustaining plant performance under urban pollution: integrated management strategies

The long-term capacity of urban vegetation to mitigate atmospheric pollution depends on maintaining structural integrity, metabolic continuity, and physiological stability under chronic environmental stress (Gaur & Singh, 2024). While plants possess inherent antioxidative and repair mechanisms, continuous exposure to ozone, nitrogen oxides, particulate deposition, elevated temperatures, and soil degradation can progressively weaken photosynthetic ability and canopy structure. Strategic management interventions are therefore essential to preserve pollutant removal efficiency over time (Chaudhuri & Roy, 2025).

9.7.1 Soil quality as the foundational determinant

Urban soils are frequently compacted, low in organic carbon, poorly structured, and contaminated with construction residues or heavy metals (Yang & Zhang, 2015). Such conditions restrict root expansion, limit microbial diversity, and impair water and nutrient acquisition (Zandybay et al., 2024). Because stomatal conductance, antioxidant enzyme synthesis, and biomass production are closely linked to root health, soil conditions fundamentally determine atmospheric mitigation capacity (Yadav et al., 2022). Amendments such as compost, vermicompost, and structured organic substrates improve aggregation, porosity, and water-holding capacity. Enhanced rhizosphere conditions support root respiration, nutrient cycling, and sustained canopy development, thereby maintaining leaf area available for particulate interception and gaseous uptake (Singh et al., 2022).

9.7.2 Biochar and rhizosphere redox regulation

Biochar represents a particularly promising soil amendment in urban systems. Its high surface area, porosity, and cation exchange capacity enhance nutrient retention and moisture stability. Improved aeration supports root proliferation and microbial colonization (Senadheera et al., 2024). Beyond structural improvements, biochar influences rhizosphere redox processes. By stabilizing nutrient availability and moderating heavy metal mobility, biochar reduces cumulative root stress. Improved root-zone function supports balanced stomatal regulation and antioxidant metabolism in aerial tissues (Etesami et al., 2025). Stabilized redox homeostasis contributes indirectly to better management of reactive oxygen species generated during pollutant exposure, preserving chlorophyll integrity and photosynthetic efficiency (Rao et al., 2025).

9.7.3 Microbial inoculation and plant–microbe synergy

The rhizosphere and phyllosphere microbiomes contribute significantly to plant performance in polluted environments (Zhu et al., 2022). Plant growth-promoting rhizobacteria and mycorrhizal fungi enhance nutrient uptake efficiency, improve water acquisition, and modulate hormonal signaling pathways involved in stress responses (Chieb & Gachomo, 2023). Certain microbial consortia can degrade organic contaminants, immobilize heavy metals, and influence nitrogen cycling, thereby reducing cumulative stress burden. Strengthening beneficial microbial communities enhances root vitality and supports sustained gas exchange and pollutant uptake processes (Chaudhary et al., 2023).

9.7.4 Hydrological management and stomatal function

Water availability directly regulates stomatal conductance, which controls both photosynthetic carbon assimilation and gaseous pollutant uptake (Lawson et al., 2010). Under drought conditions, stomatal closure limits ozone and nitrogen dioxide entry but simultaneously reduces carbon fixation and antioxidant regeneration capacity (Pandey & Ghosh, 2026). In highly impervious urban landscapes, regulated irrigation helps maintain physiological continuity without promoting excessive vegetative growth. Hydrological stability preserves leaf area, antioxidant activity, and gas exchange dynamics required for long-term mitigation (Hilaire et al., 2008).

9.7.5 Species selection and functional screening

Species variability strongly influences both tolerance to pollution and removal efficiency. Selection criteria should integrate structural traits (leaf morphology, canopy density, deposition velocity) with physiological indicators of pollutant tolerance (Rodríguez-Santamaría et al., 2022). Indices such as the air pollution tolerance index, incorporating chlorophyll content, relative water content, leaf extract pH, and ascorbic acid concentration, provide quantitative frameworks for screening species (Kour & Adak, 2023). Integrating tolerance metrics with functional performance parameters enhances the durability of green infrastructure systems and reduces long-term maintenance costs (Wang et al., 2023).

9.8 Conclusion

Urban air pollution remains a growing challenge in rapidly expanding cities, where emissions from transportation, industry, construction, and energy production continue to degrade atmospheric quality. Urban vegetation offers an effective nature-based solution for mitigating this problem by removing pollutants through particulate interception on leaf surfaces, stomatal uptake of gaseous pollutants, biochemical assimilation, and rhizosphere-mediated degradation. Through these processes, plants function as a dynamic biological interface between the urban atmosphere and terrestrial ecosystems, significantly contributing to the regulation of urban air quality.

However, the interaction between plants and polluted environments is inherently bidirectional. While vegetation helps remove atmospheric pollutants, exposure to ozone, nitrogen oxides, sulfur dioxide, and PM simultaneously imposes physiological and biochemical stress on plants. These pollutants disrupt photosynthesis, alter stomatal functioning, trigger oxidative stress through reactive oxygen species generation, and lead to visible foliar injury and reduced growth. Over time, such stress can weaken plant health and diminish the efficiency with which vegetation captures and assimilates pollutants. The effectiveness of vegetation-based pollution mitigation therefore depends on multiple factors, including plant functional traits, canopy architecture, and plant–microbe interactions. Leaf surface characteristics such as trichomes, surface roughness, and epicuticular waxes influence particulate retention, while stomatal behavior regulates gaseous pollutant uptake. At larger scales, vegetation density, canopy structure, and species composition determine aerodynamic interactions and deposition dynamics. Equally critical are the ecological and management conditions that sustain plant performance in polluted environments. Soil quality, nutrient availability, water management, microbial associations, and amendments such as biochar play essential roles in maintaining root health, antioxidant defense systems, and photosynthetic stability. Strategic species selection based on tolerance indices and functional traits further enhances the long-term efficiency of urban green infrastructure.

Ultimately, the success of vegetation-based air pollution mitigation depends not only on increasing urban green cover but also on maintaining the physiological resilience of the plants themselves. Urban sustainability therefore requires viewing vegetation as living ecological infrastructure supported by healthy soils, functional microbial communities, and informed management practices to ensure that plants can continue to perform their essential role in improving urban air quality.

References

Adjei, F. A., & Afriyie, A. (2025). Air pollution as a global health risk: Emerging evidence, interventions and policy insights. *World Journal of Advanced Research and Reviews*, *27*(01), 1925–1940.

Aduse-Poku, M. (2025). *Facade greening: Investigating the potential of climbing plants for air pollution mitigation and water-stress-dependent evapotranspiration* (Doctoral dissertation, Universität zu Köln).

Ahmad, G., Nishat, Y., Haris, M., Danish, M., & Hussain, T. (2019). Efficiency of soil, plant and microbes for healthy plant immunity and sustainable agricultural system. In *Plant microbe interface.* Cham: Springer International Publishing, 325–346.

Akbari, H., Pomerantz, M., & Taha, H. (2001). Cool surfaces and shade trees to reduce energy use and improve air quality in urban areas. *Solar energy, 70*(3), 295–310.

Ali, M. U., Liu, G., Yousaf, B., Ullah, H., Abbas, Q., & Munir, M. A. M. (2019). A systematic review on global pollution status of particulate matter-associated potential toxic elements and health perspectives in urban environment. *Environmental Geochemistry and Health, 41*(3), 1131–1162.

Alvarado, O.A. (2019). *Measuring the benefits of urban nature-based solutions through quantitative assessment tools* (Master's thesis).

Ankaya, F. (2025). Relationships between leaf structural and functional traits of urban landscape plant species: Implications for sustainable landscape planning. *BMC Plant Biology, 25*(1), 1300.

Antenozio, M. L., Caissutti, C., Caporusso, F. M., Marzi, D., & Brunetti, P. (2024). Urban air pollution and plant tolerance: Omics responses to ozone, nitrogen oxides, and particulate matter. *Plants, 13*(15), 2027.

Ashinze, U. K., Edeigba, B. A., Umoh, A. A., Biu, P. W., & Daraojimba, A. I. (2024). Urban green infrastructure and its role in sustainable cities: A comprehensive review. *World Journal of Advanced Research and Reviews, 21*(2), 928–936.

Baloch, F. B., Zeng, N., Gong, H., Zhang, Z., Zhang, N., Baloch, S. B., Ali, S., & Li, B. (2024). Rhizobacterial volatile organic compounds: Implications for agricultural ecosystems' nutrient cycling and soil health. *Heliyon, 10*(23), e40522. https://doi.org/10.1016/j.heliyon.2024.e40522.

Barima, Y. S. S., Angaman, D. M., N'gouran, K. P., Koffi, N. G. A., Bi, F. Z. T., & Samson, R. (2016). Involvement of leaf characteristics and wettability in retaining air particulate matter from tropical plant species. *Environmental Engineering Research, 21*(2), 121–131.

Barwise, Y., & Kumar, P. (2020). Designing vegetation barriers for urban air pollution abatement: A practical review for appropriate plant species selection. *Npj Climate and Atmospheric Science, 3*(1), 12.

Bernhard, G. H., Bais, A. F., Aucamp, P. J., Klekociuk, A. R., Liley, J. B., & McKenzie, R. L. (2023). Stratospheric ozone, UV radiation, and climate interactions. *Photochemical & Photobiological Sciences, 22*(5), 937–989.

Bonan, G. (2019). *Climate change and terrestrial ecosystem modeling.* Cambridge University Press.

Boontanom, J., & Tafti, D. K. (2022). A Micro-Scale Computational Investigation on the Effects of Leaf Hair on PM2. 5 Deposition. In *Proceedings of the ASME 2022 Fluids Engineering Division Summer Meeting. Volume 1: Fluid Applications and Systems (FASTC); Fluid Measurement and Instrumentation (FMITC); Fluid Mechanics (FMTC). Toronto, Ontario, Canada. August 3–5, 2022.* American Society of Mechanical Engineers. https://doi.org/10.1115/FEDSM2022-87200.

Buraerah, M. F., Patandjengi, B., Suryani, S., Hamzah, A., & Demmalino, E. B. (2023). The effect of vegetation in reducing air pollution in an urban environment: A review. *In IOP Conference Series: Earth and Environmental Science, Vol. 1253*(No. 1), 012105 IOP Publishing.

Cai, M., Xin, Z., & Yu, X. (2017). Spatio-temporal variations in PM leaf deposition: A meta-analysis. *Environmental Pollution, 231*, 207–218.

Chaudhary, P., Xu, M., Ahamad, L., Chaudhary, A., Kumar, G., Adeleke, B. S., Verma, K. K., Hu, D. M., Širić, I., Kumar, P., & Popescu, S. M. (2023). Application of synthetic consortia for improvement of soil fertility, pollution remediation, and agricultural productivity: A review. *Agronomy, 13*(3), 643.

Chaudhuri, S., & Roy, M. (2025). Ambient air quality: Trees could help… but what goes with the trees? reflections, future challenges, potential opportunities. *Environmental Quality Management, 35*(1), e70161.

Chen, D., Yan, J., Sun, N., Sun, W., Zhang, W., Long, Y., & Yin, S. (2024). Selective capture of PM2. 5 by urban trees: The role of leaf wax composition and physiological traits in air quality enhancement. *Journal of Hazardous Materials, 478*, 135428.

Chieb, M., & Gachomo, E. W. (2023). The role of plant growth promoting rhizobacteria in plant drought stress responses. *BMC Plant Biology, 23*(1), 407.

Coutts, C., & Hahn, M. (2015). Green infrastructure, ecosystem services, and human health. *International Journal of Environmental Research and Public Health, 12*(8), 9768–9798.

Cui, S., & Adams, M. (2025). Dry deposition of fine particulate matter by city-owned street trees in a city defined by urban sprawl. *Land, 14*(10), 1969.

del Carmen Redondo-Bermúdez, M., Gulenc, I. T., Cameron, R. W., & Inkson, B. J. (2021). Green barriers' for air pollutant capture: Leaf micromorphology as a mechanism to explain plants capacity to capture particulate matter. *Environmental pollution, 288*, 117809.

Denisova, K. O., Ilyin, A. A., Rumyantsev, R. N., Ilyin, A. P., & Volkova, A. V. (2019). Nitrous oxide: Production, application, and protection of the environment. *Russian Journal of General Chemistry, 89*(6), 1338–1346.

Dewan, S., & Lakhani, A. (2022). Tropospheric ozone and its natural precursors impacted by climatic changes in emission and dynamics. *Frontiers in Environmental Science, 10*, 1007942.

Dhir, B. (2016). Air pollutants and photosynthetic efficiency of plants. In *Plant responses to air pollution.* Singapore: Springer Singapore, 71–84.

Etesami, H., Hejazi Mahabadi, S., Otabek, U., Zahro, B., Tulkin, O., & Muhammad, A. (2025). Biochar as a rhizosphere modulator: Enhancing root resilience to multifaceted environmental stresses for sustainable agriculture. *Journal of Soil Science and Plant Nutrition, 25*, 10330–10363. https://doi.org/10.1007/s42729-025-02796-z.

Feng, H., Fan, X., Miller, A. J., & Xu, G. (2020). Plant nitrogen uptake and assimilation: Regulation of cellular pH homeostasis. *Journal of Experimental Botany, 71*(15), 4380–4392.

Ferrini, F., Fini, A., Mori, J., & Gori, A. (2020). Role of vegetation as a mitigating factor in the urban context. *Sustainability, 12*(10), 4247.

Gaur, R., & Singh, H. (2024). Urban tree responses to climate change and environmental pollution. In *Urban forests, climate change and environmental pollution: Physio-biochemical and molecular perspectives to enhance urban resilience.* Cham: Springer Nature Switzerland, 279–305.

Gawrońska, H., & Bakera, B. (2015). Phytoremediation of particulate matter from indoor air by *Chlorophytum comosum* L. plants. *Air Quality. Atmosphere & Health, 8*(3), 265–272.

Ghosh, S., Dutta, R., & Mukhopadhyay, S. (2023). A review on seasonal changes in particulate matter accumulation by plant bioindicators: Effects on leaf traits. *Water, Air, & Soil Pollution, 234*(8), 529.

Gligorovski, S., Strekowski, R., Barbati, S., & Vione, D. (2015). Environmental implications of hydroxyl radicals (• OH). *Chemical Reviews, 115*(24), 13051–13092.

Grulke, N. E., & Heath, R. L. (2020). Ozone effects on plants in natural ecosystems. *Plant Biology, 22*, 12–37.

Guo, J., Zhang, G., Hu, R., Xie, P., Hu, C., Cai, H., Wang, Y., Tang, H., Li, X., Huang, X., & Liu, X. (2024). Local radical chemistry driven ozone pollution in a megacity: A case study. *Atmospheric Environment, 318*, 120227.

Guo, Z., Gao, Y., Yuan, X., Yuan, M., Huang, L., Wang, S., Liu, C. E., & Duan, C. (2023). Effects of heavy metals on stomata in plants: A review. *International Journal of Molecular Sciences, 24*(11), 9302.

Gupta, V. (2019). Vehicle-generated heavy metal pollution in an urban environment and its distribution into various environmental components. In *Environmental concerns and sustainable development: Volume 1: air, water and energy resources*. Singapore: Springer Singapore, 113–127.

Hasanuzzaman, M., Bhuyan, M. B., Zulfiqar, F., Raza, A., Mohsin, S. M., Mahmud, J. A., Fujita, M., & Fotopoulos, V. (2020). Reactive oxygen species and antioxidant defense in plants under abiotic stress: Revisiting the crucial role of a universal defense regulator. *Antioxidants, 9*(8), 681.

Herath, P., & Bai, X. (2024). Benefits and co-benefits of urban green infrastructure for sustainable cities: Six current and emerging themes. *Sustainability Science, 19*(3), 1039–1063.

Hewitt, C. N., Ashworth, K., & MacKenzie, A. R. (2020). Using green infrastructure to improve urban air quality (GI4AQ). *Ambio, 49*(1), 62–73.

Hicks, B. B., Saylor, R. D., & Baker, B. D. (2016). Dry deposition of particles to canopies—A look back and the road forward. *Journal of Geophysical Research: Atmospheres, 121*(24), 14–691.

Hilaire, R. S., Arnold, M. A., Wilkerson, D. C., Devitt, D. A., Hurd, B. H., Lesikar, B. J., Lohr, V. I., Martin, C. A., McDonald, G. V., Morris, R. L., & Pittenger, D. R. (2008). Efficient water use in residential urban landscapes. *HortScience a Publication of the American Society for Horticultural Science, 43*(7), 2081–2092.

Hung, Y-T., & Ashner, I. (2020). Chapter 18: Sulfur dioxide emission and mitigation. In Y-T. Hung, L. W. Wang, & N. K. Shammas (Eds.). *Handbook of environment and waste management: acid rain and greenhouse gas pollution control* (pp. 627–658). World Scientific Publishing Co Pte Ltd.. https://doi.org/10.1142/9789811207136_0018.

Husen, A. (Ed.). (2022). *Plants and their interaction to environmental pollution: Damage detection, adaptation, tolerance, physiological and molecular responses*. Elsevier.

Husen, A. (2021). Morpho-anatomical, physiological, biochemical and molecular responses of plants to air pollution. In *Harsh environment and plant resilience: Molecular and functional aspects*. Cham: Springer International Publishing, 203–234.

Idris, A. B., Hossain, M. M., Islam, M. A., Hossain, M. Z., & Islam, M. T. (2025). Green infrastructure and urban air quality: A semi-systematic review of multiscale evidence, methodologies, and policy-relevant insights. *Water, Air, & Soil Pollution, 236*(15), 993.

Illmann, N. (2025). A perspective on the reactions of organic peroxy radicals with HO_2. *Environmental Science: Atmospheres, 5*(7), 737–755.

Irfeey, A. M. M., Chau, H. W., Sumaiya, M. M. F., Wai, C. Y., Muttil, N., & Jamei, E. (2023). Sustainable mitigation strategies for urban heat island effects in urban areas. *Sustainability, 15*(14), 10767.

Isaksen, I. S., Berntsen, T. K., Dalsøren, S. B., Eleftheratos, K., Orsolini, Y., Rognerud, B., Stordal, F., Søvde, O. A., Zerefos, C., & Holmes, C. D. (2014). Atmospheric ozone and methane in a changing climate. *Atmosphere, 5*(3), 518–535.

Jayasooriya, V. M., Ng, A. W. M., Muthukumaran, S., & Perera, B. J. C. (2017). Green infrastructure practices for improvement of urban air quality. *Urban Forestry & Urban Greening, 21*, 34–47.

Jochner, S., Markevych, I., Beck, I., Traidl-Hoffmann, C., Heinrich, J., & Menzel, A. (2015). The effects of short-and long-term air pollutants on plant phenology and leaf characteristics. *Environmental Pollution, 206*, 382–389.

Joshi, N., Gupta, C. K., Mangla, Y., & Chowdhuri, A. (2023). Green plants as a sustainable solution to air pollution. *International Journal of Plant and Environment, 9*(02), 102–112.

Joshi, R. K., Gupta, R., Mishra, A., & Garkoti, S. C. (2024). Seasonal variations of leaf ecophysiological traits and strategies of co-occurring evergreen and deciduous trees in white oak forest in the central Himalaya. *Environmental Monitoring and Assessment, 196*(7), 634.

Karmakar, S. P., Das, A. B., Gurung, C., & Ghosh, C. (2022). Effects of ozone on plant health and environment: A mini review. *Research Journal of Agricultural Sciences, 13*, 612–619.

Karri, R. R., Vera, T., Hassan, S. K. M., Khoder, M. I., Dehghani, M. H., Mubarak, N. M., & Ravindran, G. (2024). Chapter 1 - Classification, sources, and occurrence of outdoor air pollutants: A comprehensive overview. In M. H. Dehghani, R. R. Karri, T. Vera, & S. K. M. Hassan (Eds.). *Health and Environmental Effects of Ambient Air Pollution: Volume 1: Air Pollution, Human Health, and the Environment* (pp. 1–34). Academic Press. https://doi.org/10.1016/B978-0-443-16088-2.00002-8.

Kim, J., Kim, J., Kim, Y., Go, T., & Lee, S. J. (2023). Accelerated settling velocity of airborne particulate matter on hairy plant leaves. *Journal of Environmental Management, 332*, 117313.

Kim, J. J., Park, J., Jung, S. Y., & Lee, S. J. (2020). Effect of trichome structure of Tillandsia usneoides on deposition of particulate matter under flow conditions. *Journal of Hazardous Materials, 393*, 122401.

Kim, K. J., Khalekuzzaman, M., Suh, J. N., Kim, H. J., Shagol, C., Kim, H. H., & Kim, H. J. (2018). Phytoremediation of volatile organic compounds by indoor plants: A review. *Horticulture, Environment, and Biotechnology, 59*(2), 143–157.

Koppmann, R. (2020). Chemistry of volatile organic compounds in the atmosphere. In *Hydrocarbons, oils and lipids: Diversity, origin, chemistry and fate*. Cham: Springer International Publishing, 811–822.

Korkou, M., Tarigan, A. K., & Hanslin, H. M. (2023). The multifunctionality concept in urban green infrastructure planning: A systematic literature review. *Urban Forestry & Urban Greening, 85*, 127975.

Kour, N., & Adak, P. (2023). Role of air pollution tolerance index (APTI) method for green belt development: A review. *Environmental Monitoring and Assessment, 195*(7), 856.

Kumar, A., Ekka, P., Upreti, M., Shilky, & Saikia, P. (2023). Urban green spaces for environmental sustainability and climate resilience. In *The Palgrave handbook of Socio-ecological resilience in the face of climate change: Contexts from a developing country*. Singapore: Springer, 389–409.

Kumari, R., Bhatnagar, S., & Kalra, C. (2022). Nitrogen assimilation in plants. In *Advances in plant nitrogen metabolism*. CRC Press, 38–54.

Kuttippurath, J., Patel, V. K., Pathak, M., & Singh, A. (2022). Improvements in SO_2 pollution in India: Role of technology and environmental regulations. *Environmental Science and Pollution Research, 29*(52), 78637–78649.

Lawson, T., von Caemmerer, S., & Baroli, I. Lüttge, U., Beyschlag, W., Büdel, B., Francis, D.. (2010). Photosynthesis and stomatal behaviour. In U. Lüttge, W. Beyschlag, B. Büdel, & D. Francis (Eds.). *Progress in Botany 72* (pp. 265–304). Berlin, Heidelberg: Springer. https://doi.org/10.1007/978-3-642-13145-5_11.

Lelieveld, J., Evans, J. S., Fnais, M., Giannadaki, D., & Pozzer, A. (2015). The contribution of outdoor air pollution sources to premature mortality on a global scale. *Nature, 525*(7569), 367–371.

Li, X., Zhang, T., Sun, F., Song, X., Zhang, Y., Huang, F., Yuan, C., Yu, H., Zhang, G., Qi, F., & Shao, F. (2021). The relationship between particulate matter retention capacity and leaf surface micromorphology of ten tree species in Hangzhou, China. *Science of the Total Environment, 771*, 144812.

Li, Y., Wang, S., & Chen, Q. (2019). Potential of thirteen urban greening plants to capture particulate matter on leaf surfaces across three levels of ambient atmospheric pollution. *International Journal of Environmental Research and Public Health, 16*(3), 402.

Li, Y., Zhang, X., Li, M., Yin, S., Zhang, Z., Zhang, T., Meng, H., Gong, J., & Zhang, W. (2022). Particle resuspension from leaf surfaces: Effect of species, leaf traits and wind speed. *Urban Forestry & Urban Greening, 77*, 127740.

Liang, L., Wang, Z., & Li, J. (2019). The effect of urbanization on environmental pollution in rapidly developing urban agglomerations. *Journal of cleaner production, 237*, 117649.

Lin, M., Horowitz, L. W., Payton, R., Fiore, A. M., & Tonnesen, G. (2017). US surface ozone trends and extremes from 1980 to 2014: Quantifying the roles of rising Asian emissions, domestic controls, wildfires, and climate. *Atmospheric Chemistry and Physics, 17*(4), 2943–2970.

Lovell, S. T., & Taylor, J. R. (2013). Supplying urban ecosystem services through multifunctional green infrastructure in the United States. *Landscape Ecology, 28*(8), 1447–1463.

Madheshiya, P., Gupta, G. S., Sahoo, A., & Tiwari, S. (2023). Role of elevated ozone on development and metabolite contents of lemongrass [*Cymbopogon flexuosus* (Steud.) (Wats. *Metabolites, 13*(5), 597.

Malik, A., & Singh, H. (2026). Urban green spaces and blue-green infrastructure: Pathways to carbon neutrality and building sustainable cities. In *Forestry for a carbon-neutral and sustainable future: A scientific perspective*. Cham: Springer Nature Switzerland, 433–452.

Meo, S. A., Salih, M. A., Alkhalifah, J. M., Alsomali, A. H., & Almushawah, A. A. (2024). Environmental pollutants particulate matter (PM2. 5, PM10), carbon monoxide (CO), nitrogen dioxide (NO_2), sulfur dioxide (SO_2), and ozone (O_3) impact on lung functions. *Journal of King Saud University-Science, 36*(7), 103280.

Molina, L., & Segura, A. (2021). Biochemical and metabolic plant responses toward polycyclic aromatic hydrocarbons and heavy metals present in atmospheric pollution. *Plants, 10*(11), 2305.

Monaco, P., Baldoni, A., Naclerio, G., Scippa, G. S., & Bucci, A. (2024). Impact of plant–microbe interactions with a focus on poorly investigated urban ecosystems—A review. *Microorganisms, 12*(7), 1276.

Montaluisa-Mantilla, M. S., Lebrero, R., García-Encina, P. A., & Muñoz, R. (2026). Elucidation of the mechanisms of VOC removal in botanical filters during indoor air treatment in a test chamber. *Building and Environment, 287*, 113775. https://doi.org/10.1016/j.buildenv.2025.113775.

Mulenga, C., Clarke, C., & Meincken, M. (2020). Physiological and growth responses to pollutant-induced biochemical changes in plants: A review. *Pollution, 6*(4), 827–848. 10.22059/poll.2020.303151.821.

Nargis, K., & Singh, H. (2024). Tree species suitability for adapting to urban air pollution and climate change. In *Urban forests, climate change and environmental pollution: Physio-biochemical and molecular perspectives to enhance urban resilience*. Cham: Springer Nature Switzerland, 121–142.

Nations, U., 2018. The World's cities in 2018. The World's Cities in, p.34.

Ngagine, S.H. (2022). Chemical heterogeneity of atmospheric particles in urbanized and industrialized environments during pollution events. Université du Littoral Côte d'Opale.

Nowak, D. J., Crane, D. E., Stevens, J. C., Hoehn, R. E. (2003). The urban forest effects (UFORE) model: Field data collection manual. *US Department of Agriculture Forest Service, Northeastern Research Station: Syracuse, NY, USA*, pp.4-11.

Nowroz, F., Hasanuzzaman, M., Siddika, A., Parvin, K., Caparros, P. G., Nahar, K., & Prasad, P. V. (2024). Elevated tropospheric ozone and crop production: Potential negative effects and plant defense mechanisms. *Frontiers in plant science, 14*, 1244515.

Ochoa-Alvarado, L., Garzón-Gil, J., Castro-Alzate, S., Zafra-Mejía, C. A., & Rondón-Quintana, H. A. (2025). UFORE-D modeling of urban tree influence on particulate matter concentrations in a high-altitude Latin American megacity. *Earth, 6*(2), 36.

Ogwu, M. C., & Kosoe, E. A. (2025). Integrating green infrastructure into sustainable agriculture to enhance soil health, biodiversity, and microclimate resilience. *Sustainability, 17*(9), 3838.

Ogwu, M. C., Imarhiagbe, O., Ikhajiagbe, B., & Osawaru, M. E. (2024). Toward understanding the impacts of air pollution. In *Sustainable strategies for air pollution mitigation: Development, economics, and technologies*. Cham: Springer Nature Switzerland, 3–43.

Ossola, R., & Farmer, D. (2024). The chemical landscape of leaf surfaces and its interaction with the atmosphere. *Chemical Reviews, 124*(9), 5764–5794.

Pace, R., & Grote, R. (2020). Deposition and resuspension mechanisms into and from tree canopies: A study modeling particle removal of conifers and broadleaves in different cities. *Frontiers in Forests and Global Change, 3*, 26.

Pandey, B., & Ghosh, A. (2026). Combined effect of drought stress and high tropospheric ozone conditions on crop yield. In *Tropospheric ozone and food security: Interactions of ozone with other major abiotic stresses and its impact on food security*. Cham: Springer Nature Switzerland, 197–227.

Patel, A., Tiwari, S., Khandelwal, A., Singh, C., Pandey, N., Tiwari, A., Yadav, S., & Prasad, S. M. (2024). Plants as biomarkers for monitoring environmental pollution. In *Biomarkers in environmental and human health biomonitoring*. Academic Press, 169–184.

Paull, N. (2020). *Green wall technology for sustainably improving environmental quality: Investigations into green wall plant health and particulate deposition*. Australia: University of Technology Sydney.

Prithviraj, P. S., & Reshmi, G. R. (2025). Breathing plants: Analyzing air pollution's effect on leaf and stomatal morphology. *Journal of Stress Physiology & Biochemistry, 21*(2), 216–228.

Rafie-Rad, Z., Raza, T., Eash, N. S., Moradi-Khajevand, M., & Moradkhani, M. (2024). Effects of outdoor air pollutants on plants and agricultural productivity. In *Health and environmental effects of ambient air pollution*. Academic Press, 71–90.

Rai, A., Kulshreshtha, K., Srivastava, P. K., & Mohanty, C. S. (2010). Leaf surface structure alterations due to particulate pollution in some common plants. *The Environmentalist, 30*(1), 18–23.

Rao, M. J., Duan, M., Ikram, M., & Zheng, B. (2025). ROS regulation and antioxidant responses in plants under air pollution: Molecular signaling, metabolic adaptation, and biotechnological solutions. *Antioxidants, 14*(8), 907.

Roy, A., Mandal, M., Popek, R., Przybysz, A., & Sarkar, A. (2025). Decoding leaf micro- and macro-morphology: A path to effective particulate matter phytoremediation. *International Journal of Phytoremediation*, 1–18. 10.1080/15226514.2025.2523938.

Saloglu, D., Basile, A., & Cassano, A. (2025). Atmospheric chemistry. In *Air pollution, air quality, and climate change*. Elsevier, 23–49.

Santos, G., Daraboina, N., & Sarica, C. (2021). Dynamic microscopic study of wax deposition: Particulate deposition. *Energy & Fuels, 35*(15), 12065–12074.

Senadheera, S. S., Withana, P. A., Lim, J. Y., You, S., Chang, S. X., Wang, F., Rhee, J. H., & Ok, Y. S. (2024). Carbon negative biochar systems contribute to sustainable urban green infrastructure: A critical review. *Green Chemistry, 26*(21), 10634–10660.

Shahbazi, H., Abolmaali, A. M., Alizadeh, H., Salavati, H., Zokaei, H., Zandavi, R., Torbatian, S., Yazgi, D., & Hosseini, V. (2021). Development of high-resolution

emission inventory to study the relative contribution of a local power plant to criteria air pollutants and Greenhouse gases. *Urban Climate, 38*, 100897.
Shahsavari, E., Schwarz, A., Aburto-Medina, A., & Ball, A. S. (2019). Biological degradation of polycyclic aromatic compounds (PAHs) in soil: A current perspective. *Current Pollution Reports, 5*(3), 84–92.
Shao, F., Wang, L., Sun, F., Li, G., Yu, L., Wang, Y., Zeng, X., Yan, H., Dong, L., & Bao, Z. (2019). Study on different particulate matter retention capacities of the leaf surfaces of eight common garden plants in Hangzhou, China. *Science of the total environment, 652*, 939–951.
Sharma, R., Kurmi, O. P., Hariprasad, P., & Tyagi, S. K. (2024). Health implications due to exposure to fine and ultra-fine particulate matters: A short review. *International Journal of Ambient Energy, 45*(1), 2314256.
Sharma, U., Bekturova, A., Ventura, Y., & Sagi, M. (2020). Sulfite oxidase activity level determines the sulfite toxicity effect in leaves and fruits of tomato plants. *Agronomy, 10*(5), 694.
Singh, D., & Verma, S. K. (2023). Impacts of particulate matter pollution on plants. In *New frontiers in plant-environment interactions: Innovative technologies and developments*. Cham: Springer Nature Switzerland, 483–500.
Singh, H., Bandyopadhyay, S., Kumar, A., Singh, M., Malik, A., Kumari, G., Raturi, A., Yadav, S. K., Kumar, M., Barthwal, S., & Thakur, A. (2022). Understanding the physiological and biophysical response of urban roadside plantations for assessing adaptation and mitigation mechanisms toward vehicular emissions. *Urban Climate, 44*, 101183.
Singh, R., & Singh, H. (2025). Analysing the impact of atmospheric particulate matter deposition on major soil and plants attributes: A review. *Water, Air, & Soil Pollution, 236*(12), 814.
Singh, V. K., Malhi, G. S., Kaur, M., Singh, G., & Jatav, H. S. (2022). Use of organic soil amendments for improving soil ecosystem health and crop productivity. *Ecosystem Services, 12*, 45.
Smith, P., Ashmore, M. R., Black, H. I., Burgess, P. J., Evans, C. D., Quine, T. A., Thomson, A. M., Hicks, K., & Orr, H. G. (2013). The role of ecosystems and their management in regulating climate, and soil, water and air quality. *Journal of Applied Ecology, 50*(4), 812–829.
Southerland, V. A., Brauer, M., Mohegh, A., Hammer, M. S., Van Donkelaar, A., Martin, R. V., Apte, J. S., & Anenberg, S. C. (2022). Global urban temporal trends in fine particulate matter ($PM_{2.5}$) and attributable health burdens: Estimates from global datasets. *The Lancet Planetary Health, 6*(2), e139–e146.
Tallis, M. J., Amorim, J. H., Calfapietra, C., Freer-Smith, P., Grimmond, S., Kotthaus, S., Sinnett, D., Smith, N., & Burgess, S. (2015). The impacts of green infrastructure on air quality and temperature. In D. Sinnett, N. Smith, & S. Burgess (Eds.). *Handbook on green infrastructure: Planning, design and implementation* (pp. 30–49). Cheltenham: Edward Elgar Publishing.
Tiwari, S., & Agrawal, M. (2018). Effect of ozone on physiological and biochemical processes of plants. In *Tropospheric ozone and its impacts on crop plants: A threat to future global food security*. Cham: Springer International Publishing, 65–113.
Tiwari, S., & Agrawal, M. (2018). *Tropospheric ozone and its impacts on crop plants*. Cham: Springer International Publishing [Online]. Available at: doi, 10, pp.978-3.
Upadhyay, S. K. (2025). Relevance of cross talk between root exudates, hormones, and root-associated microbes in developing sustainable phytoremediation strategies: A comprehensive review. *Physiology and Molecular Biology of Plants, 31*(10), 1629–1649.

Vongelis, P., Koulouris, N. G., Bakakos, P., & Rovina, N. (2025). Air pollution and effects of tropospheric Ozone (O_3) on public health. *International Journal of Environmental Research and Public Health, 22*(5), 709.

Vos, T., Lim, S. S., Abbafati, C., Abbas, K. M., Abbasi, M., Abbasifard, M., Abbasi-Kangevari, M., Abbastabar, H., Abd-Allah, F., Abdelalim, A., & Abdollahi, M. (2020). Global burden of 369 diseases and injuries in 204 countries and territories, 1990–2019: A systematic analysis for the Global Burden of Disease Study 2019. *The Lancet, 396*(10258), 1204–1222.

Wang, H., Maher, B. A., Ahmed, I. A., & Davison, B. (2019). Efficient removal of ultrafine particles from diesel exhaust by selected tree species: Implications for roadside planting for improving the quality of urban air. *Environmental Science & Technology, 53*(12), 6906–6916.

Wang, H., Shi, H., Li, Y., Yu, Y., & Zhang, J. (2013). Seasonal variations in leaf capturing of particulate matter, surface wettability and micromorphology in urban tree species. *Frontiers of Environmental Science & Engineering,* 7(4), 579–588.

Wang, M., Zhong, X., Sun, C., Chen, T., Su, J., & Li, J. (2023). Comprehensive performance of green infrastructure through a life-cycle perspective: A review. *Sustainability, 15*(14), 10857.

Wang, P., Chen, Y., Hu, J., Zhang, H., & Ying, Q. (2018). Attribution of tropospheric ozone to NO x and VOC emissions: Considering ozone formation in the transition regime. *Environmental Science & Technology, 53*(3), 1404–1412.

Wang, X., Xiang, Y., Peng, C., Teng, M., Ma, B., Zhou, Z., & Peng, C. (2024). Impact of tree growth form on temporal and spatial patterns of particulate matter with various particle sizes in urban street canyons. *Landscape Ecology, 40*(1), 5.

Wasikoyo, E., Zsembeli, J., Gashi, N., Gumisiriya, C., & Csaba, J. (2026). Plant–soil microbe interactions' effects on CO_2 emissions, soil organic carbon and nutrients under different tillage systems. *Agriculture, 16*(4), 429.

Wedow, J. M., Ainsworth, E. A., & Li, S. (2021). Plant biochemistry influences tropospheric ozone formation, destruction, deposition, and response. *Trends in Biochemical Sciences, 46*(12), 992–1002.

Wei, X., Lyu, S., Yu, Y., Wang, Z., Liu, H., Pan, D., & Chen, J. (2017). Phylloremediation of air pollutants: Exploiting the potential of plant leaves and leaf-associated microbes. *Frontiers in Plant Science, 8*, 1318.

Weyens, N., Thijs, S., Popek, R., Witters, N., Przybysz, A., Espenshade, J., Gawronska, H., Vangronsveld, J., & Gawronski, S. W. (2015). The role of plant–microbe interactions and their exploitation for phytoremediation of air pollutants. *International Journal of Molecular Sciences, 16*(10), 25576–25604.

Xu, X., Zhang, Z., Bao, L., Mo, L., Yu, X., Fan, D., & Lun, X. (2017). Influence of rainfall duration and intensity on particulate matter removal from plant leaves. *Science of the Total Environment, 609*, 11–16.

Yadav, P., Usha, K., & Singh, B. (2022). Air pollution mitigation and global dimming: A challenge to agriculture under changing climate. In *Climate change and crop stress.* Academic Press, 271–298.

Yang, J. L., & Zhang, G. L. (2015). Formation, characteristics and eco-environmental implications of urban soils–A review. *Soil Science and Plant Nutrition, 61*(sup1), 30–46.

Zandybay, A., Saspugayeva, G., Khussainov, M., Karelkhan, N., Kydyrova, A., & Adylbek, Z. (2024). Assessing the impact of urban development on soil health and nutrient cycling across urban areas. *Journal of Ecological Engineering, 25*(12), 106–123.

Zappitelli, I., Conte, A., Alivernini, A., Finardi, S., & Fares, S. (2023). Species-specific contribution to atmospheric carbon and pollutant removal: Case studies in two Italian municipalities. *Atmosphere, 14*(2), 285.

Zayed, E. M., Imran, M., Saba, M., & Abo-Elyousr, K. A. (2025). Rhizosphere and remediation soil, water, and air pollution. In *Sustainable remediation for pollution and climate resilience*. Singapore: Springer, 443–465.
Zhang, X., Lyu, J., Han, Y., Sun, N., Sun, W., Li, J., Liu, C., & Yin, S. (2020). Effects of the leaf functional traits of coniferous and broadleaved trees in subtropical monsoon regions on PM2. 5 dry deposition velocities. *Environmental Pollution, 265*, 114845.
Zhang, X., Lyu, J., Zeng, Y., Sun, N., Liu, C., & Yin, S. (2021). Individual effects of trichomes and leaf morphology on PM2. 5 dry deposition velocity: A variable-control approach using species from the same family or genus. *Environmental Pollution, 272*, 116385.
Zheng, G., & Li, P. (2019). Resuspension of settled atmospheric particulate matter on plant leaves determined by wind and leaf surface characteristics. *Environmental Science and Pollution Research, 26*(19), 19606–19614.
Zhu, Y. G., Xiong, C., Wei, Z., Chen, Q. L., Ma, B., Zhou, S. Y. D., Tan, J., Zhang, L. M., Cui, H. L., & Duan, G. L. (2022). Impacts of global change on the phyllosphere microbiome. *New Phytologist, 234*(6), 1977–1986.

CHAPTER TEN

Urban greening for sustainable urban planning and air pollution mitigation

Aditya Abha Singh[1], Naushad Ahamad[1], and Richa Rai[2]
[1]Department of Botany, University of Lucknow, Lucknow, Uttar Pradesh, India,
[2]Plant Ecology and Climate Change Science Division, CSIR-National Botanical Research Institute, Lucknow, Uttar Pradesh, India

10.1 Introduction

Urban planning with a focus on sustainability involves creating and structuring city environments to address present and future environmental, societal, and economic requirements while preserving the Earth's ecological equilibrium. This approach aims to develop urban spaces that are both functional for current needs and adaptable to future challenges, all while minimizing negative impacts on the planet's ecosystems (UN-Habitat, 2020). The 21st century is witnessing urbanization as a major demographic shift. By 2018, 55% of the global population lived in urban areas; by 2050, this percentage is predicted to rise to 68%. The majority of this urban expansion is anticipated to occur in Asian and African regions (United Nations Department of Economic and Social Affairs Population Division, 2018). The process of urbanization involves the shift of populations from dispersed rural areas, where farming is the primary economic driver, to more densely populated urban centers characterized by industrial and service-based economies. This demographic transition results in a concentration of people in larger, more compact city environments (United Nations Department of Economic and Social Affairs Population Division, 2015). Urbanization is influenced by various elements, including demographic shifts, economic growth, and governmental policies. The expansion of urban populations can be attributed to three main factors: natural increase (when births outnumber deaths), rural to urban migration, and the redefinition of rural settlements as urban. Several push factors contribute to the exodus from rural regions or countries, such as economic hardship,

Urban Air Pollution and Plant Responses
DOI: https://doi.org/10.1016/B978-0-443-40320-0.00002-4

social unrest due to conflicts, loss or seizure of land, severe weather conditions, and unemployment (potentially exacerbated by climate change). Concurrently, urban areas attract people through pull factors such as enhanced economic prospects, better availability of services, and contemporary conveniences. Additionally, migration policies play a direct role in shaping population distribution, including strategies aimed at reducing the flow of people from rural to urban regions (United Nations Department of Economic and Social Affairs Population Division, 2016). Worldwide, this has resulted in unsustainable urban growth as cities struggle to accommodate rapidly expanding populations and meet the increasing societal demands. These demands include access to potable water, adequate sanitation, unpolluted air, efficient transportation networks, and new infrastructure for housing, education, and leisure activities (Shao et al., 2020).

Densely populated cities tackle the challenges of human and environmental sustainability in an era of rapid population growth and urbanization. As the urban environment expands, it is accompanied by the harmful effects of pollution, specifically air pollution. Skyscraper buildings have contributed to the pollutant's accumulation because of their long structure and large surface areas in urban ecosystems (Hamling et al., 2024). The cities face challenges like air pollution, excessive noise, natural resource consumption, biodiversity loss, climate change, waste production and disposal, water availability, land use, etc. The social and economic difficulties in guaranteeing the necessary standard of living and equal opportunities for all city dwellers are linked to these environmental problems (Michalina et al., 2021; Zimmermann, 2016). These critical issues require creative solutions to reduce pollution and improve the quality of urban life (Hamling et al., 2024).

Several examples have emerged, including the nature-based solutions (NBS) assessment framework proposed by Raymond et al. (2017), the Urban Integrated Assessment Framework suggested by Ford et al. (2019), conceptual framework for water net-positive buildings given by Joustra and Yeh (2014), sustainability framework for ecosystem services trade-offs suggested by Cavender-Bares et al. (2015), and conceptual framework for urban water sustainability given by Yang et al. (2016), and Mellino et al. (2014), combined spatial planning assessment with Geographic Information System (GIS) software. However, there is limited research on comprehensive, systems-thinking frameworks that incorporate the planning system process and align urban design solutions with organizational-level decision-making. Bozovic et al. (2017) emphasized a highly structured preplanning phase in their Blue Green Systems approach but did not provide detailed guidance on its implementation.

In order to address these and other urban challenges, NBSs are being taken up as a sustainable solution that can provide a number of advantages (Costanza et al., 1998; Gascon et al., 2015; Gómez-Baggethun & Barton, 2013) like ecosystem services (Dick et al., 2019), including flood/runoff control, microclimate regulation (Langergraber et al., 2020), and enhance life, human health, and well-being (Haase et al., 2014). In order to promote more resilient, sustainable, and livable cities (Ahern et al., 2014), it is crucial to implement NBS for providing various ecosystem services (Bona et al., 2023). Therefore the European Green Deal recently released by the European policy framework could highly increase the recognition and function of NBS and SUP as powerful drivers of inclusive and sustainable urban transition (de Luca et al., 2021).

Sustainable urban design originated as an important strategy for tackling the issues of rapidly increasing climate change and urbanization (Gil-Ozoudeh et al., 2023). The goal of sustainable urban planning and design is to decrease water consumption, enhance renewable energy, reduce urban heat island effect, air pollution, provide sustainable life, good health, and develop resilient communities (Ghosh, 2017). Adopting sustainable activities in green buildings reduces the carbon footprint (Ahram & Zakaria, 2023). Incorporation of green building infrastructure provides positive physio-ecological effects and long-term urban environments (Min, 2025). High-density cities deal with the problem of high urban temperature caused by urban heat island effects (Hamling et al., 2024). Vertical garden plays a vital role in reducing surface temperature by providing shade and it is a sustainable method to reduce the urban heat island (UHI) effects, improve air quality, enhance biodiversity, and improve mental health (Yadav et al., 2024). This chapter highlights the significant role of urban greening in air pollution mitigation pollution.

10.2 Need for sustainable green planning development in urban areas

Urbanization often leads to higher emissions from vehicles, industries, and residential activities, which worsens air quality. Green planning that is sustainable has become crucial in tackling these issues. Vegetation plays a key role in reducing the negative effects of pollution through natural filtration and absorption processes (Barwise & Kumar, 2020), producing cooling effects to mitigate the UHI phenomenon (Kumar et al., 2024), and improving mental and physical health (Barwise & Kumar, 2020; Kardan et al., 2015). Singapore's urban greening initiatives, which have greatly enhanced the air quality and urban esthetics, are successful examples of sustainable green planning (Yuen, 2011). However, due to rapidly expanding populations and rising societal demands, ecological imbalance needs theurban planning (Musakwa & Niekerk, 2014; Pickett et al., 2013). Green infrastructure, which includes green roofs, green walls, vegetation barriers, community gardens, and street trees, is the most sustainable solution to urban planning.

10.3 Types of green infrastructure

10.3.1 Green roofs

Green roofs are typically defined as building roofs that are entirely or partially covered in vegetation of any kind. Other names for green roofs include living roofs, eco-roofs, and roof gardens (Berardi et al., 2014; Khotbehsaara et al., 2019). A green roof is made up of vegetation, a substrate for growth, protective coating like root barriers and insulation (Vijayaraghavan, 2016), drainage, and waterproofing (Department of Planning Local Government, 2010; GSA, 2011).

Green roofs are of three types, which are extensive, semiintensive, and intensive, depending on the thickness of the substrate and the vegetation they support (Berardi et al., 2014; GSA, 2011; Vijayaraghavan, 2016). Grass and shrubs are supported by extensive green roofs, which have substrates that are less than 15 cm thick (Carter & Butler, 2008). Semiintensive green roofs (20–30 cm) support shrubs (Vacek et al., 2017), while intensive green roofs (>30 cm to 100 cm) support shrubs and small trees (Jim & Tsang, 2011). Green roofs are usually low-maintenance, affordable solutions that control runoff, lessen air pollution, and reduce urban heat islands. Using green roofs can help transform a rooftop's sealed & solar-heated surfaces into multipurpose ecological areas (Bianchini & Hewage, 2012). It is especially true of large systems with thin substrates, as larger plants require thicker substrates, intensive systems that are more costly and require more upkeep, but they also provide greater ecological advantages. Systems that are semiintensive offer a balance between moderate maintenance and cost.

Green roofs improve ecological preservation, lower noise pollution, and increase energy efficiency (Abhijith et al., 2017; Berardi et al., 2014; Oberndorfer et al., 2007; Vijayaraghavan, 2016). The first variety (extensive) is capable of supporting grass and shrubs, while the second variety (intensive) can support low trees and shrubs.

Green roofs improve ecosystem services while lowering the effect of UHI, energy consumption, and emission of greenhouse gases (Bianchini & Hewage, 2012; Semeraro et al., 2019), enhance the sustainability of buildings, and are frequently designed to partially offset the loss of vegetation brought on by construction of buildings (Berardi et al., 2014; Ferrini et al., 2020; Rowe, 2011). Compared to extensive green roofs, intensive green roofs offer more advantages, such as enhancing public areas and promoting human health, but they also come with higher installation costs, more maintenance requirements, and a greater structural support requirement (Williams et al., 2016). However, the extensive green roof is most widely used because it is lighter, doesn't require irrigation, reduces the effect of flooding, and costs less to install and maintain (Shafique et al., 2018). A green roof's cooling effect and its effect on street canyon air quality showed that a cooling temperature of 2°C at breathing level could reduce pollutant concentrations by 32% because of higher vertical dispersion and more canyon vortices caused by cool air moving downwards (Baik et al., 2012). According to Pugh et al. (2012), green roofs had little effect on removing pollutants as compared to green canyon, even when their cooling advantages were taken into consideration. According to Speak et al. (2012), roofs close to traffic corridors greatly enhanced the quality of the air, decreasing particulate matter from automobile emissions (Kostadinović et al., 2023). Seasonal variations, plant species, wind speed, and the roof's location all affect how quickly the pollutants are removed by green roofs (Currie & Bass, 2008; Li et al., 2010; Speak et al., 2012; Yang et al., 2008). The removal of pollutants can further be increased by intensive green roofs by approximately 2046.89 Mt as compared to the extensive type, which could be up to 1405.50 Mt (Yang et al., 2008). While green roofs improve urban biodiversity (Mayrand & Clergeau, 2018), the use of native species can be challenging due to the artificial environment created by green roofs having high radiation and temperature (Semeraro et al., 2021). Green roof may not always correlate with native species conservation, while plant species such as aromatic, medicinal, and fruit-bearing plants, which are used to create urban gardens, serve as community hubs and social attractions (Varela-Stasinopoulou et al., 2023).

By converting urban areas from single-purpose spaces into ones that integrate multiple functions, green roofs offer multifunctionality and improve social and ecological well-being (Semeraro et al., 2021). Finding ways to lower installation costs while taking roof weight restrictions into account and putting in place efficient management techniques is crucial for integrating green roofs into global urban strategies (Whittinghill & Rowe, 2012).

Mitigation of air pollution is one of the possible advantages of green roofs (Berardi et al., 2014; Francis & Jensen, 2017; Rowe, 2011; Yang et al., 2008). However, only a small number of studies have looked at how air quality can be enhanced by green roofs (Francis & Jensen, 2017). Most of these have only shown performance at the rooftop level or have been carried out at the city level. Yang et al. (2008), indicated that green roofs effects on air quality are contingent upon numerous variables. These include the environmental conditions of the site (Speak et al., 2012), location of the installation of green roofs (Li et al., 2010), direction of ambient air (Baik et al., 2012), wind velocity (Li et al., 2010), seasonal changes (Currie & Bass, 2008; Yang et al., 2008), meteorological factors, concentrations of airborne pollutants, rate of pollutant dry deposition (Yang et al., 2008), selection of plant species (Currie & Bass, 2008; Speak et al., 2012), and morphological attributes of the vegetation (Yang et al., 2008). Green roofs, when strategically implemented with consideration for their surroundings, offer an additional method of incorporating vegetation to mitigate noise pollution exposure at street level (Yang et al., 2012). Furthermore, these rooftop green spaces have the capacity to boost air quality in urban streets (Pugh et al., 2012), provided that relevant considerations are appropriately addressed during their design and implementation.

10.3.2 Community gardens

Urban community gardens are an open area managed by the community associated with cultivating crops in urban and suburban regions (Faoun, 2007; Martellozzo et al., 2014; Pearsall et al., 2016). These gardens not only contribute to the food production but also contribute to medicinal, aromatic, and ornamental plant cultivation (Faoun, 2007). Urban agriculture practices are changeable at particular urban spaces, which can be executed in various shapes and places like roof, brownfield and greenfield areas, that is, gardens and parks (Egerer & Fairbairn, 2018; Koopmans et al., 2017). Agricultural schemes can illustrate multioperation land use in urban areas and can integrate agricultural activities with ecological and social goals (Duzi et al., 2014).

Due to urbanization, which caused food scarcity, biodiversity loss, and social isolation, community gardens provide an essential space to the residents, where they can grow food and reconnect with nature in a social ecosystem. This garden creates habitual space and a new form of collaboration and community interaction (Egerer & Fairbairn, 2018; Okvat & Zautra, 2011; Taylor & Lovell, 2013). Particularly, these contribute to enhancing environmental sustainability through agri-environment, which can reduce the water runoff, soil erosion, and enhance soil quality (Martin et al., 2016). They also affect the urban climate and microclimate changes, whereas promoting the organic waste recycling from urban spaces, and reducing their footprint (De Zeeuw, 2011). Moreover, these have provided various economic and social benefits (Levkoe, 2006). They enhance community bond strength, fresh produce, local economies revitalization, and convert vacant, neglected, and disturbed areas into productive spaces. Community gardens can also improve ecological services related to education and culture through enhancing people's practical gardening skills and fostering a connection to biodiversity, natural processes, and the earth's cycles (Ellis, 2010; Miedema et al., 2013). Community gardens play a role in empowering people's food and flower production in urban areas (Ohmer et al., 2009). Through teamwork and independence, community gardens are considered "participatory landscapes" that oppose marginalization and racism (Gray et al., 2013; Mares & Pena, 2011; White, 2011).

10.3.3 Green walls

Green walls are also known as vertical gardens, bio wall, or living wall (Balany et al., 2020; Tomson et al., 2021). Green walls are classified into living walls and green facades. The difference between living walls and green facades is that living walls have planting materials supported by various plants, whereas green facades have the climbing plants (Perez et al., 2014). Green facades are subsequently classified into direct and indirect green facades (Bustami et al., 2018). In direct green facades, plants are attached directly from the wall, which are directly rooted to the ground or pots (Feng & Hewage, 2014), whereas indirect green facades utilize vertical supporting systems like meshes, wire, cables, ropes, or modular trellis for climbing plants, which secure vegetation away from the wall (Tomson et al., 2021). Green walls provide various benefits, such as building insulation by direct shading of the wall surface and providing attractive designs on walls. They improve air quality, produce a cooler microclimate, and allow for the growth of plants in areas that usually cannot support vegetation. Green walls include a variety of plants like small shrubs and herbaceous species (Balany et al., 2020). Depending on the source of pollution, green facades enhance the air quality from background and local emission sources. Living walls have the ability to reduce air pollution in urban environments and enhance air quality (Feng & Hewage, 2014; Joshi & Ghosh, 2014). Litschike and Kuttler (2008) suggested green walls are a potent planting concept to reduce particulate matter by deposition, unmodified air exchange between the street canyon and the atmosphere (Litschike & Kuttler, 2008). Several researches have reported that green walls help in reducing pollutants along the street canyon, open roads and footpaths. Furthermore, research has found a collection of pollutants through the green wall vegetation (Joshi & Ghosh, 2014; Pugh et al., 2012; Tong et al., 2016).

A study conducted at the city level revealed that the green wall significantly improved the quality of the air, but the influence of trees outweighed the reductions (Jayasooriya et al., 2016). A green wall produced dispersion patterns resembling those of a solid wall in conditions of an open road, with pollution reduction in a high concentration zone in front of the barrier (on the road) and behind the green walls (Morakinyo et al., 2016; Tong et al., 2016). Furthermore, the wall's vegetation cover eliminated pollution by the way of deposition (Joshi & Ghosh, 2014; Morakinyo et al., 2016). A computational fluid dynamics (CFD) study conducted by Pugh et al. (2012) examined the impact of green walls (*Hedera helix*) in street canyons with height-to-width ratios of 1 and 2 (H/W 1 & 2), reducing NO_2 concentrations by up to 40% and PM_{10} concentrations by 60% (Pugh et al., 2012). It was discovered that common climbing plants like Lianas and ivy species would work well regarding the green walls (Chen et al., 2016; Sternberg et al., 2010). Leaf Area Index, humidity, wind velocity, and street canyon geometry were found to have an impact on the potential for pollutant removal with a green wall (Joshi & Ghosh, 2014; Pugh et al., 2012). Particle depositions at various heights on the green wall close to a traffic corridor did not differ from one another. An enormous amount of wall space used as surface greenery has various benefits in modern cities, like biodiversity conservation, cooling surface, and reducing energy consumption (Feng & Hewage, 2014; He et al., 2017; Louis & Jim, 2017; Perez et al., 2014).

10.3.3.1 Types of green walls

Green facades usually feature climbing plants, and living walls support a wider variety of plants by utilizing technology and plant material. Green walls can be categorized into two main types, each with distinct characteristics. Green facades, one of these types, are further divided into direct and indirect varieties (Bustami et al., 2018). Direct green facades, traditionally known, feature plants rooted in soil or containers at various heights, adhering to walls via self-clinging climbers (Feng & Hewage, 2014). In contrast, indirect green facades, also termed double-skin green facades or green curtains, employ vertical support structures such as continuous guides or modular trellises for climbing plants, forming a separate green layer from the wall (Perez et al., 2011). The other main type, living walls, consists of panels, geotextile felts, or modules affixed to walls to support vegetation (Tomson et al., 2021). These can be either continuous or modular systems. Continuous living walls, also known as felt-layer systems or vertical gardens, utilize permeable screens for individual plant placement (Palermo & Turco, 2020). Modular green walls, or planter box systems, comprise components attached directly to buildings or supported by additional structures, containing growing media in various forms (Ottelé et al., 2011). An alternative classification by Perez et al. (2011) separates green walls into intensive (living walls) and extensive (green facades) systems, based on their complexity and maintenance requirements. Additionally, active green walls, also referred to as functional green walls or active botanical biofilters, represent a distinct category used indoors for air purification (Pettit et al., 2019).

10.3.4 Vertical green screens

The idea of green screens, which are made of plastic or steel mesh with climbing plant growing on it, is newer than, say, the utilization of hedges or trees (Tomson et al., 2021). They are made to offer a variety of advantages, such as absorbing air pollution, promoting biodiversity, lowering noise levels, and reducing rainfall. Additionally, green screens can be used for fencing, noise barrier greening, privacy and tranquility (Dover & Phillips, 2015; Kateshastabadan et al., 2025). Green screens can be erected in place of hedges and trees in urban areas, where planting space is usually scarce. Additionally, they could be placed in backyards, indoor spaces, or open spaces like playgrounds or schools. Compared to green walls, green screens are easier to install, less expensive, and require less upkeep (Tomson et al., 2021). Before being planted on location, climbers can be pregrown or planted in greenhouse nurseries for almost 2 years in panels of metal mesh that are held up through poles or other facilities (Dover & Phillips, 2015), or they are sown right there and kept there until they are about two-and-a-half years old (Tremper & Green, 2018). Only a small number of studies have evaluated green screens' ability to reduce air pollution, and none of them were conducted in street canyon settings. Green screens are useful at reducing PM_{10}, according to a 2015 study using *Hedera helix* to capture PM_{10} on a busy street in Birmingham, UK (Dover & Phillips, 2015). In contrast to pristine leaves gathered from a background location (nursery), the research discovered that the quantity of particles recorded by green screens increased fourfold. Despite baseline information (for example, gradient in NO_2 concentrations before installation), according to this study, NO_2 levels had decreased through 53% in both outside the school and the playground, indicating that ivy screens have the potential to greatly lower the exposure to traffic pollution (Hedges Direct, 2023). Tremper and Green (2018) carried out research to evaluate how well an ivy screen works to stop vehicle emissions from entering a nearby schoolyard. The level of NO_2 was measured on the two sides during school hours and ivy screens toward the playground side showed a mean decrease in concentrations of up to 23%, with a mean daily pollution concentration reduction of 22%. Ivy screens also help in reducing daily average concentrations of pollutants on the playground side of the screen by 24% and 38% for NO_2 and PM_{10}, respectively. During school hours, concentration reduced by up to 36% and 41% of NO_2 and PM_{10}, respectively (Tremper et al., 2015). Based on limited field research, green screens could be a useful green infrastructure (GI) option for reducing air pollution. The key benefits of green screens are required for minimum ground space, which can create a better green screen for a more suitable green infrastructure option in a street canyon than rows of trees or hedges. But further studies are required to determine the efficiency of green screens specially which involved the plant species other than *Hedera helix*, and in the creation of various street canyon environments (Tomson et al., 2021).

10.3.5 Hedges

The hedges have the ability to capture emissions at the local level and improve air quality because they are frequently found close to emission sources (Abhijith et al., 2017). In open road areas, hedges can create a boundary between pollution receptors and sources. In street canyons, hedges can mitigate exposure of pedestrians by creating local vortices and pollutant diverting form footpaths, but this mitigation depends on aspect ratios (Barwise & Kumar, 2020; Gromke et al., 2016; Kumar et al., 2019). Hedges significantly reduced the traffic-emitted gaseous pollutants (carbon monoxide) as observed by Li et al. (2016) in a CFD modeling study.

Additionally, Hedges may vary in height (1.5 or 2.25 m), length (continuous or discontinuous), and permeability (1.67 or 3.34/m). Studies explored the spread of gaseous pollutants throughout the street canyons by wind tunnel experiments. It observed two different configurations of hedges (1) with an eccentric hedgerow on both sides of the main road, (2) with one hedgerow in central traffic lane. Both sides of discontinuous hedgerows of the canyon reduced pollution concentration up to 3%–19% higher than without hedgerows, and the central hedgerow demonstrates a higher reduction of pollutants up to 61%, which is greater than compared to the parallel hedges. Greater pollutants reduced in higher and less permeable hedges (Gromke et al., 2016). Kumar et al. (2019) also advised growing the hedges continuously, without any spaces or gaps. On the other hand, Vos et al. (2013) found that, when taking into account various vegetation scenarios, in street canyon, hedges negatively affect the overall air quality. The ideal hedge size varies depending on the canyon's aspect ratio. For moderately deep canyons, Li et al. (2016) found that the optimal height for hedges was 1.1 meter, while hedges of about 2 m have been found to work well in shallow canyon (Abhijith et al., 2017; Kumar et al., 2019). Although it may be generalized that most studies to date have discovered that hedges enhance the quality of air in street canyon, hence careful and site-specific hedge configuration is essential. Pilot modeling studies ought to be conducted whenever feasible for every street canyon when planting trees or hedges is suggested (Barwise & Kumar, 2020; Hewitt et al., 2020). A diagrammatic representation of the types and benefits of green infrastructure in urban areas has been provided in Fig. 10.1.

10.4 Urban green spaces and air quality improvement

As massive sinks, plants are the primary absorbers of various pollutants and contribute to atmospheric purification. On the surface of their leaves or bark, plants have the ability to adsorb and accumulate pollutants (Park et al., 2025). Since leaves are the most vulnerable to air pollution, their degree of sensitivity can cause a variety of noticeable changes (Chavan & Sonwane, 2012). Particulate matter and dust are either

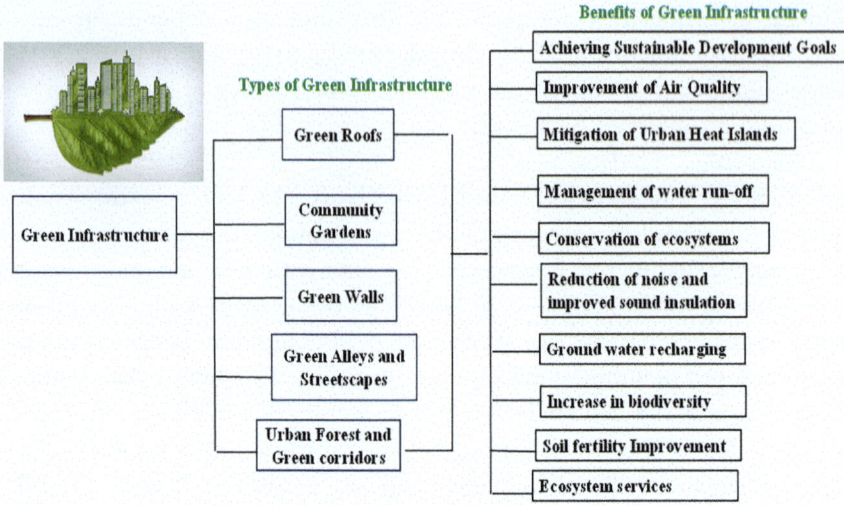

Figure 10.1 Types and benefits of green infrastructure in urban areas.

absorbed by the leaf through stomatal pores or adsorbed on the leaf surface (Farmer, 2002). An area's pollution level can be investigated by looking at the types of plants that grow there. These days, roadside plants are thought to be among the best systems for monitoring air pollution because they can serve as both bio-monitor and bio-indicator of air pollution (Esfahani et al., 2013; Holt & Miller, 2011; Kaler et al., 2017). Plants react differently to prolonged exposure to vehicle pollution; some can withstand high pollution loads, while others become sensitive to it and exhibit a range of symptoms, such as reduced level of chlorophyll and consequently slower growth of the plants (Davison & Blakemore, 1976; Joshi & Swami, 2009). Plant species selection is a crucial consideration in the development of green cover, which can reduce air pollution to a certain extent through the plantation of tolerant species. Plants can be chosen based on their tolerance level using the Air pollution Tolerance Index, one of the popular indices. Concerning air pollution, plants that exhibit exceptional performance can be chosen using the Anticipated Performance Index (Joshi & Swami, 2009; Kaur & Nagpal, 2017; Rai et al., 2013).

Various green infrastructure practices have been designed for air pollution reduction. These include both active and passive practices, uptake filtering devices like commercial City Tree are examples of active, and green lamppost or green walls are passive solutions (Nemitz et al., 2020;

Sänger & Splittgerber, 2016). These are primarily developed for esthetic purpose but presently used for both esthetic and improve urban environment by reducing pollution that is sustainable for the future (Hamling et al., 2024). Green roofs and green walls have various benefits for the environment, like reduced UHI effects by reducing temperature, air pollution, noise, energy consumption, managing storm water, improving air quality, and urban biodiversity. Numerous studies have provided ample evidence of these advantages (Berardi et al., 2014; Hunter et al., 2014; Perez et al., 2014; Vijayaraghavan, 2016; Wood et al., 2018). Different types of green infrastructures and their efficacy in improving air quality is listed in Table 10.1.

10.4.1 Absorbing and removing pollutants

Urban vegetation is essential for enhancing air quality because it absorbs and eliminates pollutants through stomatal uptake and deposition onto plant surfaces (Wei et al., 2017). Deposition entails filtering air pollutants through bark or leaves and, in the case of reactive contaminants such as ozone (O_3), their breakdown in the gas phase is brought on by substances released by the plants (Janhall, 2015; Wei et al., 2017). The absorption capacity of plant surfaces, movement of air in the crown space, and transmission via the boundary layer next to all surfaces affect deposition rates. These factors are impacted by stomatal conductance and vegetation characteristics like canopy structure, leaf shape, and surface features (Grote et al., 2016). Compared to species like *Platanus*, trees with intricate leaf structures, hairs, or waxes like *Tilia* have a greater capacity for particle deposition; waxes have been demonstrated to almost double the removal of particulate matter (Dzierżanowski et al., 2011). During the growing season, particulates may become immobilized within wax layers, and pollutants like SO_2 and NO_2 can dissolve in water films on leaf surfaces (Altimir et al., 2006; Grecia et al., 2025). To balance photosynthesis and water loss, plants can absorb gaseous pollutants like CO_2 through stomata. With varying deposition efficiency across particle sizes, urban trees greatly increase the rates of deposition of gaseous pollutants (NO_2, SO_2, CO, O_3) and particulate matter (PM_{10}, $PM_{2.5}$, ultrafine particles) (Janhall, 2015; Kumar et al., 2019). Urban vegetation helps lower health risks related to respiratory and cardiovascular diseases by reducing air pollution (EEA, 2017).

Table 10.1 The table explores various green infrastructure strategies and their effectiveness in improving air quality.

GIs strategies	Description	Benefits	References
Green roofs	Installing vegetation layers, such as soil layers to promote plant growth, waterproofing, and drainage, on building rooftops. There are two types of green roofs: intensive (heavier with deeper soil) and extensive (lightweight with shallow soil)	**Mitigating air pollution:** Vegetation on rooftop gardens captures airborne contaminants, enhancing the quality of the atmosphere **Climate control:** These installations act as insulators, minimizing energy use for indoor temperature management **Management of rainfall:** Green rooftops help in the capture of precipitation, reducing water runoff and reducing pressure on urban drainage system **Ecological diversity:** Green roof act as environment for various organism and encouraging urban biodiversity	Tomson et al. (2021), Chaudhuri and Kumar (2022), Semeraro et al. (2021)
Green walls (vertical gardens)	Vegetation covers vertical structures, either in the form of modular panels with pregrown plants or climbing plants. These can be erected as freestanding structure or on the exterior of buildings	**Compact vertical gardening:** Suitable for metropolitan regions due to land availability **Pollution removal:** Vegetation remove airborne contaminants and function as natural air purifier **Temperature regulation:** Green walls function as natural insulator, specially reducing surface and internal temperature **Esthetic improvement:** Enhance visual appeal of urban spaces	Tang (2023), Oquendo-Di Cosola et al. (2022), Semeraro et al. (2021)
Green barriers	Installation of vegetation barrier along road sides such as tree lines and hedges, to filter emissions from moving vehicles	**Air purification:** Green barriers purify noxious fumes and suspended matter of vehicle exhaust **Noise mitigation:** Green barriers can reduce intensity of path noise **Scenic enhancement:** Boosts the visual attractiveness of thoroughfares and urban installations	Vitaliano et al. (2024), -Di Cosola et al. (2022), Tremper and Green (2018)

Urban forests and green corridors	Establishing parks, street trees, green belt and other connected green spaces within urban areas, as well as dense tree plantations	**Absorption of pollutants:** Arboreal vegetation captures carbon dioxide and other airborne contaminants, improve environmental conditions **Heat mitigation:** Foliage coverage of green corridors reduces urban temperature **Social engagement venues:** Furnishes locations for physical activities and communal interactions, promoting societal health **Fauna sanctuaries:** metropolitan biodiversity cultivation provide space for various animal populations	Li et al. (2016), Badach et al. (2020), Lin and Li (2025)
Green alleys and streetscapes	Revitalizing streets and alleys with sustainable drainage system, vegetation, and permeable pavements to turn unused urban areas into green useful places	**Water runoff control:** Porous surface promotes water recharge and reduce water runoff **Temperature mitigation:** Plants decrease environmental and surface temperature **Improved visual appeal:** Enhances the appearance of city structures **Public interaction:** For community and social gathering, create a welcoming space in urban areas	Dai et al. (2020), Newell et al. (2013), Ni et al. (2025)
Parks and green spaces	Urban lungs are open green spaces with a variety of vegetation	Increases carbon sequestration, boosts well-being, and improves air quality	Escobedo et al. (2019)
Bio-filtration systems	Microorganisms and plant roots are used as green filters to purify water and air	Improve biodiversity and remove aerosols, particulate matter, and CO_2	Gonzalez-Meler et al. (2013), Tomson et al. (2025)

10.4.2 Filtering particulate matter deposition

To reduce urban air pollution, researchers have looked into both artificial and natural solutions. While artificial materials like nanofiber air filters and selective catalysts efficiently eliminate pollutants through filtration or adsorption (Zhang et al., 2016), increase in green infrastructure, which include grasslands, trees and hedges, are examples of nature-based solutions that provide a passive, sustainable way to lower exposure levels and concentrations of air pollutants (Abhijith et al., 2017; Ottosen & Kumar, 2020; Tallis et al., 2011). Airborne particulate matter ($PM_{2.5}$ and PM_{10}) captured on the rough surface stems, bark, and leaves of plants serves as a natural filter preventing the particles from resuspending in the environment (Park et al., 2025). Due to foliar variation throughout the year, trees are categorized into deciduous (broadleaf) and coniferous (evergreen), having various levels of efficiency to remove pollutants (Tomson et al., 2021). For example, green infrastructure affects the concentration of air pollution by changing environmental turbulence apart from deposition. This aerodynamic effect enhances turbulence in open areas and reduced under tree canopy, which has reduced pollutant concentration respectively (Salmond et al., 2013). According to Barnes et al. (2014), enhancing the roughness of the surface decreased the ground level of air pollutants by using operational dispersion modeling (ADMS [Atmospheric Dispersion Modelling System]-Urban, UK). Green infrastructure also gives extra environmental benefits like management of urban surface water runoff and reduces the urban heat island effect (Grey et al., 2018; Loughner et al., 2012). Green infrastructure enhances the air quality by combining with the deposition and dispersion effects on air pollutants. These quantify individual as well as combined effects (dispersed or deposition, dispersed and deposition) on the reduction of concentration and pollutant under 8 green infrastructure scenarios from the past (2015) to the future (2039). A case study of Guildford (UK), green infrastructure planning reduces the concentration of various traffic-level air pollutants, such as $PM_{2.5}$, PM_{10}, and NOx under green infrastructure scenarios. Annual pollutant reduction estimated by green infrastructure influenced the air quality modeling and estimated the pollutant deposition concentration by deposition effect subtracted from the aerodynamic effect (Tiwari & Kumar, 2020).

10.5 Technological innovations

Technological innovations like urban modeling, remote sensing and GIS enable planners to efficiently pinpoint possible locations for green infrastructure development (Bai et al., 2022). Innovation in green technology is greatly aided by financial support. Since the development of green technology necessitates assured, steady, and adequate financial resources, financial support is essential to encourage innovation in green businesses (Lv et al., 2021; Owen et al., 2018). R&D investment and corporate managers' financial literacy are two examples of factors that can impact a company's performance in green innovation (Delgado-Ceballos et al., 2012; García-Pérez-de-Lema et al., 2021; Murillo-Luna et al., 2008). Nevertheless, market mechanisms alone cannot support corporate green innovation due to the high sunk costs of due diligence and high regulatory costs brought on by information asymmetry. Instead, the limitations of traditional financial systems in promoting green technology innovation call for complementary fiscal and monetary policies. These include tax exemptions, reimbursements, and government-sponsored equity investments such as venture capital and crowdfunding initiatives, which are crucial for channeling funds to enterprises (Criscuolo & Menon, 2015; Olmos et al., 2012; Owen et al., 2018). The research investigated the empirical foundation of financing constraints theory by examining how to integrate digital technologies in financial sectors, which affects the support of green technological innovations. Moreover, it elucidates the pathways by which digital finance impacts the innovation of green technologies, focused on basically three areas, such as financial constraints, industrial restructuring, and manufacturing sector development. These approaches provide a comprehensive perspective of their interconnectedness from a microeconomic and macroeconomic view. The research also investigates how environmental decentralization enhances the impact of digital finance on green technologies innovation that utter important for promoting reforms in decentralized environmental systems. Such inquiry is crucial for balanced distribution of environmental responsibilities between local and national governments, which is essential for improving environmental activity (Feng et al., 2022).

10.6 Public-private partnerships

Public-private partnerships (PPPs) offer both financial and technological assistance for extensive urban greening projects. As a component of their Corporate Social Responsibility programs, corporations frequently back these kinds of initiatives. PPP refers to any form of collaboration between the public and private sectors that is intended to promote public welfare or that is seen in the context of major projects that are intended to meet public expectations and develop public services. PPP projects' primary goal is to maximize public service performance (Koscielniak & Gorka, 2016; Leung & Hui, 2005). The following reasons make it feasible;

Through the utilization of private sector resources and expertise, PPPs facilitate the successful execution of projects. Public partners profit from the integrated approach and risk-sharing mechanisms, concentrating on the end result rather than budgetary limitations. In order to ensure efficiency and accountability, PPP success depends on allocating risks to the party most qualified to handle them (Kanakoudis et al., 2007).

There are a variety of partnership models, from full-service to accommodation-only agreements, with government supervision in franchises that generate income. By paying payments to performance, private financing improves infrastructure efficiency and service quality. Failure in the private sector may occur due to improper risk allocation, highlighting the necessity of strategic assignment (Anwar et al., 2017; Bielenberg et al, 2016).

Contractual tasks and deliverable process of PPP vary with different phases of the project, like funding, construction, and assembly over the agreement period. Upon completion of the tenure, the infrastructure or service is transferred to the government. After the design and construction are complete, the private sectors can operate and deliver the facilities according to government norms. It is a long-term method with public and private sector cooperation in project integration aimed at public service provision (Anwar et al., 2017; Koscielniak & Gorka, 2016).

Urban finance should be green-oriented increment of property (residential, multiresidential, farm, commercial, and industrial property), transportation, and water taxes that stimulate density, reduce vehicle traffic, and water consumption. Cities can apply to attract private finance systems, such as bonds, loans, and the carbon market are used to attract the private sector for investment in new projects of green infrastructure development and involved in partnership. The government can benefit from private skills and experience in green infrastructure (Merk et al., 2012). Creating multipurpose urban areas that integrate basic features of green and smart city projects, like good communication and public transportation, application of electric transportation, renewable energy, modern housing, environmental protection, power consumption, waste recycling management, rational use of space, and public safety monitoring. Cities should be free from traffic jams, with extensive green space and modern amenities; the existing solutions require redesigning. Each element can be integrated separately or within a combined project (Koscielniak & Gorka, 2016). The project should be focused on accessibility, flexibility, and green sustainability (Shor, 2022). Strong collaboration between the public and private sectors is essential to the success of these projects, both during the planning and execution stages (Kang et al., 2019). Public-private partnerships are thought to be a promising way to deal with these issues and guarantee that green city projects are carried out successfully (Almarri & Boussabaine, 2025). Sports, tourism, and leisure are the main areas of investment. Urban renewal comes next, and transportation and communication come in third. Science, technology, environmental protection, and housing all see lower investment (Chen et al., 2020; Koscielniak & Gorka, 2016). Some of these projects meet the basic needs of the local population, while others address sustainability, transportation, and contemporary housing in accordance with the principles of green cities (Fell & Mattsson, 2021; Selim & Saeed, 2021). Protection of the environment, public transportation, sports, tourism, housing, regeneration, and technology are important areas. Stakeholder engagement, thorough planning, and the creation of policies that strike a balance between environmental sustainability and urban growth are all necessary to address these issues (Mersal, 2016; Reed, 2008).

10.7 Conclusion and future perspectives

Sustainable urban planning and green infrastructure approaches develop sustainable, resilient, and livable cities as well as mitigate environmental challenges such as air pollution, biodiversity degradation, urban heat island, waste disposal, and climate change. The integration of green roofs, green walls, community gardens, hedges, and vertical gardens helps in improving air quality, temperature regulation, biodiversity conservation, and ecological advantages. The financial resources and private sector skills are leveraged by a public-private partnership, which makes sure long-term feasibility of urban green projects. However, resource allocation and planning regulation enhance technological innovation like urban modeling, remote sensing, geographical information system, and digital finance. The future of urban sustainability utilizes smart innovative technologies with green infrastructure and urban planning to succeed in smart and sustainable development. Innovative techniques like bioengineered flora, AI-based urban design, and Internet of Things-based environmental monitoring, enhance green space effectiveness. Climate adaptability and environmental restoration can be increased by nature-based solutions such as vertical gardens, green corridors, green roofs, green walls, and community gardens. Financial schemes like sustainable investment funds, loans, bonds, and carbon market systems encourage the private sector's involvement in sustainable urban projects. A flexible urban planning framework is necessary for maintaining a balance between high-density urban areas and sustainable green areas. Cities should be focused on smart and sustainable green spaces, ensuring more resilience, a climate-adaptive future, and an urban population. This can be done by policy implementation, multisector partnership, and increased research investment, which implements economic growth with environmental conservation. By integration of these strategies, cities can enhance public health and well-being, and also mitigate environmental challenges.

Conflict of interest

Authors declare no conflict of interest.

References

Abhijith, K. V., Kumar, P., Gallagher, J., McNabola, A., Baldauf, R., Pilla, F., Broderick, B., Di Sabatino, S., & Pulvirenti, B. (2017). Air pollution abatement performances of green infrastructure in open road and built-up street canyon environments; A review. *Atmospheric Environment, 162*, 71–86.

Ahern, J., Cilliers, S., & Niemelä, J. (2014). The concept of ecosystem services in adaptive urban planning and design; A framework for supporting innovation. *Landscape and Urban Planning, 125*, 254–259.

Ahram, F. H., & Zakaria, S. A. (2023). The contribution of green building in reducing the carbon footprint and attaining SDG 13. *IOP Conference Series: Earth and Environmental Science, 1205* Article 012034.

Almarri, K., & Boussabaine, H. (2025). Critical success factors for public–private partnerships in smart city infrastructure projects. *Construction Innovation: Information Process Management, 25*(2), 224–247.

Altimir, N., Kolari, P., Tuovinen, J.-P., et al. (2006). Foliage surface ozone deposition; A role for surface moisture? *Biogeosciences, 3*, 209–228.

Anwar, B., Xiao, Z., Akter, S., & Rehman, R.-U. (2017). Sustainable urbanization and development goals; Strategy through public–private partnerships in a South-Asian metropolis. *Sustainability, 9*(11), 1940.

Badach, J., Dymnicka, M., & Baranowski, A. (2020). Urban vegetation in air quality management: A review and policy framework. *Sustainability, 12*, 1258.

Bai, H., Li, Z., Guo, H., Chen, H., & Luo, P. (2022). Urban green space planning based on remote sensing and geographic information systems. *Remote Sensing, 14*(17) Article 4213.

Baik, J., Kwak, K.-H., Park, S.-B., & Ryu, Y.-H. (2012). Effects of building roof greening on air quality in street canyons. *Atmospheric Environment, 61*, 48–55.

Balany, F., Ng, A. W. M., Muttil, N., Muthukumaran, S., & Wong, M. S. (2020). Green infrastructure as an urban heat island mitigation strategy—A review. *Water, 12*(12), 3577.

Barnes, M. J., Brade, T. K., MacKenzie, A. R., Whyatt, J. D., Carruthers, D. J., Stocker, J., Cai, X., & Hewitt, C. N. (2014). Spatially-varying surface roughness and ground-level air quality in an operational dispersion model. *Environmental Pollution, 185*, 44–51.

Barwise, Y., & Kumar, P. (2020). Designing vegetation barriers for urban air pollution abatement; A practical review for appropriate plant species selection. *Climate and Atmospheric Science, 3*, 12.

Berardi, U., GhaffarianHoseini, A., & GhaffarianHoseini, A. (2014). State-of-the-art analysis of the environmental benefits of green roofs. *Applied Energy, 115*, 411–428.

Bianchini, F., & Hewage, K. (2012). How "green" are the green roofs? Life cycle analysis of green roof materials. *Building and Environment, 48*, 57–65.

Bielenberg, A., Kerlin, M., Oppenheim, J., & Roberts, M. (2016). *Financing change: How to mobilize private-sector financing for sustainable infrastructure*. McKinsey Center for Business and Environment.

Bona, S., Silva-Afonso, A., Gomes, R., Matos, R., & Rodrigues, F. (2023). Nature-based solutions in urban areas: A European analysis. *Applied Sciences, 13*(1) Article 168.

Bozovic, R., Maksimovic, C., Mijic, A., Suter, I., & Van Reeuwijk, M. (2017). *Blue green solutions; A systems approach to sustainable, resilient and cost-efficient urban development*. Imperial College London.

Bustami, R. A., Belusko, M., Ward, J., & Beecham, S. (2018). Vertical greenery systems: A systematic review of research trends. *Building and Environment, 146*, 226–237.

Carter, T., & Butler, C. (2008). Ecological impacts of replacing traditional roofs with green roofs in two urban areas. *Cities and the Environment, 1*(1), 17.

Cavender-Bares, J., Polasky, S., King, E., & Balvanera, P. (2015). A sustainability framework for assessing trade-offs in ecosystem services. *Ecology and Society, 20*(1), 17.

Chaudhuri, S., & Kumar, A. (2022). Urban greenery for air pollution control: A meta-analysis of current practice, progress, and challenges. *Environmental Monitoring and Assessment, 194*, 235.

Chavan, B. L., & Sonwane, N. S. (2012). Anticipated performance index (API) of some tree species grown in Aurangabad city. *International Journal of Environmental Rehabilitation and Conservation, 3*(2), 9–13.

Chen, L., Liu, C., Zou, R., Yang, M., & Zhang, Z. (2016). Experimental examination of effectiveness of vegetation as a bio-filter of particulate matters in the urban environment. *Environmental Pollution, 208*, 198–208.

Chen, Z., Zhao, Y., Zhou, X., & Zhang, L. (2020). Investigating critical factors that encourage private partners to participate in sports and leisure characteristic town public-private partnerships: Evidence from China. *Sustainability, 12*(8).

Costanza, R., D'Arge, R., De Groot, R., Farber, S., Grasso, M., Hannon, B., Limburg, K., Naeem, S., O'Neill, R., Paruelo, J., et al. (1998). The value of the world's ecosystem services and natural capital. *Ecological Economics, 387*, 3–15.

Criscuolo, C., & Menon, C. (2015). Environmental policies and risk finance in the green sector; Cross-country evidence. *Energy Policy, 83*(8), 38–56.

Currie, B. A., & Bass, B. (2008). Estimates of air pollution mitigation with green plants and green roofs using the UFORE model. *Urban Ecosystems, 11*(4), 409–422.

Dai, Y., Jiang, J., Gu, X., Zhao, Y., & Ni, F. (2020). Sustainable urban street comprising permeable pavement and bioretention facilities: A practice. *Sustainability, 12*, 8288.

Davison, A. W., & Blakemore, J. (1976). Effects of air pollution on plants. In T. A. Mansfield (Ed.). *Effects of air pollution on plants* (pp. 209). Cambridge University Press.

de Luca, C., Naumann, S., Davis, M., & Tondelli, S. (2021). Nature-based solutions and sustainable urban planning in the European environmental policy framework; Analysis of the state of the art and recommendations for future development. *Sustainability, 13*(9), 5021.

De Zeeuw, H. (2011). Cities, climate change and urban agriculture. *Urban Agriculture Magazine, 25*, 39–42.

Delgado-Ceballos, J., Aragon-Correa, J. A., Ortiz-De-Mandojana, N., & Rueda-Manzanares, A. (2012). The effect of internal barriers on the connection between stakeholder integration and proactive environmental strategies. *Journal of Business Ethics, 107*(3), 281–293.

Department of Planning and Local Government. (2010). Rain gardens, green roofs, and infiltration systems. In Water Sensitive Urban Design Technical Manual, Government of South Australia, pp. 12–20. https://www.watersensitivesa.com/wp-content/uploads/WSUD_chapter_6.pdf.

Dick, J., Miller, J. D., Carruthers-Jones, J., Dobel, A. J., Carver, S., Garbutt, A., Hester, A., Hails, R., Magreehan, V., & Quinn, M. (2019). How are nature-based solutions contributing to priority societal challenges surrounding human well-being in the United Kingdom: A systematic map protocol. *Environmental Evidence, 8* Article 37.

Dover, J., & Phillips, S. (2015). Particulate pollution capture by green screens along the A38 Bristol Street in Birmingham. *Mobilane Report, 44*. Retrieved from www.mobilane.co.uk.

Duzi, B., Tóth, A., Bihunová, M., & Stojanov, R. (2014). Challenges of urban agriculture; Highlights on the Czech and Slovak Republic Specifics. In *Current Challenges of Central Europe; Society and Environment*; Vávra, J., Lapka, M., Cudlínová, E., Eds.; Univerzita Karlova v Praze, Filozofická Fakulta; Praha, Czech, 82–107.

Dzierżanowski, K., Popek, R., Gawrońska, H., Saebø, A., & Gawroński, S. W. (2011). Deposition of particulate matter of different size fractions on leaf surfaces and in waxes of urban forest species. *International Journal of Phytoremediation, 13*, 1037–1046.

EEA. (2017). Air quality in Europe. In; EEA Report No 13/2017, pp. 80. https://www.eea.europa.eu/publications/air-quality-in-europe-2017.

Egerer, M., & Fairbairn, M. (2018). Gated gardens; Effects of urbanization on community formation and commons management in community gardens. *Geoforum; Journal of Physical, Human, and Regional Geosciences, 96*, 61–69.

Ellis, R. (2010). Sowing the seeds of democracy; Community gardening in Parkdale, Toronto. Master's Thesis, University of Western Ontario, Toronto, ON, Canada.

Escobedo, F. J., Giannico, V., Jim, C. Y., Sanesi, G., & Lafortezza, R. (2019). Urban forests, ecosystem services, green infrastructure and nature-based solutions: Nexus or evolving metaphors? *Urban Forestry & Urban Greening, 37*, 3–12.

Esfahani, A. A., Amini, H. N., Kar, S. S., Hoodaji, M., Shirvani, M., & Porsakhi, K. (2013). Assessment of air pollution tolerance index of higher plants suitable for green belt development in East of Esfahan City, Iran. *Journal of Ornamental Horticultural Plants, 3*(2), 87–94.

Farmer, A. (2002). Effects of particulates. In J. N. B. Bell, & M. Treshow (Eds.). *Air pollution and plant life* (pp. 187–199). (2nd ed). Wiley.

Fell, T., & Mattsson, J. (2021). The role of public-private partnerships in housing as a potential contributor to sustainable cities and communities: A systematic review. *Sustainability, 13*, 7783.

Feng, H., & Hewage, K. (2014). Lifecycle assessment of living walls; Air purification and energy performance. *Journal of Cleaner Production, 69*, 91–99.

Feng, S., Zhang, R., & Li, G. (2022). Environmental decentralization, digital finance and green technology innovation. *Structural Change and Economic Dynamics, 61*, 70–83.

Ferrini, F., Fini, A., Mori, J., & Gori, A. (2020). Role of vegetation as a mitigating factor in the urban context. *Sustainability, 12*(10), 4247 Article.

Food and Agriculture Organization of the United Nations (FAOUN) (2007). *Profitability and sustainability of urban and peri-urban agriculture*. Rome, Italy: FAO.

Ford, A., Barr, S., Dawson, R., Virgo, J., Batty, M., & Hall, J. (2019). A multi-scale urban integrated assessment framework for climate change studies; A flooding application. *Computers, Environment and Urban Systems, 75*, 229–243.

Francis, L. F. M., & Jensen, M. B. (2017). Benefits of green roofs; A systematic review of the evidence for three ecosystem services. *Urban Forestry and Urban Greening, 28*, 167–176.

García-Pérez-de-Lema, D., Ruiz-Palomo, D., & Diéguez-Soto, J. (2021). Analyzing the roles of CEO's financial literacy and financial constraints on Spanish SMEs' technological innovation. *Technology in Society, 64*(2) Article 101519.

Gascon, M., Mas, M. T., Martínez, D., Dadvand, P., Forns, J., Plasència, A., & Nieuwenhuijsen, M. J. (2015). Mental health benefits of long-term exposure to residential green and blue spaces; A systematic review. *International Journal of Environmental Research and Public Health, 12*, 4354–4379.

Ghosh, S. (2017). Sustainable urban design and planning for precincts. In M. A. Abraham (Ed.). *Encyclopedia of sustainable technologies* (pp. 13–30). Elsevier.

Gil-Ozoudeh, I., Iwuanyanwu, O., Okwandu, A. C., & Ike, C. S. (2023). Sustainable urban design; The role of green buildings in shaping resilient cities. *International Journal of Applied Research in Social Sciences, 5*(10), 674–692.

Gómez-Baggethun, E., & Barton, D. N. (2013). Classifying and valuing ecosystem services for urban planning. *Ecological Economics, 86*, 235–245.

Gonzalez-Meler, M. A., Cotner, L. A., Massey, D. A., Zellner, M. L., & Minor, E. S. (2013). The environmental and ecological benefits of green infrastructure for stormwater runoff in urban areas. *JSM Environmental Science & Ecology, 1*(2), 1007.

Gray, L., Guzman, P., Glowa, K. M., & Drevno, A. G. (2013). Can home gardens scale up into movements for social change? The role of home gardens in providing food security and community change in San Jose, California. *Local Environment. The International Journal of Justice and Sustainability, 18*(7), 772–786.

Grecia, D. A.-M., Arturo, S. T.-S., & Marlenne, G.-R. (2025). Implications of air pollution on trees located in urban areas: A review. *Earth, 6* Article 38.

Grey, V., Livesley, S. J., Fletcher, T. D., & Szota, C. (2018). Tree pits to help mitigate runoff in dense urban areas. *Journal of Hydrology, 565*, 400–410.

Gromke, C., Jamarkattel, N., & Ruck, B. (2016). Influence of roadside hedgerows on air quality in urban street canyons. *Atmospheric Environment, 139*, 75–86.

Grote, R., Samson, R., Alonso, R., Amorim, J. H., Cariñanos, P., Churkina, G., Fares, S., Thiec, D. L., Niinemets, Ü., Mikkelsen, T. N., Paoletti, E., Tiwary, A., & Calfapietra, C. (2016). Functional traits of urban trees; Air pollution mitigation potential. *Frontiers in Ecology and the Environment, 14*(10), 543–550.

GSA. (2011). The benefits and challenges of green roofs on public and commercial buildings; A report of the United States General Service Administration. Retrieved from https://www.gsa.gov/cdnstatic/The_Benefits_and_Challenges_of_Green_Roofs_on_Public_and_Commercial_Buildings.pdf.

Haase, D., Larondelle, N., Andersson, E., Artmann, M., Borgström, S., Breuste, J., Gomez-Baggethun, E., Gren, Å., Hamstead, Z., Hansen, R., et al. (2014). A quantitative review of urban ecosystem service assessments; Concepts, models, and implementation. *Ambio, 43*, 413–433.

Hamling, A., Van Dam, C., & Dubbeldee, V. (2024). Vertical green landscape facades as a reducer of pollution in high-density cities; A review. *Urban Planning and Construction, 2*(1).

He, Y., Yu, H., Ozaki, A., Dong, N., & Zheng, S. (2017). An investigation on the thermal and energy performance of living wall system in Shanghai area. *Energy and Buildings, 140*, 324–335.

Hedges Direct. (2023). *How hedging can reduce air pollution*. https://www.hedgesdirect.co.uk/advice/hedging-reduce-air-pollution.

Hewitt, C. N., Ashworth, K., & MacKenzie, A. R. (2020). Using green infrastructure to improve urban air quality (GI4AQ). *Ambio, 49*(1), 62–73.

Holt, E. A., & Miller, S. W. (2011). Bioindicators; Using organisms to measure environmental impacts. *Nature Education Knowledge, 2*(2), 1–10.

Hunter, A. M., Williams, N. S. G., Rayner, J. P., Aye, L., Hes, D., & Livesley, S. J. (2014). Quantifying the thermal performance of green façades; A critical review. *Ecological Engineering, 63*, 102–113.

Janhall, S. (2015). Review on urban vegetation and particle air pollution—Deposition and dispersion. *Atmospheric Environment, 105*, 130–137.

Jayasooriya, V. M., Ng, A. W., Muthukumaran, S., & Perera, B. J. (2016). Green infrastructure practices for improvement of urban air quality. *Urban Forestry and Urban Greening, 21*, 34–47.

Jim, C. Y., & Tsang, S. W. (2011). Biophysical properties and thermal performance of an intensive green roof. *Building and Environment, 46*, 1263–1274.

Joshi, P. C., & Swami, A. (2009). Air pollution-induced changes in the photosynthetic pigment of selected plant species. *Journal of Environmental Biology, 30*(2), 295–298.

Joshi, S. V., & Ghosh, S. (2014). On the air cleansing efficiency of an extended green wall; A CFD analysis of mechanistic details of transport processes. *Journal of Theoretical Biology, 361*, 101–110.

Joustra, C. M., & Yeh, D. H. (2014). Framework for net-zero and net-positive building water cycle management. *Building Research and Information, 43*(1), 121–132.

Kaler, N. S., Kashyap, P., Prasad, H., & Singh, T. J. (2017). Air pollution tolerance index (APTI) of tree species: A review. *International Journal of Chemical Studies, 5*(4), 716–720.

Kanakoudis, V., Papotis, A., Sanopoulos, A., & Gkoutzios, V. (2007). Crucial parameters for PPP projects' successful planning and implementation. In M. Schrenk, V. V. Popovich, & J. Benedikt (Eds.), *REAL CORP 2007 proceedings: Tagungsband*.

Kang, S., Mulaphong, D., Hwang, E., & Chang, C.-K. (2019). Public-private partnerships in developing countries: Factors for successful adoption and implementation. *International Journal of Public Sector Management, 32*(4), 334–356.

Kardan, O., Gozdyra, P., Misic, B., Moola, F., Palmer, L. J., Paus, T., & Berman, M. G. (2015). Neighborhood greenspace and health in a large urban center. *Scientific Reports, 5* Article 11610.

Kateshastabadan, M. A., Azmoodeh, M., & Mohammadhosseini, B. (2025). Experimental evaluation of a vertical green system for indoor noise reduction. *Building Acoustics, 32*(3), 377–395.

Kaur, M., & Nagpal, A. K. (2017). Evaluation of air pollution tolerance index and anticipated performance index of plants and their application in development of green space along the urban areas. *Environmental Science and Pollution Research, 24*, 18881–18895.

Khotbehsaara, E. M., Baghaei Daemei, A., & Asadi Malekjahan, F. (2019). Simulation study of the eco green roof in order to reduce heat transfer in four different climatic zones. *Results in Engineering, 2*, 100010 Article.

Koopmans, M., Keech, D., Sovova, L., & Reed, M. (2017). Urban agriculture and place-making; Narratives about place and space in Ghent, Brno and Bristol. *Moravian Geographical Reports, 25*(2), 154–165.

Koscielniak, H., & Gorka, A. (2016). PPP as a method of financing sustainable urban development. *Transportation Research Procedia, 16*, 227–235.
Kostadinović, D., Jovanović, M., Bakić, V., & Stepanić, N. (2023). Mitigation of urban particulate pollution using lightweight green roof system. *Energy & Buildings, 293*, 1–13 113203.
Kumar, P., Debele, S. E., Khalili, S., Halios, C. H., Sahani, J., Aghamohammadi, N., & Jones, L. (2024). Urban heat mitigation by green and blue infrastructure: Drivers, effectiveness, and future needs. *The Innovation, 5*(2).
Kumar, P., Druckman, A., Gallagher, J., Gatersleben, B., Allison, S., Eisenman, T. S., Hoang, U., Hama, S., Tiwari, A., Sharma, A., Abhijith, K. V., Adlakha, D., McNabola, A., Astell-Burt, T., Feng, X., Skeldon, A. C., de Lusignan, S., & Morawska, L. (2019). The nexus between air pollution, green infrastructure, and human health. *Environment International, 133*, 105181.
Langergraber, G., Pucher, B., Simperler, L., Kisser, J., Katsou, E., Buehler, D., Mateo, M. C. G., & Atanasova, N. (2020). Implementing nature-based solutions for creating a resourceful circular city. *Blue-Green Systems, 2*(2), 173–185.
Leung, B. Y. P., & Hui, E. C. M. (2005). Evaluation approach on public–private partnership (PPP) urban redevelopments. *International Journal of Strategic Property Management, 9*(1), 1–16.
Levkoe, C. Z. (2006). Learning democracy through food justice movements. *Agriculture and Human Values, 23*(1), 89–98.
Li, J. J., Wai, O. W. H., Li, Y. S., Zhan, J., Ho, Y. A., Li, J. J., & Lam, E. (2010). Effect of green roof on ambient CO2 concentration. *Building and Environment, 45*(12), 2644–2651.
Li, X.-B., Lu, Q.-C., Lu, S.-J., He, H.-D., Peng, Z.-R., Gao, Y., & Wang, Z.-Y. (2016). The impacts of roadside vegetation barriers on the dispersion of gaseous traffic pollution in urban street canyons. *Urban Forestry and Urban Greening, 17*, 80–91.
Lin, H., & Li, X. (2025). The role of urban green spaces in mitigating the urban heat island effect: A systematic review from the perspective of types and mechanisms. *Sustainability, 17*(13), 6132.
Litschike, T., & Kuttler, W. (2008). On the reduction of urban particle concentration by vegetation – A review. *Meteorologische Zeitschrift, 17*(3), 229–240.
Loughner, C. P., Allen, D. J., Zhang, D. L., Pickering, K. E., Dickerson, R. R., & Landry, L. (2012). Roles of urban tree canopy and buildings in urban heat island effects; Parameterization and preliminary results. *Journal of Applied Meteorology and Climatology, 51*, 1775–1793.
Louis, S. H., & Jim, C. Y. (2017). Subtropical summer thermal effects of wirerope climber green walls with different air-gap depths. *Building and Environment, 126*, 1–12.
Lv, C., Shao, C., & Lee, C. C. (2021). Green technology innovation and financial development; Do environmental regulation and innovation output matter? *Energy Economics, 98* Article 105237.
Mares, T. M., & Pena, D. G. (2011). Environmental and food justice. In *Cultivating food justice; race, class, and sustainability*. Boston, MA, USA: MIT Press, 197–219.
Martellozzo, F., Landry, J. S., Plouffe, D., Seufert, V., Rowhani, P., & Ramankutty, N. (2014). Urban agriculture; A global analysis of the space constraint to meet urban vegetable demand. *Environmental Research Letters, 9*(6), 1–8.
Martin, G., Clift, R., & Christie, I. (2016). Urban cultivation and its contributions to sustainability; nibbles of food but oodles of social capital. *Sustainability, 8*(4), 409.
Mayrand, F., & Clergeau, P. (2018). Green roofs and green walls for biodiversity conservation: A contribution to urban connectivity. *Sustainability, 10*(4) Article 985.
Mellino, S., Ripa, M., Zucaro, A., & Ulgiati, S. (2014). An emergy–GIS approach to the evaluation of renewable resource flows; A case study of Campania Region, Italy. *Ecological Modelling, 271*, 103–112.

Merk, O., et al. (2012). Financing green urban infrastructure. In *OECD regional development working papers*. OECD Publishing, 19.

Mersal, A. (2016). Sustainable urban futures: Environmental planning for sustainable urban development. *Procedia Environmental Sciences, 34*, 49–61.

Michalina, D., Mederly, P., Diefenbacher, H., & Held, B. (2021). Sustainable urban development; A review of urban sustainability indicator frameworks. *Sustainability, 13*, 9348.

Miedema, J.M., Desjardins, E., & Marshall, K. (2013). 'Not Just a Passing Fancy'; How community gardens contribute to healthy and inclusive neighbourhoods. *AgriUrban; Quèbec, QC, Canada*. Retrieved February 20, 2025 from; http://chd.region.waterloo.on.ca/en/researchResourcesPublications/resources/Community_Gardening_Storytelling_Project.pdf

Min, K. (2025). Exploring research fields in green buildings and urban green spaces for carbon-neutral city development. *Buildings, 15* Article 1463.

Morakinyo, T. E., Lam, Y. F., & Hao, S. (2016). Evaluating the role of green infrastructures on near-road pollutant dispersion and removal; Modelling and measurement. *Journal of Environmental Management, 182*, 595–605.

Murillo-Luna, J. L., Garcés-Ayerbe, C., & Rivera-Torres, P. (2008). Why do patterns of environmental response differ? A stakeholders' pressure approach. *Strategic Management Journal, 29*(11), 1225–1240.

Musakwa, W., & Niekerk, A. V. (2014). Earth observation for sustainable urban planning in developing countries: Needs, trends, and future directions. *Journal of Planning Literature, 29*(1), 3–21.

Nemitz, E., Vieno, M., Carnell, E., Fitch, A., Steadman, C., Cryle, P., Holland, M., Morton, R. D., Hall, J., Mills, G., Hayes, F., Dickie, I., Carruthers, D., Fowler, D., Reis, S., & Jones, L. (2020). Potential and limitation of air pollution mitigation by vegetation and uncertainties of deposition-based evaluations. *Philosophical Transactions of the Royal Society A: Mathematical, Physical and Engineering Sciences, 378*(2183), 20190320. https://doi.org/10.1098/rsta.2019.0320.

Newell, J. P., Seymour, M., Yee, T., Renteria, J., Longcore, T., Wolch, J. R., & Shishkovsky, A. (2013). Green alley programs: Planning for a sustainable urban infrastructure? *Cities (London, England), 31*, 144–155.

Ni, Y., Lin, L., Xia, H., & Wang, X. (2025). Urban street greening and resident comfort: An integrated approach based on high-precision shadow distribution and facade visual assessment. *Sustainability, 17*(3), 1026.

Oberndorfer, E., Lundholm, J., Bass, B., Coffman, R. R., Doshi, H., Dunnett, N., Gaffin, S., Kohler, M., Liu, K. K. Y., & Rowe, B. (2007). Green roofs as urban ecosystems; Ecological structures, functions, and services. *Bioscience, 57*(10), 823–833.

Ohmer, M. L., Meadowcroft, P., Freed, K., & Lewis, E. (2009). Community gardening and community development: Individual, social, and community benefits of a community conservation program. *Journal of Community Practice, 17*(4), 377–399.

Okvat, H. A., & Zautra, A. J. (2011). Community gardening: A parsimonious path to individual, community, and environmental resilience. *American Journal of Community Psychology, 47*(3–4), 374–387.

Oquendo-Di Cosola, V., Olivieri, F., & Ruiz-García, L. (2022). A systematic review of the impact of green walls on urban comfort: temperature reduction and noise attenuation. *Renewable and Sustainable Energy Reviews, 162*, 112463. https://doi.org/10.1016/j.rser.2022.112463.

Olmos, L., Ruester, S., & Liong, S. J. (2012). On the selection of financing instruments to push the development of new technologies; Application to clean energy technologies. *Energy Policy, 43*(4), 252–266.

Ottelé, M., Perini, K., Fraaij, A. L. A., Haas, E. M., & Raiteri, R. (2011). Comparative life cycle analysis for green façades and living wall systems. *Energy and Buildings, 43*(12), 3419–3429.

Ottosen, T. B., & Kumar, P. (2020). The influence of the vegetation cycle on the mitigation of air pollution by a deciduous roadside hedge. *Sustainable Cities and Society, 53*, 101919.

Owen, R., Brennan, G., & Lyon, F. (2018). Enabling investment for the transition to a low carbon economy; Government policy to finance early-stage green innovation. *Current Opinion in Environmental Sustainability, 31*(2), 137–145.

Palermo, S. A., & Turco, M. (2020). Green wall systems: Where do we stand? *IOP Conference Series: Earth and Environmental Science, 410* Article 012013.

Park, J., Bui, H.-T., Lee, E., Lim, H.-S., Lim, H.-B., & Park, B.-J. (2025). Accumulation of particulate matter, heavy metals, and air pollution tolerance index of 10 species of urban forest plants. *Water, Air, & Soil Pollution, 236* 233, 1-17.

Pearsall, H., Gachuz, S., Rodriguez Sosa, M., Schmook, B., van der Wal, H., & Gracia, M. A. (2016). Urban community garden agrobiodiversity and cultural identity in Philadelphia, Pennsylvania. *U.S. Geographical Review, 106*(1), 86–104.

Perez, G., Coma, J., Martorell, I., & Cabeza, L. F. (2014). Vertical Greenery Systems (VGS) for energy saving in buildings; A review. *Renewable and Sustainable Energy Reviews, 39*, 139–165.

Perez, G., Rincon, L., Vila, A., Gonzalez, J. M., & Cabeza, L. F. (2011). Green vertical systems for buildings as passive systems for energy savings. *Applied Energy, 88*, 4854–4859.

Pettit, T., Irga, P. J., & Torpy, F. R. (2019). The in-situ pilot-scale phytoremediation of airborne VOCs and particulate matter with an active green wall. *Air Quality, Atmosphere and Health, 12*, 33–44.

Pickett, S. T. A., Boone, C. G., McGrath, B. P., Cadenasso, M. L., Childers, D. L., Ogden, L. A., McHale, M., & Grove, J. M. (2013). Ecological science and transformation to the sustainable city. *Cities (London, England), 32*, S10–S20.

Pugh, M. T. A., MacKenzie, A. R., Whyatt, J. D., & Hewitt, C. N. (2012). Effectiveness of green infrastructure for improvement of air quality in urban street canyons. *Environmental Science and Technology, 46*(13), 7692–7699.

Rai, P. J., Panda, L. L. S., Chutia, B. M., & Singh, M. M. (2013). Comparative assessment of Air Pollution Tolerance Index (APTI) in the industrial (Rourkela) and non-industrial area (Aizawl) of India; An ecomanagement approach. *African Journal of Environmental Science and Technology*, 7(10), 944–948.

Raymond, C. M., et al. (2017). A framework for assessing and implementing the co-benefits of nature-based solutions in urban areas. *Environmental Science and Policy, 77*, 15–24.

Reed, M. S. (2008). Stakeholder participation for environmental management: A literature review. *Biological Conservation, 141*(10), 2417–2431.

Rowe, D. B. (2011). Green roofs as a means of pollution abatement. *Environmental Pollution, 159*(8-9), 2100–2110.

Salmond, J. A., Williams, D. E., Laing, G., Kingham, S., Dirks, K., Longley, I., & Henshaw, G. S. (2013). The influence of vegetation on the horizontal and vertical distribution of pollutants in a street canyon. *Science of the Total Environment, 443*, 287–298.

Sänger, P., & Splittgerber, V. (2016). The CityTree: A vertical plant filter for enhanced temperature management. In W. Leal Filho (Ed.). *Innovation in climate change adaptation. Climate change management.* Cham: Springer.

Selim, A. M., & Saeed, D. M. (2021). Infrastructure projects for green cities between implementation challenges and efficiency indicators. *Transport, 5*, 6.

Semeraro, T., Aretano, R., & Pomes, A. (2019). Green roof technology as a sustainable strategy to improve urban water availability. *IOP Conference Series: Materials Science and Engineering, 471* Article 092065.

Semeraro, T., Scarano, A., Buccolieri, R., Santino, A., & Aarrevaara, E. (2021). Planning of urban green spaces; An ecological perspective on human benefits. *Land, 10*, 105.

Shafique, M., Kim, R., & Rafiq, M. (2018). Green roof benefits, opportunities and challenges–A review. *Renewable and Sustainable Energy Reviews, 90*, 757–773.

Shao, Z., Sumari, N. S., Portnov, A., Ujoh, F., Musakwa, W., & Mandela, P. J. (2020). Urban sprawl and its impact on sustainable urban development: A combination of remote sensing and social media data. *Geo-spatial Information Science, 24*(2), 241–255. https://doi.org/10.1080/10095020.2020.1787800.

Shor, I. M. (2022). Modern aspects of urban infrastructure development by means of public and private partnership. *Regionalnaya Ekonomika. Yug Rossii,* 46–55.

Speak, A. F., Rothwell, J. J., Lindley, S. J., & Smith, C. L. (2012). Urban particulate pollution reduction by four species of green roof vegetation in a UK city. *Atmospheric Environment, 61*, 283–293.

Sternberg, T., Viles, H., Cathersides, A., & Edwards, M. (2010). Dust particulate absorption by ivy (Hedera helix L.) on historic walls in urban environments. *Science of the Total Environment, 409*(1), 162–168.

Tallis, M., Taylor, G., Sinnett, D., & Freer-Smith, P. (2011). Estimating the removal of atmospheric particulate pollution by the urban tree canopy of London, under current and future environments. *Landscape and Urban Planning, 103*(2), 129–138.

Tang, K. H. D. (2023). Green walls as mitigation of urban air pollution: A review of their effectiveness. *Research in Ecology, 5*(2), 1–13. https://doi.org/10.30564/re.v5i2.5710.

Taylor, J. R., & Lovell, S. T. (2013). Urban home food gardens in the Global North; Research traditions and future directions. *Agriculture and Human Values, 31*(2), 285–305.

Tiwari, A., & Kumar, P. (2020). Integrated dispersion-deposition modelling for air pollutant reduction via green infrastructure at an urban scale. *Science of the Total Environment, 723*, 138078.

Tomson, M., Kumar, P., Barwise, Y., Perez, P., Forehead, H., French, K., Morawska, L., & Watts, J. F. (2021). Green infrastructure for air quality improvement in street canyons. *Environment International, 146*, 106288.

Tomson, N., Michael, R. N., & Agranovski, I. E. (2025). Classic theory of aerosol filtration for application to urban green infrastructure. *Water, Air, & Soil Pollution, 236*, 188.

Tong, Z., Baldauf, R. W., Isakov, V., Deshmukh, P., & Max Zhang, K. (2016). Roadside vegetation barrier designs to mitigate near-road air pollution impacts. *Science of the Total Environment, 541*, 920–927.

Tremper, A. H., & Green, D. C. (2018). The impact of a green screen on concentrations of nitrogen dioxide at Bowes Primary School, Enfield. Retrieved June 5, 2020, from https://www.londonair.org.uk/london/reports/Green_Screen_Enfield_Report_final.pdf.

Tremper, A.H., Green, D.C., Chatter-Singh, D., & Eleftheriou-Vaus, K. (2015). Impact of green screens on concentrations of particulate matter and oxides of nitrogen in near-road environments. Retrieved June 8, 2020, from http://www.londonair.org.uk/london/reports/GreenScreen_Report.pdf.

UN-Habitat. (2020). World cities report 2020; The value of sustainable urbanization. United Nations Human Settlements Programme (UN-Habitat).

United Nations Department of Economic and Social Affairs Population Division. (2015). World Urbanization Prospects; the 2014 Revision, (ST/ESA/SER.A/366).

United Nations Department of Economic and Social Affairs Population Division. (2016). Policies on Spatial Distribution and Urbanization. Policies Spat. Distrib. Urban. Bookl. https//www.un.org/en/development/desa/population/publications/pdf/policy/DataBookletUrbanizationPolicies.pdf.

United Nations Department of Economic and Social Affairs Population Division. (2018). World Urbanization Prospects; The 2018 Revision (Online Edition).

Vacek, P., Struhala, K., & Matějka, L. (2017). Life-cycle study on semi-intensive green roofs. *Journal of Cleaner Production, 154*, 203–213.

Varela-Stasinopoulou, D. S., Nektarios, P. A., Ntoulas, N., Trigas, P., & Roukounakis, G. I. (2023). Sustainable growth of medicinal and aromatic Mediterranean plants growing as communities in shallow substrate urban green roof systems. *Sustainability, 15*(7) Article 5940.

Vijayaraghavan, K. (2016). Green roofs; A critical review on the role of components, benefits, limitations, and trends. *Renewable and Sustainable Energy Reviews, 57*, 740–752.

Vitaliano, S., Cascone, S., & D'Urso, P. R. (2024). Mitigating built environment air pollution by green systems: An in-depth review. *Applied Sciences, 14*(15), 6487.

Vos, P. E. J., Maiheu, B., Vankerkom, J., & Janssen, S. (2013). Improving local air quality in cities; To tree or not to tree? *Environmental Pollution, 183*, 113–122.

Wei, X., Lyu, S., Yu, Y., Wang, Z., Liu, H., Pan, D., & Chen, J. (2017). Phylloremediation of air pollutants: Exploiting the potential of plant leaves and leaf-associated microbes. *Frontiers in Plant Science, 8* Article 1318.

White, M. M. (2011). D-town farm; African American resistance to food insecurity and the transformation of Detroit. *Environmental Practice, 13*(4), 406–417.

Whittinghill, L. J., & Rowe, D. B. (2012). The role of green roof technology in urban agriculture. *Renewable Agriculture and Food Systems, 27*(4), 314–322.

Williams, R., Goodwell, A., Richardson, M., Le, P. V. V., Kumara, P., & Stillwell, A. S. (2016). An environmental cost–benefit analysis of alternative green roofing strategies. *Ecological Engineering, 95*, 1–9.

Wood, E., Harsant, A., Dallimer, M., Cronin de Chavez, A., McEachan, R. R. C., & Hassall, C. (2018). Not all green space is created equal: Biodiversity predicts psychological restorative benefits from urban green space. *Frontiers in Psychology, 9* Article 2320.

Yadav, L., Kumar, S., & Kumar, K. (2024). Vertical gardening (Chapter 2). In *Gardening*. Emerald Publishing House.

Yang, H. S., Kang, J., & Choi, M. S. (2012). Acoustic effects of green roof systems on a low-profiled structure at street level. *Building and Environment, 50*, 44–55.

Yang, J., Yu, Q., & Gong, P. (2008). Quantifying air pollution removal by green roofs in Chicago. *Atmospheric Environment, 42*(34), 7266–7273.

Yang, W., Hyndman, D. W., Winkler, J. A., Viña, A., Deines, J. M., Lupi, F., Luo, L., Li, Y., Basso, B., Zheng, C., Ma, D., Li, S., Liu, X., Zheng, H., Cao, G., Meng, Q., Ouyang, Z., & Liu, J. (2016). Urban water sustainability: Framework and application. *Ecology and Society, 21*(4), 4. http://dx.doi.org/10.5751/ES-08685-210404.

Yuen, B. (2011). Urban planning in Southeast Asia: Perspective from Singapore. *Town Planning Review, 82*(2), 145–167.

Zhang, R., Liu, C., Hsu, P. C., Zhang, C., Liu, N., Zhang, J., Lee, H. R., Lu, Y., Qiu, Y., Chu, S., & Cui, Y. (2016). Nanofiber air filters with high-temperature stability for efficient PM2.5 removal from the pollution sources. *Nano Letters, 16*, 3642–3649.

Zimmermann, F. (2016). *Nachhaltigkeit wofür? Von Chancen und Herausforderungen für eine nachhaltige Zukunft*. Berlin, Germany: Springer 288.

CHAPTER ELEVEN

Integrating urban heat island mitigation and climate-responsive planning: policy and management strategies for sustainable urban landscapes

Syed Zaki Ahmed and Tejaswini Singh
Department of Ecology and Environmental Sciences, School of Life Sciences, Pondicherry University, Puducherry, India

11.1 Introduction

Urbanization represents one of the most profound anthropogenic transformations of the Earth's surface, reshaping natural landscapes into complex socio-ecological systems. Over the past few decades, the global urban population has grown rapidly, with projections indicating that nearly 68% of the world's population will reside in urban areas by 2050 (IPCC, 2022). This accelerated urban expansion, particularly in developing countries, has led to extensive land-use changes, increased impervious surfaces, and significant alterations in local and regional climates. Urban areas are now recognized as critical hotspots of climate change impacts and vulnerabilities, requiring urgent integration of mitigation and adaptation strategies within urban planning frameworks (Seto et al., 2014).

One of the most prominent manifestations of urban-induced climatic alteration is the urban heat island (UHI) effect, wherein urban areas exhibit higher temperatures than their rural surroundings (Arnfield, 2003; Oke, 1982). This phenomenon arises due to multiple factors, including reduced vegetation cover, increased heat-absorbing materials such as concrete and asphalt, and anthropogenic heat emissions. According to the Intergovernmental Panel on Climate Change Sixth Assessment Report (AR6), urban centers are consistently warmer than surrounding rural areas, and this temperature difference is expected to intensify with continued urbanization (IPCC, 2021). Furthermore, there is very high confidence that future urban

Urban Air Pollution and Plant Responses
DOI: https://doi.org/10.1016/B978-0-443-40320-0.00010-3

growth will amplify warming trends, particularly during nighttime, where heat retention by built surfaces leads to elevated nocturnal temperatures (IPCC, 2021). This intensification is especially critical in tropical and subtropical cities, where baseline temperatures are already high.

The interaction between urbanization and global climate change further exacerbates thermal stress in cities. The IPCC highlights that human-induced climate change has already increased the frequency and intensity of heatwaves since the mid-20th century, and these trends are projected to continue (IPCC, 2021). When combined with the UHI effect, heatwaves in urban areas become more severe, longer-lasting, and spatially heterogeneous. Indeed, future projections suggest that extreme heat risks could affect a substantial proportion of the urban population, particularly in the Global South (IPCC, 2022). Such compounded risks underscore the urgent need to address urban thermal environments through climate-responsive planning.

Urban climate dynamics are also intricately linked with air pollution, forming a complex feedback system that influences both environmental quality and human health. High temperatures accelerate photochemical reactions, leading to increased formation of secondary pollutants such as ozone (Jacob & Winner, 2009). Conversely, certain air pollutants, including black carbon and aerosols, can influence radiative forcing and atmospheric stability, thereby modifying local climate conditions. Recent studies indicate that simultaneous exposure to elevated temperatures and air pollution significantly increases health risks, with mortality rising substantially under combined stress conditions (Li et al., 2019; Schnell & Prather, 2017). This interaction highlights the need for integrated approaches that address both urban climate and air quality management, rather than treating them as separate domains.

The concept of sustainable urban landscapes has emerged as a critical response to these challenges. Sustainable urban planning emphasizes the integration of ecological principles, energy efficiency, and climate resilience into urban design. Nature-based solutions, such as urban greening, green roofs, and permeable surfaces, play a vital role in mitigating UHI effects by enhancing evapotranspiration, increasing albedo, and reducing surface temperatures. Urban greening has been widely recognized as an effective strategy to cool cities and improve environmental quality (Bowler et al., 2010; Kabisch et al., 2017). The IPCC further notes that such land-use modifications contribute significantly to both mitigation and adaptation, offering cobenefits such as improved air quality and enhanced biodiversity

(IPCC, 2022). These strategies are particularly relevant in rapidly urbanizing regions, where traditional gray infrastructure alone is insufficient to address climate challenges.

Within this broader framework, thermal comfort has gained increasing attention as a key indicator of urban livability and human well-being. Thermal comfort is influenced by multiple factors, including air temperature, humidity, wind speed, and radiation, all of which are affected by urban morphology and land-use patterns (Arnfield, 2003). The intensification of UHI and increasing frequency of heatwaves directly compromise outdoor and indoor thermal comfort, leading to increased energy demand for cooling and heightened vulnerability among urban populations. From a planning perspective, incorporating thermal comfort indices into urban design can facilitate the development of climate-sensitive infrastructure and public spaces.

The implications of urban climate change extend beyond discomfort to significant environmental and public health impacts. Prolonged exposure to high temperatures is associated with heat stress, cardiovascular and respiratory illnesses, and increased mortality, particularly among vulnerable groups such as the elderly, children, and low-income populations (IPCC, 2022). The IPCC emphasizes that climate change poses a serious threat to human well-being and planetary health, with urban populations facing disproportionate risks due to high exposure and limited adaptive capacity (IPCC, 2022). Additionally, increased temperatures can exacerbate energy consumption, strain urban infrastructure, and intensify socio-economic inequalities.

In this context, the integration of UHI mitigation strategies with climate-responsive planning becomes essential for achieving sustainable urban development. Traditional urban planning approaches, often characterized by sectoral and fragmented decision-making, are inadequate to address the interconnected challenges of urbanization, climate change, and environmental degradation. Instead, there is a growing need for holistic, multiscale, and interdisciplinary approaches that incorporate climate science into policy and planning processes (Creutzig et al., 2016; Rosenzweig et al., 2018).

This chapter aims to explore the intersection of UHI mitigation and climate-responsive planning, focusing on policy and management strategies that promote sustainable urban landscapes. By synthesizing insights from IPCC reports and contemporary urban climate literature, the chapter highlights the importance of integrating scientific knowledge into urban governance frameworks. It emphasizes the role of adaptive planning, nature-based solutions, and evidence-based policy interventions in enhancing urban resilience and reducing climate risks.

Ultimately, addressing urban heat and climate challenges requires a paradigm shift towards climate-resilient urbanism, where cities are designed not only as centers of economic growth but also as sustainable and liveable environments. The integration of environmental, social, and climatic considerations into urban planning is crucial for ensuring long-term sustainability and improving the quality of life for urban residents.

11.2 Urban heat island and urban climate dynamics

The UHI effect represents a localized climatic modification driven by the transformation of natural landscapes into built environments. While commonly described as the temperature difference between urban and rural areas, UHI is more accurately understood as a complex alteration of the urban energy balance and atmospheric processes. It manifests across multiple spatial and vertical scales, including the surface urban heat island, which reflects land surface temperature differences, and the atmospheric urban heat island, which includes both canopy layer and boundary layer dynamics (Oke, 1987; Voogt & Oke, 2003). These distinctions are critical for understanding how different components of the urban system contribute to thermal variability.

At its core, UHI is governed by changes in the surface energy balance, where the partitioning of incoming solar radiation into sensible heat, latent heat, and storage heat is significantly altered in urban areas. Natural landscapes typically dissipate a large proportion of energy through evapotranspiration, whereas urban surfaces, characterized by impervious materials, reduce latent heat flux and enhance sensible heat flux (Arnfield, 2003). This shift results in higher surface and near-surface air temperatures. Additionally, urban materials such as concrete, asphalt, and brick possess high thermal inertia, enabling them to absorb and store heat during the day and release it gradually at night, thereby intensifying nocturnal warming (Oke, 1987).

A key driver of UHI dynamics is anthropogenic heat release, which originates from energy consumption in buildings, transportation systems, and industrial activities. Unlike natural heat fluxes, anthropogenic heat is directly injected into the urban atmosphere, often coinciding with peak human activity. In densely populated cities, particularly in developing regions, anthropogenic heat can significantly contribute to the urban energy budget, especially during nighttime when other heat sources diminish (Sailor, 2011). The increasing reliance on air conditioning further exacerbates this phenomenon, as waste heat from cooling systems is released into the surrounding environment, creating a feedback loop that reinforces urban warming.

Land cover transformation is another fundamental factor influencing urban climate dynamics. The replacement of vegetated surfaces with impervious structures not only alters thermal properties but also disrupts hydrological processes. Vegetation plays a crucial role in cooling through evapotranspiration and shading, while also maintaining soil moisture and regulating microclimatic conditions. The loss of green spaces reduces these cooling mechanisms, leading to increased surface temperatures and decreased humidity regulation (Zhou et al., 2014). Moreover, impervious surfaces enhance runoff and reduce water infiltration, further limiting the availability of moisture for evaporative cooling.

Urban climate is also strongly shaped by urban morphology, which determines the spatial arrangement and geometric configuration of buildings and infrastructure. Parameters such as building height, density, street orientation, and sky view factor influence radiative exchange, airflow, and heat storage within urban canyons. Dense urban forms with high aspect ratios tend to trap longwave radiation and restrict convective heat loss, leading to elevated temperatures, particularly at night (Oke, 1987). Reduced sky view factor limits the ability of surfaces to radiate heat back to the atmosphere, while complex building geometries enhance multiple reflections of solar radiation, increasing heat absorption.

The built environment further modulates urban climate through material properties and surface characteristics. Low-albedo materials absorb a greater fraction of solar radiation, while high thermal conductivity facilitates heat transfer into deeper layers, prolonging heat retention. Conversely, reflective and permeable materials can mitigate these effects by enhancing radiative cooling and reducing heat storage (Santamouris, 2015). The widespread use of dark, impervious materials in urban construction has therefore been identified as a key contributor to UHI intensity.

Urban climate dynamics are not static but exhibit significant spatio-temporal variability. UHI intensity varies across different land-use types, with commercial and industrial areas typically experiencing higher temperatures than residential or peri-urban zones. Temporal variations are influenced by diurnal cycles, seasonal changes, and synoptic weather conditions. For instance, UHI intensity is generally stronger during calm and clear nights, when radiative cooling differences between urban and rural areas are most pronounced (Oke, 1987). Seasonal variations also play a role, with higher intensities observed during summer months due to increased solar radiation and reduced soil moisture.

Another important dimension of urban climate dynamics is its interaction with the urban atmospheric boundary layer. The UHI effect can enhance vertical mixing during daytime, leading to increased boundary layer height, while at night, stable atmospheric conditions can trap heat and pollutants near the surface. This has important implications for air quality and pollutant dispersion, as elevated temperatures can accelerate chemical reactions and influence the formation of secondary pollutants such as ozone (Jacob & Winner, 2009). The coupling between UHI and air pollution thus represents a critical component of urban environmental systems, linking thermal processes with atmospheric chemistry and public health outcomes.

Emerging research has further advanced the understanding of intraurban thermal variability through the concept of Local Climate Zones (LCZs). The LCZ framework enables a shift from generalized urban–rural comparisons to spatially explicit, intraurban thermal analysis (Stewart & Oke, 2012). It classifies urban and natural landscapes into 17 standardized categories, including 10 built types (LCZ 1–10) and seven land cover types (LCZ A–G), based on parameters such as building density, surface cover, thermal properties, and anthropogenic heat emissions.

The built LCZ classes range from compact, high-density urban cores to sparsely developed suburban areas. For example, LCZ 1 (compact high-rise) is characterized by dense built-up structures, low sky view factor, and high heat storage, often corresponding to areas of intense UHI. In contrast, LCZ 6 (open low-rise) and LCZ 9 (sparsely built) exhibit greater vegetation cover and improved ventilation, resulting in comparatively lower temperatures. Industrial zones, classified as LCZ 10, often experience elevated temperatures due to significant anthropogenic heat emissions and extensive impervious surfaces.

The land cover LCZ classes (A–G), including dense vegetation (LCZ A), scattered vegetation (LCZ B), and water bodies (LCZ G), typically exhibit lower temperatures due to higher evapotranspiration, shading, and moisture availability. The contrast between built and natural LCZs provides a robust framework for identifying urban thermal hotspots and understanding the cooling potential of vegetation.

One of the key strengths of the LCZ framework lies in its universality and comparability across cities, enabling standardized assessment of urban thermal environments. By linking urban form with temperature patterns, LCZ classification facilitates more targeted and spatially explicit mitigation strategies. Advances in remote sensing and GIS-based mapping have further enabled large-scale application of LCZ analysis across diverse urban settings (Bechtel et al., 2015). In rapidly urbanizing regions, including Indian cities, LCZ-based studies have proven particularly useful for identifying vulnerable zones and informing climate-sensitive planning interventions (Kotharkar et al., 2018).

In addition to physical drivers, socio-economic factors also influence urban climate dynamics by shaping land-use patterns, infrastructure development, and energy consumption. Rapid and often unplanned urbanization in developing countries leads to heterogeneous urban forms, informal settlements, and inadequate infrastructure, all of which can exacerbate thermal stress. Vulnerable populations, particularly those in high-density and low-income areas, are disproportionately affected by UHI due to limited access to cooling resources and green spaces.

From a broader perspective, UHI is increasingly recognized as a critical component of urban climate systems, interacting with regional climate processes and global climate change. The amplification of heat in urban areas not only affects local environments but can also influence mesoscale weather patterns, including precipitation and wind circulation. As cities continue to expand, these interactions are expected to intensify, highlighting the need for integrated approaches to urban climate management.

These interconnected processes demonstrate that urban heat dynamics are closely linked with atmospheric processes and land surface characteristics, while also influencing air pollutant formation and dispersion. This highlights the importance of examining the interactions between UHI, air pollution, and urban vegetation as interconnected components of urban environmental systems.

11.3 Interaction between urban heat island, air pollution and vegetation

Urban environments are strongly influenced by the UHI effect, wherein impervious surfaces absorb and retain heat, significantly increasing ambient temperatures and suppressing evaporative cooling processes (Castillo et al., 2021; Tan et al., 2021). Elevated urban temperatures accelerate photochemical reactions, leading to the formation of secondary pollutants such as ground-level ozone, thereby exacerbating urban air quality issues (Wu et al., 2024). This interaction is further intensified by pollutants such as nitrogen oxides and particulate matter, which can absorb and retain heat, alter surface radiative properties, and contribute to localized warming (Rahaman et al., 2024). Consequently, a feedback loop emerges in which increased temperatures enhance pollutant formation, while pollutants further modify urban thermal conditions, reinforcing both UHI intensity and the formation of "urban pollution islands" (Ernst et al., 2022; Wajeetongratana, 2023).

Vegetation plays a crucial mediating role in this interaction by influencing both thermal and atmospheric processes. Trees and other plant species can absorb and metabolize gaseous pollutants such as volatile organic compounds, with broadleaf species being particularly effective due to their larger surface area and waxy leaf structures that facilitate pollutant deposition (Edeigba et al., 2024). In addition, vegetation contributes to particulate matter removal through mechanisms such as interception, impaction, and gravitational settling, thereby improving urban air quality (Schaefer, 2022). However, the effectiveness of vegetation in pollutant removal is strongly influenced by environmental stressors, including elevated temperature and drought conditions. Such stress can alter physiological processes such as stomatal conductance, photosynthetic activity, and transpiration rates, ultimately reducing the capacity of plants to absorb pollutants (Arsenov et al., 2025; Yang et al., 2025).

Beyond pollutant removal, vegetation significantly modifies urban microclimates through shading and evapotranspiration. These processes reduce surface and air temperatures by converting sensible heat into latent heat flux, thereby mitigating UHI intensity (Yuan, 2024; Zhang et al., 2021). The cooling potential of vegetation varies depending on its structure and density, with tree-dominated systems exhibiting significantly greater cooling capacity compared to treeless green spaces (Joshi et al., 2024). Moreover, the integration of trees and water bodies within urban environments has been shown to enhance cooling efficiency and promote more stable microclimatic conditions (Miles et al., 2023).

The interaction between vegetation and urban climate is further shaped by urban form and design. The configuration of green infrastructure, including green roofs, vertical greening systems, and urban parks, can influence airflow, radiation balance, and pollutant dispersion (Wu et al., 2024). While well-designed vegetation can enhance ventilation and reduce pollutant concentrations, poorly planned green infrastructure—particularly in dense street canyons—may obstruct airflow and lead to localized increases in particulate matter concentrations (Schaefer, 2022). Additionally, certain plant species emit biogenic volatile organic compounds, which can contribute to secondary pollutant formation under high-temperature conditions, further highlighting the complexity of vegetation–atmosphere interactions (Picone et al., 2023).

A comprehensive understanding of the interplay between urban design, vegetation characteristics, and climatic conditions is therefore essential for developing integrated strategies that address both thermal comfort and air quality in urban environments (Costa et al., 2023; Yadav & Singh, 2024). Such strategies leverage the biophysical functions of vegetation—including evapotranspiration, shading, and pollutant uptake—to enhance urban resilience and environmental quality (Linhares et al., 2021). Urban green spaces have been shown to significantly reduce concentrations of key pollutants, including ozone, sulfur dioxide, nitrogen dioxide, carbon monoxide, and particulate matter (Singh et al., 2022). However, their effectiveness depends on appropriate species selection, spatial configuration, and maintenance, particularly under conditions of climatic stress (Salvalai et al., 2023).

These interactions highlight the need for holistic urban design approaches that integrate heat mitigation and air pollution control rather than addressing them as isolated challenges (Roy & Roy, 2024). Evidence suggests that strategically designed green infrastructure can improve urban ventilation, reduce energy demand during heatwaves, and enhance public health outcomes by lowering exposure to extreme temperatures and air pollutants (Joshi et al., 2024; Massaro et al., 2023). Moreover, green infrastructure plays a vital role in regulating short-lived climate pollutants such as ozone and black carbon, which simultaneously influence temperature and air quality (Menon & Sharma, 2021).

The interactions between urban heat, air pollution, and vegetation are illustrated in Fig. 11.1.

Overall, the interdependence of UHI, air pollution, and vegetation underscores the importance of nature-based and design-oriented interventions that can simultaneously regulate urban temperature and atmospheric quality. These insights provide a strong foundation for exploring urban vegetation and nature-based solutions as key strategies for sustainable and climate-responsive urban landscapes.

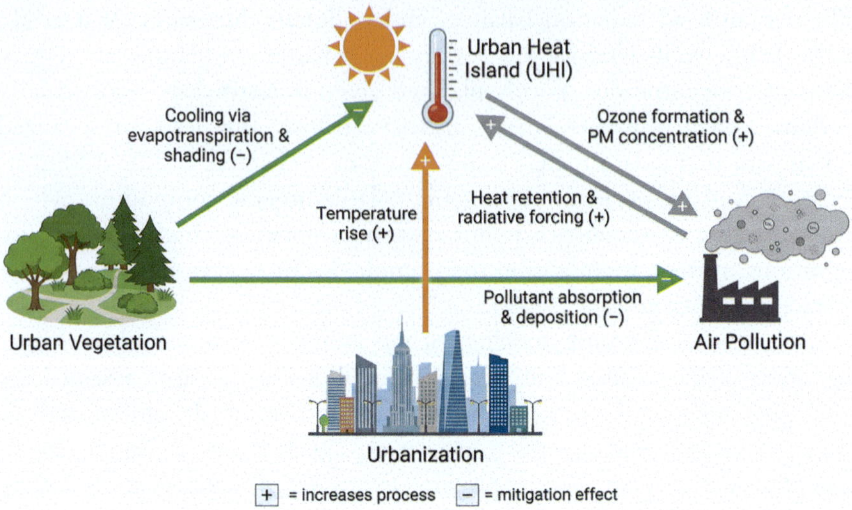

Figure 11.1 Conceptual framework illustrating the interactions between urbanization, urban heat island (UHI) intensity, air pollution, and urban vegetation, highlighting reinforcing (+) and mitigating (–) processes.

11.4 Urban vegetation and nature-based solutions

Urban vegetation and nature-based solutions (NbS) have emerged as key strategies for addressing the interconnected challenges of urban heat, air pollution, and environmental degradation. Urban vegetation plays a critical role in regulating microclimates through shading, evapotranspiration, and modification of surface albedo, thereby reducing urban heat stress while simultaneously improving air quality (Hussain et al., 2024; Menon & Sharma, 2021). A comprehensive understanding of the relationship among urban design, vegetation types, and climatic conditions is therefore essential for developing strategies that address both thermal comfort and atmospheric pollution (Costa et al., 2023; Yadav & Singh, 2024).

Nature-based solutions are defined as cost-effective and sustainable interventions derived from natural systems that deliver environmental, social, and economic benefits while enhancing urban resilience (Menon & Sharma, 2021). The planned incorporation of green infrastructure—including urban forests, street trees, green roofs, green walls, and urban parks—represents a key NbS strategy to mitigate urban heat island effects and improve air quality (Adélaïde et al., 2024; Wu et al., 2024). Such interventions are particularly important in rapidly urbanizing regions, including Indian cities, where urban expansion has led to a significant reduction in green cover and increased vulnerability to heat stress (Ahmad & Hassan, 2025).

Urban vegetation functions as a biological filter for air pollutants through two primary mechanisms: dry deposition for particulate matter and stomatal absorption for gaseous pollutants (Barwise & Kumar, 2020; Gong et al., 2023). Gaseous pollutants are absorbed through leaf stomata during gas exchange, a process influenced by plant physiology and environmental conditions (Xian et al., 2022). Large green spaces, such as urban forests and parks, generally provide greater pollutant removal due to higher leaf area index and canopy density (Gong et al., 2023). However, in certain contexts, smaller and well-distributed green spaces may be more effective, as dense canopies can sometimes restrict airflow and trap pollutants at ground level (Xian et al., 2022). Consequently, the effectiveness of vegetation in air pollution mitigation depends on careful species selection, canopy structure, and spatial configuration (Menon & Sharma, 2021; Wanner et al., 2022).

In addition to air purification, urban vegetation significantly contributes to thermal regulation. Through shading and evapotranspiration, vegetation reduces surface and air temperatures by converting sensible heat into latent heat flux (Zhang et al., 2021). Among NbS, green roofs and green walls are particularly effective in densely built environments where horizontal space is limited. Green roofs improve building insulation, reduce surface temperatures, and lower energy consumption, with studies indicating reductions in air temperature of approximately 1°C (Hussain et al., 2024; Vardhu & Sharma, 2023). Similarly, green walls reduce heat gain in buildings and contribute to localized cooling while also acting as air filters at street level (Perini et al., 2011; Pugh et al., 2012).

Urban forests and street trees represent some of the most effective NbS for mitigating UHI impacts. Trees reduce land surface and ambient temperatures through shading and evapotranspiration, thereby enhancing thermal comfort (Bowler et al., 2010; Livesley et al., 2016). They also function as natural air filters by capturing particulate matter ($PM_{2.5}$ and PM_{10}) and absorbing gaseous pollutants such as nitrogen dioxide (NO_2) and ozone (O_3) (Nowak et al., 2014). In Indian cities such as Delhi and Bengaluru, urban tree cover has been associated with improved micro-climate regulation and enhanced biodiversity (Nagendra & Gopal, 2010). Similarly, in Mumbai, the restoration of mangroves and coastal green belts has been shown to reduce local temperatures by up to 1.5°C, while the loss of green cover has contributed to significant increases in land surface temperatures (Pachouri & Kothari, 2024; Panda, 2025). In Chennai, the restoration of wetlands and green networks has also contributed to localized cooling effects of approximately 1°C (Pachouri & Kothari, 2024).

Urban parks serve as important cooling islands, often exhibiting temperatures several degrees lower than surrounding built-up areas (Spronken-Smith & Oke, 1998). In addition to providing thermal regulation, parks enhance air quality through pollutant deposition and improved air circulation (Tiwary et al., 2009), while also supporting biodiversity by serving as habitats for diverse species (Aronson et al., 2017). Prominent examples such as Central Park in New York and Lodhi Garden in New Delhi illustrate the multifunctional role of urban green spaces in improving thermal comfort, reducing pollution, and supporting ecological systems.

Beyond individual green elements, the integration of blue-green infrastructure (BGI) further enhances the effectiveness of NbS. The combination of water bodies and vegetation creates multifunctional ecological networks that regulate temperature, support biodiversity, and improve hydrological processes (Hussain et al., 2024). Initiatives such as the Blue-Green Master Plan in Delhi aim to restore the balance between built and natural environments by integrating water-sensitive urban design principles. Additionally, ecological corridors and "stepping-stone" habitats, including ponds, green roofs, and urban green patches, facilitate species movement across fragmented landscapes and contribute to long-term ecosystem resilience (Balbi et al., 2021; Beaujean et al., 2021; Perrelet et al., 2024). The multifunctional role of different NbS in regulating urban temperature and air quality is summarized in Table 11.1.

However, the implementation of NbS also involves certain trade-offs. For instance, stormwater ponds and other BGI components may accumulate pollutants, potentially affecting sensitive species (Perrelet et al., 2024). Similarly, inappropriate vegetation design may lead to reduced airflow or increased emissions of biogenic volatile organic compounds, contributing to secondary pollutant formation. Therefore, the successful implementation of NbS requires a balanced and context-specific approach that considers both ecological benefits and potential limitations.

Integrating nature-based solutions into urban planning, as demonstrated in cities such as Bengaluru, Delhi, and Kolkata, enables a transition toward more sustainable and circular urban systems that align engineering objectives with ecological processes (Pachouri & Kothari, 2024; Thyagaraj et al., 2026).

Table 11.1 Biophysical mechanisms and benefits of nature-based solutions (NbS) for UHI mitigation and air quality improvement.

Intervention	UHI mitigation mechanism	Air pollution mitigation	Quantitative evidence	References
Urban trees	Shading, evapotranspiration	PM deposition, NO_2 & O_3 uptake	Air temp ↓ 1–5°C; PM ↓ up to 10%–25%	Bowler et al. (2010), Nowak et al. (2014), Santamouris (2015)
Urban forests	High leaf area index, evapotranspiration	Large-scale pollutant removal	Cooling ~2°C–4°C	Livesley et al. (2016), Nowak et al. (2014)
Green roofs	Albedo increase, evapotranspiration, insulation	Minor PM capture	Roof temp ↓ 20°C–40°C; air temp ↓ ~1°C	Santamouris (2015), Speak et al. (2012)
Green walls	Vertical shading, evapotranspiration	Pollutant filtration	Local temp ↓ 0.5°C–2°C	Perini et al. (2011), Pugh et al. (2012)
Urban parks	Cooling island effect	Pollutant dispersion + deposition	Cooling ~1°C–3°C	Spronken-Smith & Oke (1998), Bowler et al. (2010)
Blue-green infrastructure	Evaporative cooling, vegetation synergy	Indirect improvement	Cooling ~1°C–2°C	Kabisch et al. (2017), Gunawardena et al. (2017)
Permeable/ vegetated surfaces	Soil moisture, reduced heat storage	Reduced dust resuspension	Surface temp ↓ 5°C–15°C	Santamouris (2015), Li et al. (2019)

UHI, urban heat island.

Overall, urban vegetation and NbS offer a multifunctional approach to simultaneously address urban heat, air pollution, and biodiversity challenges. Their effective implementation, however, depends on strategic planning, appropriate design, and integration within broader urban development frameworks, which are further explored in the context of climate-responsive urban planning in the following section.

11.5 Climate-responsive urban planning

Climate-responsive urban planning has emerged as a critical approach to mitigate UHI effects and enhance urban thermal environments. It focuses on designing urban form, infrastructure, and land-use patterns in a manner that optimizes energy balance, promotes natural ventilation, and minimizes heat accumulation. Unlike conventional planning approaches, which often prioritize density and economic efficiency, climate-responsive planning integrates principles of urban climatology into spatial design and governance frameworks (Arnfield, 2003; IPCC, 2022; Oke, 1982). Importantly, this approach also complements nature-based solutions by embedding vegetation and ecological functions within urban design to simultaneously address thermal stress and air quality challenges.

Urban geometry, defined by building height, spacing, and street configuration, plays a fundamental role in regulating solar radiation, heat storage, and airflow within cities. The formation of urban canyons—resulting from closely spaced buildings—significantly alters microclimatic conditions by trapping radiation and reducing longwave heat loss, particularly during nighttime (Oke, 1982). The height-to-width ratio of streets determines shading patterns and radiative exchange, with higher ratios often increasing heat retention. Building orientation further influences solar exposure and wind flow, where aligning structures to minimize direct solar gain and maximize prevailing wind directions can enhance passive cooling and reduce energy demand (Krüger et al., 2010). These strategies are particularly critical in tropical cities, where excessive solar radiation contributes significantly to thermal stress.

Ventilation corridors represent another essential component of climate-responsive planning. These open spaces facilitate air movement across urban areas, allowing heat dissipation and improving pollutant dispersion. When aligned with prevailing wind directions or natural landscape features such as river systems, ventilation corridors can significantly reduce heat stagnation in dense urban environments. Conversely, high-density and unplanned construction often obstructs airflow, leading to reduced convective cooling and intensified UHI effects (Ng et al., 2011). Integrating such corridors into urban master plans requires careful regulation of building height, density, and spatial layout to maintain effective airflow pathways.

Material-based interventions, including cool roofs and cool pavements, provide additional mechanisms to reduce urban heat. Cool roofs, characterized by high solar reflectance and thermal emittance, reflect a substantial portion of incoming solar radiation, thereby reducing heat absorption. Similarly, cool pavements utilize reflective or permeable materials to lower surface temperatures and enhance evaporative cooling. These interventions have been shown to reduce surface temperatures significantly and contribute to overall urban cooling, while also decreasing building energy demand (Santamouris, 2015).

The relationship between urban form and thermal performance is inherently complex. Compact urban forms, characterized by high density and reduced land consumption, can enhance shading and reduce transportation-related emissions. However, without adequate ventilation and green infrastructure, such forms may lead to increased heat accumulation. In contrast, dispersed urban development allows for improved airflow and greater integration of green spaces but may result in urban sprawl and higher energy consumption. Climate-responsive planning, therefore, advocates for an optimized urban form that balances density with ventilation, green infrastructure, and spatial configuration (Creutzig et al., 2016; Seto et al., 2014).

Reducing heat accumulation in urban environments requires an integrated approach that combines design, materials, and ecological strategies. Enhancing urban green cover through parks, street trees, and green roofs promotes evapotranspiration and shading, thereby lowering surface and air temperatures (Bowler et al., 2010; Kabisch et al., 2017). Increasing surface albedo through reflective materials further reduces heat absorption, while the protection and restoration of blue infrastructure—such as lakes and urban water bodies—contributes to microclimate regulation. Additionally, low-impact development techniques improve soil moisture retention and reduce surface heating, while the incorporation of thermal comfort indices into planning standards enables more human-centric urban design. These approaches collectively align with the broader concept of nature-based solutions, which provide cobenefits including improved air quality, biodiversity conservation, and enhanced climate resilience (IPCC, 2022).

Overall, climate-responsive urban planning provides a scientifically grounded framework for mitigating UHI effects while simultaneously addressing air quality and ecological considerations. By integrating urban design, material innovation, and nature-based strategies, cities can reduce heat accumulation and improve thermal comfort in a sustainable manner. Importantly, the successful implementation of these strategies depends on their integration into policy frameworks and governance systems, thereby reinforcing the need for supportive policy and governance mechanisms.

11.6 Policy and governance strategies for climate-responsive urban landscapes

Addressing the challenges of UHI effects and broader urban climate risks requires robust policy and governance frameworks that integrate climate considerations into urban planning and development. As cities expand and climate change intensifies, traditional sectoral approaches are increasingly inadequate. There is a growing emphasis on multilevel governance, wherein national policies, state-level initiatives, and local urban planning mechanisms operate in coordination to promote sustainable and climate-resilient urban landscapes (Bulkeley et al., 2014; IPCC, 2022).

Urban climate governance operates across multiple scales, encompassing international commitments, national missions, and local implementation strategies. At the global level, frameworks such as the Paris Agreement emphasize the role of cities in climate mitigation and adaptation (IPCC, 2022). Translating these commitments into actionable strategies requires strong institutional mechanisms at both national and local levels. In India, urban climate policy is embedded within broader environmental and developmental frameworks involving multiple stakeholders, including central ministries, state governments, urban local bodies, and planning authorities. However, fragmented institutional responsibilities, limited technical capacity, and inadequate integration of scientific data continue to hinder effective implementation (Rosenzweig et al., 2018). Strengthening coordination among these actors is therefore essential for mainstreaming climate considerations into urban governance.

Planning regulations form a critical interface between policy intent and on-ground implementation, shaping land-use patterns, building density, and urban morphology (Seto et al., 2014). Instruments such as master plans, zoning regulations, and building bye-laws have significant implications for urban thermal environments, air quality, and the distribution of green infrastructure. Integrating climate-responsive principles into these frameworks enables cities to enhance green cover, promote the adoption of cool roofs and high-albedo materials (Santamouris, 2015), regulate building heights and orientations to improve ventilation (Oke, 1982), and protect blue infrastructure such as lakes and urban water bodies. Despite these opportunities, climate considerations remain insufficiently embedded within statutory planning processes in many Indian cities, highlighting the need for a transition towards climate-informed urban planning (Creutzig et al., 2016).

The integration of climate mitigation strategies into city planning requires a systems-based approach that links land use, transport, energy, and environmental management (Creutzig et al., 2016). Strategies such as transit-oriented development can reduce vehicular emissions, while energy-efficient building codes contribute to lowering urban energy demand. Expanding urban forests and green corridors enhances evapotranspiration and pollutant removal, and the implementation of NbS provides additional cooling and resilience benefits (Bowler et al., 2010; Kabisch et al., 2017). Such integrated approaches not only mitigate UHI effects but also deliver cobenefits, including improved air quality, reduced energy consumption, and enhanced urban livability (IPCC, 2022).

In the Indian context, several national and city-level initiatives demonstrate evolving efforts to address urban climate and environmental challenges. The National Clean Air Programme, launched in 2019, represents a comprehensive national strategy to reduce particulate matter ($PM_{2.5}$ and PM_{10}) concentrations in urban areas (Ministry of Environment, Forest, & Climate Change [MoEFCC], 2019). Although primarily focused on air quality, its measures—such as increasing green cover, reducing vehicular emissions, and improving urban infrastructure—also contribute indirectly to UHI mitigation and climate resilience. Similarly, the Smart Cities Mission emphasizes sustainable urban development through technology-driven solutions, improved infrastructure, and enhanced governance (Ministry of Housing & Urban Affairs, MoHUA, 2015). While not explicitly framed as a climate program, it provides a platform for integrating climate-responsive planning through green infrastructure, smart mobility, and energy-efficient systems (Creutzig et al., 2016).

Heat Action Plans (HAPs) have emerged as critical tools for climate adaptation, particularly in response to increasing heatwave events. Cities such as Ahmedabad have pioneered these plans, which incorporate early warning systems, public awareness campaigns, and institutional preparedness to reduce heat-related mortality (Gupta et al., 2020). These interventions represent a shift towards risk-based governance approaches. However, their long-term effectiveness depends on integration with structural urban planning measures, including land-use regulation, urban greening, and infrastructure design (Dasgupta et al., 2025).

Urban greening policies and nature-based solutions are increasingly recognized as cost-effective strategies for addressing both UHI and air pollution. Initiatives promoting urban forests, green belts, green roofs, and vertical gardens contribute to temperature reduction, pollutant removal, and biodiversity enhancement (Bowler et al., 2010; Kabisch et al., 2017). These interventions align closely with climate mitigation and adaptation objectives, offering multiple cobenefits. However, their implementation in dense urban areas is often constrained by land availability, maintenance challenges, and competing development priorities (IPCC, 2022).

Despite the presence of multiple policy initiatives, a significant challenge lies in the lack of integration across sectors and governance levels. Air quality management, climate adaptation, urban planning, and public health policies are frequently developed in isolation, leading to fragmented outcomes (Rosenzweig et al., 2018). Addressing this requires a shift towards integrated governance frameworks that promote policy convergence, strengthen the role of urban local bodies, incorporate scientific data and modeling into planning processes, and encourage participatory decision-making. Such approaches can enhance the effectiveness of urban climate strategies and support sustainable urban transformation (IPCC, 2022).

Effective policy and governance frameworks must therefore move beyond sectoral approaches and adopt integrated, multilevel strategies that align urban planning, environmental management, and public health objectives. This integrated approach is essential for translating climate-responsive planning principles into actionable interventions, as reflected in the following case studies from diverse urban contexts.

11.7 Case studies

Building upon the policy and governance frameworks discussed in the previous section, this section presents selected case studies from diverse geographical contexts to illustrate the practical implementation of urban climate resilience strategies. These examples highlight how integrated planning, NbS, and policy interventions can collectively address urban heat, air pollution, and sustainability challenges.

Singapore represents a leading example of the strategic integration of green infrastructure within a highly urbanized and land-constrained environment, evolving into a "City in a Garden." Landmark projects such as Marina Bay Sands SkyPark and Gardens by the Bay exemplify the incorporation of large-scale green infrastructure into urban design (Umoh et al., 2024). In particular, the Supertree Grove and Cloud Forest Dome function as critical ecological assets that mitigate the UHI effect while providing recreational and environmental benefits (Iwuanyanwu et al., 2024). Singapore's urban forestry initiatives, including afforestation and the development of green corridors along major transport routes, further enhance carbon sequestration and contribute to urban cooling (Prasad et al., 2008).

These vegetated systems promote significant cooling through evapotranspiration and shading, with studies reporting reductions in ambient air temperature and mean radiant temperature of up to 4°C and 4.5°C, respectively (Tan & bin Abdul Hamid, 2014). Policy support has played a critical role in enabling such transformations. For instance, the "Skyrise Greenery Incentive Scheme" (2009–15) facilitated the widespread adoption of green roofs and vertical greenery, resulting in substantial expansion of vegetated surfaces across the city (Al-Zu'bi & Mansour, 2017). These integrated approaches not only mitigate UHI effects but also enhance biodiversity and improve urban air quality.

Singapore Changi Airport further demonstrates the integration of NbS within large-scale infrastructure. Features such as indoor forests, green roofs, vertical greenery, and water elements—including the Rain Vortex—contribute to cooling through evapotranspiration and shading, thereby improving microclimatic conditions and reducing dependence on mechanical cooling systems. The incorporation of biophilic design principles has been shown to enhance thermal comfort, air quality, and ecological resilience in tropical urban environments (Tan et al., 2013; Wong et al., 2010).

In the Indian context, the city of Ahmedabad provides a notable example of policy-driven climate adaptation through its HAP, developed in response to increasing heatwave risks (Kjellstrom et al., 2019). Initiated in 2013, the Ahmedabad HAP focuses on early warning systems, public awareness campaigns, and targeted interventions to reduce heat-related morbidity and mortality (Tun et al., 2025). Measures such as the dissemination of weather forecasts, establishment of cooling centers, and promotion of green infrastructure have contributed to significant reductions in heat-related mortality, with estimates suggesting the prevention of approximately 1190 deaths annually following their implementation (Witze, 2021). The success of this model has informed national-level expansion, with over 100 Indian cities developing similar plans under the guidance of the National Disaster Management Authority (Randazza et al., 2023).

Across Europe, numerous cities have implemented "Cool City" initiatives that combine advanced materials, green infrastructure, and urban design strategies to mitigate UHI effects and enhance climate resilience (Nastar, 2020). These approaches include the use of high-albedo materials, permeable pavements, urban forests, and blue infrastructure elements to regulate temperature and improve environmental performance (Aartsen et al., 2018). Additionally, innovations such as phase change materials in building envelopes have been employed to enhance thermal efficiency and reduce cooling demand (Joshi et al., 2024). These initiatives reflect a broader transition toward proactive, multiscalar climate adaptation strategies embedded within urban planning processes (Akinsanola et al., 2025; Śmietanka, 2023).

While such policy-driven and technologically advanced approaches have proven effective in developed urban systems, there is a growing need to examine context-specific, scalable solutions in rapidly urbanizing regions of the Global South. Indian cities, in particular, are increasingly adopting community-driven and nature-based interventions to address UHI effects under local socio-economic and infrastructural constraints.

One such example is the "Patchai Madi Thottam" initiative in Chennai, a rooftop greening program that promotes the conversion of unused terrace spaces into vegetated home gardens within active community neighborhoods, while also encouraging the development of vegetable gardens in municipal institutions to enhance urban greenery, reduce heat stress, and improve environmental awareness. This initiative represents a grassroots approach to urban climate adaptation, combining citizen participation with NbS.

The Chennai Urban Horticulture Initiative further extends this approach by integrating horticultural practices into residential, institutional, and public spaces to enhance biodiversity, improve air quality, and reduce urban heat (An & Dedekorkut-Howes, 2025). Studies have demonstrated that terrace gardens and green roofs in Chennai can significantly improve indoor thermal comfort and reduce surface temperatures in densely built environments (Visvanathan et al., 2024). Additionally, the Chennai Resilience Strategy (2019) outlines an ambitious target of establishing rooftop gardens across hundreds of thousands of households by 2030, reflecting a long-term commitment to integrating nature-based solutions into urban climate planning (Gloria & Gnanasekaran, 2024).

These case studies collectively demonstrate that effective urban climate mitigation requires a combination of policy support, technological innovation, and community engagement. While global cities highlight structured and large-scale interventions, emerging approaches in Indian cities emphasize locally adapted, participatory, and nature-based strategies. Together, these insights underscore the importance of integrating policy, planning, and ecological design to achieve sustainable and climate-resilient urban landscapes.

11.8 Future directions and research needs

As urban systems continue to evolve under the combined pressures of climate change, rapid urbanization, and environmental degradation, there is an increasing need to reorient research and policy towards climate-resilient and sustainability-driven urban development. Future research must move beyond isolated mitigation strategies and instead adopt integrated, policy-driven, and data-informed approaches that align urban development with long-term environmental sustainability and climate goals (Creutzig et al., 2016; IPCC, 2022).

The concept of climate-resilient cities is increasingly intertwined with broader sustainability objectives, including resource efficiency, low-carbon development, and social equity. Urban resilience must extend beyond thermal risk reduction to encompass sustainable urban metabolism, wherein energy, water, and land resources are managed in an efficient and circular manner (Rosenzweig et al., 2018). In this context, future research should focus on developing integrated resilience frameworks that simultaneously address thermal comfort, energy consumption, ecosystem services, biodiversity conservation, and social vulnerability. Furthermore, there is a pressing need to evaluate how UHI mitigation strategies contribute to global sustainability agendas, particularly the Sustainable Development Goals (SDGs), including SDG 11 (Sustainable Cities and

Communities), SDG 13 (Climate Action), and SDG 3 (Good Health and Well-being).

A critical research gap lies in the limited convergence between climate policies and urban planning frameworks. Existing initiatives—such as air quality management programs, heat action plans, and urban development missions—often operate in silos, thereby reducing their overall effectiveness. Future directions should therefore emphasize policy integration and cobenefits-based planning, wherein UHI mitigation is embedded within statutory planning instruments such as master plans, zoning regulations, and building codes. Integrating climate considerations into infrastructure policies and aligning air quality management with climate adaptation and mitigation strategies can significantly enhance urban resilience. This aligns with the growing recognition that urban infrastructure and planning decisions fundamentally shape climate outcomes (Creutzig et al., 2016). Strengthening institutional coordination across sectors and governance levels will thus be essential for enabling effective and sustainable urban transformation.

The transition toward data-driven and smart urban climate governance represents another critical frontier. Advances in geospatial technologies, urban informatics, and smart city systems have enabled real-time monitoring and predictive modeling of urban climate dynamics. Integrating multisource datasets—including remote sensing observations, ground-based measurements, and Internet of Things sensors—can support the development of urban climate dashboards and decision-support systems. Furthermore, the application of machine learning and artificial intelligence offers significant potential for predicting heat risks, optimizing urban design, and enhancing adaptive planning. Such approaches can facilitate evidence-based policymaking, improve governance transparency, and enable dynamic responses to evolving urban climate challenges (Batty, 2018).

Remote sensing and urban climate modeling are not only essential research tools but also critical enablers of climate-responsive policy and planning. Satellite-derived datasets provide spatially explicit information on land surface temperature, vegetation cover, and land-use change, which are fundamental for identifying urban heat hotspots and assessing environmental conditions (Voogt & Oke, 2003). When coupled with advanced urban climate models such as the Weather Research and Forecasting model and ENVI-met (a microclimate simulation software), these tools can simulate the impacts of various planning scenarios, evaluate the effectiveness of mitigation strategies, and support evidence-based policy formulation. Future research should therefore prioritize the development of integrated modeling frameworks that bridge the gap between scientific analysis and urban governance, ensuring that data-driven insights are translated into actionable planning interventions (Mirzaei & Haghighat, 2010).

An equally important emerging priority is the integration of equity and environmental justice into urban climate governance. The impacts of UHI and air pollution are disproportionately experienced by vulnerable populations, including low-income communities, informal settlements, and outdoor workers. These groups often have limited access to cooling infrastructure, green spaces, and adaptive resources, thereby increasing their exposure to thermal stress and health risks. Future research and policy must therefore address the distributional inequalities associated with urban climate impacts, promote inclusive and participatory planning processes, and ensure equitable access to climate adaptation measures. This shift towards just and inclusive urban transitions is essential for ensuring that climate-responsive planning contributes not only to environmental sustainability but also to social well-being and resilience (IPCC, 2022).

Overall, future research must adopt an integrated, interdisciplinary, and equity-oriented approach to urban climate challenges, linking UHI mitigation, air quality management, and urban vegetation within cohesive planning and governance frameworks. These directions provide a foundation for advancing sustainable and climate-resilient urban landscapes, as synthesized in the concluding section of this chapter.

11.9 Conclusion

The growing challenges of urban heat, air pollution, and rapid urbanization underscore the need for integrated and climate-responsive approaches to urban development. This chapter has demonstrated that the UHI effect is not an isolated phenomenon but a complex outcome of interactions between land-use change, built environment characteristics, anthropogenic heat emissions, and atmospheric processes. These dynamics are further intensified by their close linkage with air pollution and the degradation of urban vegetation, collectively shaping the environmental quality and livability of cities.

A key insight emerging from this synthesis is the critical role of urban vegetation and nature-based solutions in addressing multiple urban challenges simultaneously. Through mechanisms such as evapotranspiration, shading, and pollutant uptake, vegetation contributes to temperature regulation, air quality improvement, and biodiversity enhancement. However, the effectiveness of these interventions depends on their strategic integration within urban design, planning, and governance frameworks, rather than their isolated or fragmented implementation.

Climate-responsive urban planning provides a crucial bridge between scientific understanding and practical application. By incorporating principles of urban geometry, ventilation, material innovation, and ecosystem-based design, cities can reduce heat accumulation and improve thermal comfort. At the same time, such approaches must be aligned with broader sustainability goals, including energy efficiency, reduced emissions, and enhanced public health outcomes. The integration of thermal comfort considerations into planning processes further strengthens the human-centric dimension of urban development.

Policy and governance frameworks play a central role in enabling the successful implementation of these strategies. National and city-level initiatives, such as the National Clean Air Programme, Smart Cities Mission, and Heat Action Plans, illustrate evolving efforts to address urban climate risks in India. However, the persistence of fragmented governance structures highlights the need for greater policy convergence, institutional coordination, and incorporation of scientific evidence into decision-making processes. Strengthening the role of urban local bodies and promoting participatory governance are essential for ensuring effective and inclusive implementation.

The case studies presented in this chapter further demonstrate that successful urban climate mitigation requires a combination of policy support, technological innovation, and community engagement. While global cities showcase large-scale, infrastructure-driven solutions, emerging approaches in Indian cities emphasize locally adapted, community-based, and nature-driven interventions. This diversity of strategies highlights the importance of context-specific solutions that align with local environmental, socio-economic, and governance conditions.

Looking ahead, future research and practice must adopt interdisciplinary and data-driven approaches to urban climate management. Advances in remote sensing, modeling, and smart urban systems offer significant potential for improving decision-making and enabling adaptive governance. At the same time, integrating principles of equity and environmental justice is essential to ensure that climate-responsive planning benefits all sections of society, particularly vulnerable populations disproportionately affected by heat and pollution.

Ultimately, achieving sustainable and climate-resilient urban landscapes requires a paradigm shift toward integrated planning and governance that simultaneously addresses UHI, air pollution, and ecological systems. By aligning scientific knowledge with policy, design, and community action, cities can transition towards more resilient, livable, and environmentally sustainable futures.

References

Aartsen, M., Koop, S., Hegger, D., Goswami, B., Oost, J., & Van Leeuwen, K. (2018). Connecting water science and policy in India: Lessons from a systematic water governance assessment in the city of Ahmedabad. *Regional Environmental Change, 18*(8), 2445–2457. https://doi.org/10.1007/s10113-018-1363-1.

Adélaïde, L., Hough, I., Seyve, E., Kloog, I., Fifre, G., Launoy, G., Launay, L., Pascal, M., & Lepeule, J. (2024). Environmental and social inequities in continental France: An analysis of exposure to heat, air pollution, and lack of vegetation. *Journal of Exposure Science & Environmental Epidemiology, 34*, 962–972. https://doi.org/10.1038/s41370-024-00641-6.

Ahmad, N., & Hassan, Q. (2025). Ecosystem services linked to nature-based solutions for resilient and sustainable cities in India. *Frontiers in Water, 6*, 1504492. https://doi.org/10.3389/frwa.2024.1504492.

Akinsanola, A. A., Singhai, P., Taguela, T. N., Folorunsho, A. H., Adeyeri, O. E., Morakinyo, T. E., & Adebiyi, A. A. (2025). A review of urban resilience to weather and climate extremes. *City and Built Environment, 3*(1), 24.

Al-Zu'bi, M., & Mansour, O. (2017). Water, energy, and rooftops: Integrating green roof systems into building policies in the Arab region. *Environment and Natural Resources Research, 7*(2), 11–36.

An, S., & Dedekorkut-Howes, A. (2025). A review of adaptation strategies to increased urban temperatures and heat island effect. *Discover Cities, 2*(1), 32.

Arnfield, A. J. (2003). Two decades of urban climate research: A review of turbulence, exchanges of energy and water, and the urban heat island. *International Journal of Climatology: A Journal of the Royal Meteorological Society, 23*(1), 1–26.

Aronson, M. F., Lepczyk, C. A., Evans, K. L., Goddard, M. A., Lerman, S. B., MacIvor, J. S., & Vargo, T. (2017). Biodiversity in the city: Key challenges for urban green space management. *Frontiers in Ecology and the Environment, 15*(4), 189–196.

Arsenov, D., Borišev, M., Nikolić, N., Horák, R., & Pajević, S. (2025). Physiological and biochemical indicators of urban environmental stress in Tilia, Celtis, and Platanus: A functional trait-based approach. *Plants, 14*(22), 3451. https://doi.org/10.3390/plants14223451.

Balbi, M., Croci, S., Petit, E. J., Butet, A., Georges, R., Madec, L., & Ernoult, A. (2021). Least-cost path analysis for urban greenways planning: A test with moths and birds across two habitats and two cities. *Journal of Applied Ecology, 58*(3), 632–643. https://doi.org/10.1111/1365–2664.13800.

Barwise, Y., & Kumar, P. (2020). Designing vegetation barriers for urban air pollution abatement: A practical review for appropriate plant species selection. *Npj Climate and Atmospheric Science, 3*(1), 12.

Batty, M. (2018). Artificial intelligence and smart cities. *Environment and Planning B: Urban Analytics and City Science, 45*(1), 3–6. 10.1177/2399808317751169.

Beaujean, S., Nor, A. N. M., Brewer, T., Zamorano, J. G., Dumitriu, A. C., Harris, J., & Corstanje, R. (2021). A multistep approach to improving connectivity and co-use of spatial ecological networks in cities. *Landscape Ecology, 36*(7), 2077–2093.

Bechtel, B., Alexander, P. J., Böhner, J., Ching, J., Conrad, O., Feddema, J., et al. (2015). Mapping local climate zones for a worldwide database of the form and function of cities. *ISPRS International Journal of Geo-Information, 4*(1), 199–219. https://doi.org/10.3390/ijgi4010199.

Bowler, D. E., Buyung-Ali, L., Knight, T. M., & Pullin, A. S. (2010). Urban greening to cool towns and cities: A systematic review of the empirical evidence. *Landscape and Urban Planning, 97*(3), 147–155. https://doi.org/10.1016/j.landurbplan.2010.05.006.

Bulkeley, H., Castán Broto, V., & Edwards, G. A. S. (2014). *An urban politics of climate change: Experimentation and the governing of socio-technical transitions*. Routledge.

Castillo, M. D., Anenberg, S. C., Chafe, Z., Huxley, R., Johnson, L. S., Kheirbek, I., Malik, M., Marshall, J., Naidoo, S., Nelson, M. L., Pendleton, N. V., Sun, Y., d'Obrenan, H., van den, B., & Kinney, P. L. (2021). Quantifying the health benefits of urban climate mitigation actions: Current state of the epidemiological evidence and application in health impact assessments. *Frontiers in Sustainable Cities, 3*. https://doi.org/10.3389/frsc.2021.768227.

Costa, F. C.da, Chai, C. G., Carneiro, M. F., Júnior, C. M. A. F., & Adorno, A. M. P. (2023). Flood resilient smart cities and urban safety in a world affected by climate change. *Beijing Law Review, 14*(3), 1541. https://doi.org/10.4236/blr.2023.143083.

Creutzig, F., Baiocchi, G., Bierkandt, R., Pichler, P. P., & Seto, K. C. (2016). Urban infrastructure choices structure climate solutions. *Nature Climate Change, 6*(12), 1054–1056. https://doi.org/10.1038/nclimate3169.

Dasgupta, S., Singh, C., & Kumar, P. (2025). Heat action plans in Indian cities: Knowledge gaps and opportunities for intersectoral heat governance. *Indian Journal of Medical Research, 162*(3), 330–341. 10.25259/IJMR_1019_2025.

Edeigba, B. A., Ashinze, U. K., Umoh, A. A., Biu, P. W., & Daraojimba, A. I. (2024). Urban green spaces and their impact on environmental health: A global review. *World Journal of Advanced Research and Reviews, 21*(2), 917. https://doi.org/10.30574/wjarr.2024.21.2.0518.

Ernst, M., Mentec, S. L., Louvrier, M., Loubet, B., Personne, E., & Stella, P. (2022). Impact of urban greening on microclimate and air quality in the urban canopy layer: Identification of knowledge gaps and challenges. *Frontiers in Environmental Science, 10*. https://doi.org/10.3389/fenvs.2022.924742.

Gloria, S. J., & Gnanasekaran, S. P. (2024). Impact of urban vegetation loss on urban heat islands: A case study of Chennai Metropolitan Area. *Indian Journal of Science and Technology, 17*(2), 134–141. https://doi.org/10.17485/IJST/v17i2.1071.

Gong, C., Xian, C., Wu, T., Liu, J., & Ouyang, Z. (2023). Role of urban vegetation in air phytoremediation: Differences between scientific research and environmental management perspectives. *Npj Urban Sustainability, 3*(1), 24.

Gunawardena, K. R., Wells, M. J., & Kershaw, T. (2017). Utilising green and blue infrastructure to mitigate UHI. *Landscape and Urban Planning, 166*, 104–115.

Gupta, R., Gregg, M., & Muthukumaran, S. (2020). Heat action plans in India: Evaluating effectiveness and identifying gaps. *Climate Policy, 20*(9), 1130–1145. https://doi.org/10.1080/14693062.2020.1792634.

Hussain, S., Hussain, E., Saxena, P., Sharma, A., Thathola, P., & Sonwani, S. (2024). Navigating the impact of climate change in India: A perspective on climate action (SDG13) and sustainable cities and communities (SDG11). *Frontiers in Sustainable Cities, 5*, 1308684. https://doi.org/10.3389/frsc.2023.1308684.

IPCC (2021). Climate change 2021: The physical science basis. In V. Masson-Delmotte, (Ed.). *Contribution of Working Group I to the Sixth Assessment Report of the Intergovernmental Panel on Climate Change* Cambridge University Press. https://doi.org/10.1017/9781009157896.

IPCC (2022). Climate change 2022: Impacts, adaptation and vulnerability. In H.-O. Pörtner, (Ed.). *Contribution of Working Group II to the Sixth Assessment Report of the Intergovernmental Panel on Climate Change* Cambridge University Press. https://doi.org/10.1017/9781009325844.

Iwuanyanwu, O., Gil-Ozoudeh, I., Okwandu, A. C., & Ike, C. S. (2024). Cultural and social dimensions of green architecture: Designing for sustainability and community well-being. *International Journal of Applied Research in Social Sciences, 6*(8), 1951–1968.

Jacob, D. J., & Winner, D. A. (2009). Effect of climate change on air quality. *Atmospheric Environment, 43*(1), 51–63. https://doi.org/10.1016/j.atmosenv.2008.09.051.

Joshi, K. K., Khan, A., Anand, P., & Sen, J. (2024). Understanding the synergy between heat waves and the built environment: A three-decade systematic review informing policies for mitigating urban heat island in cities. *Sustainable Earth Reviews,* 7(1). https://doi.org/10.1186/s42055-024-00094-7.

Kabisch, N., Korn, H., Stadler, J., & Bonn, A. (2017). *Nature-based solutions to climate change adaptation in urban areas: Linkages between science, policy and practice.* Cham: Springer. https://doi.org/10.1007/978-3-319-56091-5.

Kjellstrom, T., Maître, N., Saget, C., Otto, M., & Karimova, T. (2019). *Working on a warmer planet: The impact of heat stress on labour productivity and decent work.* ILO.

Kotharkar, R., Bagade, A., & Surawar, M. (2018). Local climate zone classification for Indian cities: A case study of Nagpur. *Urban Climate, 24*, 369–392.

Krüger, E., Pearlmutter, D., & Rasia, F. (2010). Evaluating the impact of canyon geometry and orientation on cooling loads in a high-mass building in a hot dry environment. *Applied Energy, 87*(6), 2068–2078. 10.1016/j.apenergy.2009.11.034.

Li, T., Horton, R. M., Bader, D. A., Zhou, M., Liang, X., Ban, J., Sun, Q., & Kinney, P. L. (2019). Heat wave and air pollution exposure and their combined effects on human health in China. *Environmental International, 132*, 105–111. https://doi.org/10.1016/j.envint.2019.105105.

Linhares, C. S. F., Gonçalves, R., Martins, L., & Knapic, S. (2021). Structural stability of urban trees using visual and instrumental techniques: A review. *Forests, 12*(12), 1752. https://doi.org/10.3390/f12121752.

Livesley, S. J., McPherson, E. G., & Calfapietra, C. (2016). The urban forest and ecosystem services: Impacts on urban water, heat, and pollution cycles at the tree, street, and city scale. *Journal of Environmental Quality, 45*(1), 119–124. https://doi.org/10.2134/jeq2015.11.0567.

Massaro, E., Schifanella, R., Piccardo, M., Caporaso, L., Taubenböck, H., Cescatti, A., & Duveiller, G. (2023). Spatially-optimized urban greening for reduction of population exposure to land surface temperature extremes. *Nature Communications, 14*(1). https://doi.org/10.1038/s41467-023-38596-1.

Menon, J. S., & Sharma, R. (2021). Nature-based solutions for co-mitigation of air pollution and urban heat in Indian cities. *Frontiers in Sustainable Cities, 3*. https://doi.org/10.3389/frsc.2021.705185.

Miles, V., Esau, I., & Pettersson, L. H. (2023). Using web GIS to promote stakeholder understanding of scientific results in sustainable urban development: A case study in Bergen, Norway. *Sustainable Development, 32*(3), 2517. https://doi.org/10.1002/sd.2787.

Ministry of Environment, Forest and Climate Change (MoEFCC). (2019). *National Clean Air Programme (NCAP)*. Government of India. https://moef.gov.in/wp-content/uploads/2019/05/NCAP_Report.pdf.

Ministry of Housing and Urban Affairs (MoHUA). (2015). *Smart Cities Mission: Statement and guidelines*. Government of India. https://smartcities.gov.in/upload/uploadfiles/files/SmartCityGuidelines(1).pdf.

Mirzaei, P. A., & Haghighat, F. (2010). Approaches to study Urban Heat Island—Abilities and limitations. *Building and Environment, 45*(10), 2192–2201. https://doi.org/10.1016/j.buildenv.2010.04.001.

Nagendra, H., & Gopal, D. (2010). Street trees in Bangalore: Density, diversity, composition and distribution. *Urban Forestry & Urban Greening, 9*(2), 129–137. https://doi.org/10.1016/j.ufug.2009.12.005.

Nastar, M. (2020). Message sent, now what? A critical analysis of the heat action plan in Ahmedabad, India. *Urban Science, 4*(4), 53. https://doi.org/10.3390/urbansci4040053.

Ng, E., Chen, L., Wang, Y., & Yuan, C. (2011). A study on the cooling effects of greening in a high-density city: An experience from Hong Kong. *Building and Environment, 47*, 256–271.

Nowak, D. J., Hirabayashi, S., Bodine, A., & Greenfield, E. (2014). Tree and forest effects on air quality and human health in the United States. *Environmental Pollution, 193*, 119–129. https://doi.org/10.1016/j.envpol.2014.05.028.

Oke, T. R. (1982). The energetic basis of the urban heat island. *Quarterly Journal of the Royal Meteorological Society, 108*(455), 1–24.

Oke, T. R. (1987). *Boundary layer climates* (2nd ed.). Routledge.

Pachouri, V., & Kothari, P. (2024). Optimizing urban sustainability: The effects of green infrastructure and its application in Indian cities. *Research Square*. https://doi.org/10.21203/rs.3.rs-4594516/v1.

Panda, A. (2025). Parametric insurance for urban green zones: Linking NDVI triggers with climate risk in four global cities. *Research Square*. https://doi.org/10.21203/rs.3.rs-7578117/v1.

Perini, K., Ottelé, M., Fraaij, A. L. A., Haas, E. M., & Raiteri, R. (2011). Vertical greening systems and the effect on air flow and temperature on the building envelope. *Building and Environment, 46*(11), 2287–2294. https://doi.org/10.1016/j.buildenv.2011.05.009.

Perrelet, K., Moretti, M., Dietzel, A., Altermatt, F., & Cook, L. M. (2024). Engineering blue-green infrastructure for and with biodiversity in cities. *npj Urban Sustainability, 4*(1), 27.

Picone, N., Esposito, A., Emmanuel, R., & Buccolieri, R. (2023). Potential impacts of green infrastructure on NOx and PM10 in different local climate zones of Brindisi, Italy. *Sustainability, 16*(1), 229. https://doi.org/10.3390/su16010229.

Prasad, N., Ranghieri, F., Shah, F., Trohanis, Z., Kessler, E., & Sinha, R. (2008). *Climate resilient cities: A primer on reducing vulnerabilities to disasters*. The World Bank.

Pugh, T. A., MacKenzie, A. R., Whyatt, J. D., & Hewitt, C. N. (2012). Effectiveness of green infrastructure for improvement of air quality in urban street canyons. *Environmental Science & Technology, 46*(14), 7692–7699. https://doi.org/10.1021/es300826w.

Rahaman, S. N., Nelson, J. R., Ali, A. A. B., Shermin, N., Pricope, N. G., Kafy, A.– A., Sabuj, M. S., & Toa, S. S. (2024). A multivariate geostatistical framework to assess the spatio-temporal dynamics of air pollution and land surface temperature in Bangladesh. *Earth Systems and Environment, 9*, 71–91. https://doi.org/10.1007/s41748-024-00418-9.

Randazza, J. M., Hess, J. J., Bostrom, A., Hartwell, C., Adams, Q. H., Nori-Sarma, A., & Errett, N. A. (2023). Planning to reduce the health impacts of extreme heat: A content analysis of heat action plans in local United States jurisdictions. *American Journal of Public Health, 113*(5), 559–567.

Rosenzweig, C., Solecki, W., Romero-Lankao, P., Mehrotra, S., Dhakal, S., & Ibrahim, S. A. (Eds.). (2018). *Climate change and cities: Second assessment report of the Urban Climate Change Research Network (ARC3.2*Cambridge University Presshttps://doi.org/10.1017/9781316563878.

Roy, S., & Roy, R. (2024). Dynamics of per-urban land alteration in and around Siliguri City, India: A geo-spatial study. *Current World Environment, 19*(1), 454. https://doi.org/10.12944/cwe.19.1.38.

Sailor, D. J. (2011). A review of methods for estimating anthropogenic heat and moisture emissions in the urban environment. *International Journal of Climatology, 31*(2), 189–199. https://doi.org/10.1002/joc.2106.

Salvalai, G., Cadena, J. D. B., & Quagliarini, E. (2023). Greenery as a mitigation strategy to urban heat and air pollution: A comparative simulation-based study in a densely built environment. *Rivista Tema, 9*(2). https://doi.org/10.30682/tema090003.

Santamouris, M. (2015). Regulating the damaged thermostat of the cities—Status, impacts and mitigation challenges. *Energy and Buildings, 91*, 43–56. https://doi.org/10.1016/j.enbuild.2015.01.027.

Schaefer, M. (2022). Between vision and action: The predicted effects of co-designed green infrastructure solutions on environmental burdens. *Urban Ecosystems, 25*(6), 1805. https://doi.org/10.1007/s11252-022-01268-x.

Schnell, J. L., & Prather, M. J. (2017). Co-occurrence of extremes in surface ozone, particulate matter, and temperature over North America. *Nature Climate Change*, 7(9), 657–663. https://doi.org/10.1038/nclimate3321.

Seto, K. C., Dhakal, S., Bigio, A., Blanco, H., Delgado, G. C., Dewar, D., Huang, L., Inaba, A., Kansal, A., Lwasa, S., McMahon, J. E., Müller, D. B., Murakami, J., Nagendra, H., & Ramaswami, A. (2014). Human Settlements, Infrastructure and Spatial Planning. In O. Edenhofer, R. Pichs-Madruga, Y. Sokona, E. Farahani, S. Kadner, K. Seyboth, A. Adler, I. Baum, S. Brunner, P. Eickemeier, B. Kriemann, J. Savolainen, S. Schlömer, C. von Stechow, T. Zwickel, & J. C. Minx (Eds.). *Climate Change 2014: Mitigation of Climate Change. Contribution of Working Group III to the Fifth Assessment Report of the Intergovernmental Panel on Climate Change*Cambridge, United Kingdom and New York, NY, USA: Cambridge University Presshttps://www.ipcc.ch/site/assets/uploads/2018/02/ipcc_wg3_ar5_chapter12.pdf.

Singh, H., Bandyopadhyay, S., Kumar, A., Singh, M., Malik, A., Kumari, G., Raturi, A., Yadav, S. K., Kumar, M., Barthwal, S., Thakur, A., & Rawat, P. S. (2022). Understanding the physiological and biophysical response of urban roadside plantations for assessing adaptation and mitigation mechanisms toward vehicular emissions. *Urban Climate, 44*, 101183. https://doi.org/10.1016/j.uclim.2022.101183.

Śmietanka, T. (2023). The condition of climate change adaptation in polish municipalities before and after the pandemic on the basis of cso environmental indicators-management implications. *Civil and Environmental Engineering Reports, 33*(3), 19–32.

Speak, A. F., Rothwell, J. J., Lindley, S. J., & Smith, C. L. (2012). Urban particulate pollution reduction by four species of green roof vegetation in a UK city. *Atmospheric Environment, 61*, 283–293.

Spronken-Smith, R. A., & Oke, T. R. (1998). The thermal regime of urban parks in two cities with different summer climates. *International Journal of Remote Sensing, 19*(11), 2085–2104. https://doi.org/10.1080/014311698214884.

Stewart, I. D., & Oke, T. R. (2012). Local climate zones for urban temperature studies. *Bulletin of the American Meteorological Society, 93*(12), 1879–1900. https://doi.org/10.1175/BAMS-D-11-00019.1.

Tan, J. K. N., Belcher, R. N., Tan, H., Menz, S., & Schroepfer, T. (2021). The urban heat island mitigation potential of vegetation depends on local surface type and shade. *Urban Forestry & Urban Greening, 62*, 127128. https://doi.org/10.1016/j.ufug.2021.127128.

Tan, P. Y., & bin Abdul Hamid, A. R. (2014). Urban ecological research in Singapore and its relevance to the advancement of urban ecology and sustainability. *Landscape and Urban Planning, 125*, 271–289.

Tan, P. Y., Wang, J., & Sia, A. (2013). Perspectives on five decades of the urban greening of Singapore. *Cities (London, England), 32*, 24–32.

Thyagaraj, V., Hanumesh, M., Muniyappa, M., & Saravanakumar, A. (2026). Innovative green roof materials for India: Synergizing nature-based solutions and sustainable manufacturing for climate-responsive cities. *EPJ Web of Conferences, Vol. 354*, 01008. https://doi.org/10.1051/epjconf/202635401008.

Tiwary, A., Sinnett, D., Peachey, C., Chalabi, Z., Vardoulakis, S., Fletcher, T., & Hutchings, T. R. (2009). An integrated tool to assess the role of new planting in PM10 capture and the human health benefits: A case study in London. *Environmental Pollution, 157*(10), 2645–2653. https://doi.org/10.1016/j.envpol.2009.05.005.

Tun, K. Z., Pramanik, M., Mallick, S. K., Chakrabortty, R., Halder, B., Moharir, K. N., & Zhran, M. (2025). Cooling the cities: A comprehensive review of urban heat island mitigation strategies in Southeast Asia. *Human Settlements and Sustainability, 1*(2), 91–102.

Umoh, A. A., Ohenhen, P. E., Chidolue, O., Ngozichukwu, B., Fafure, A. V., & Ibekwe, K. I. (2024). Incorporating energy efficiency in urban planning: A review of policies and best practices. *Engineering Science & Technology Journal, 5*(1), 83. https://doi.org/10.51594/estj.v5i1.729.

Vardhu, V. A. K., & Sharma, A. (2023). Classification, mitigations and methods to detect UHI: A review. *International Journal of Scientific Research in Engineering and Management,* 7(02).

Visvanathan, G., Patil, K., Suryawanshi, Y., Meshram, V., & Jadhav, S. (2024). Mitigating urban heat island and enhancing indoor thermal comfort using terrace garden. *Scientific Reports, 14*(1), 9697.

Voogt, J. A., & Oke, T. R. (2003). Thermal remote sensing of urban climates. *Remote Sensing of Environment, 86*(3), 370–384. https://doi.org/10.1016/S0034–4257(03)00079-8.

Wajeetongratana, S. (2023). The role of ecological management and green infrastructure in improving air quality. *E3S Web of Conferences, 452*, 5001. https://doi.org/10.1051/e3sconf/202345205001.

Wanner, A., Pröbstl-Haider, U., & Feilhammer, M. (2022). New green infrastructure for European cities: Multiple ways for improvement and climate change adaptation. In *The sustianable City XVI.* Southampton, UK: WIT Press, 121–133.

Witze, A. (2021). Racism is magnifying the deadly impact of rising city heat. *Nature, 595*(7867), 349–351. https://doi.org/10.1038/d41586-021-01881-4.

Wong, N. H., Tan, A. Y. K., Tan, P. Y., Chiang, K., & Wong, N. C. (2010). Acoustics evaluation of vertical greenery systems for building walls. *Building and Environment, 45*(2), 411–420.

Wu, Q., Huang, Y., Irga, P. J., Kumar, P., Li, W., Wei, S., Shon, H. K., Lei, C., & Zhou, J. L. (2024). Synergistic control of urban heat island and urban pollution island effects using green infrastructure. *Journal of Environmental Management, 370*, 122985. https://doi.org/10.1016/j.jenvman.2024.122985.

Xian, C., Gong, C., Wu, T., Liu, J., & Ouyang, Z. (2022). Air pollutant removal by urban vegetation: A meta-analysis of gaps in research and environmental management. *Research Square.* https://doi.org/10.1038/s42949-023-00105-0.

Yadav, A., & Singh, J. (2024). A study on Urban Heat Island (UHI): Challenges and opportunities for mitigation. *Current World Environment, 19*(1), 436. https://doi.org/10.12944/cwe.19.1.37.

Yang, X., Li, B., Li, N., & Ma, X. (2025). A study of the cooling effect of urban trees: Influencing factors, assessment methods, planning strategies, and impacts. *Theoretical and Applied Climatology, 157*(1). https://doi.org/10.1007/s00704-025-05904-2.

Yuan, T. (2024). The role of green infrastructure in mitigating the urban heat island effect. *Open Journal of Applied Sciences, 14*(11), 3155. https://doi.org/10.4236/ojapps.2024.1411207.

Zhang, Y., Balzter, H., & Li, Y. (2021). Influence of impervious surface area and fractional vegetation cover on seasonal urban surface heating/cooling rates. *Remote Sensing, 13*(7), 1263. https://doi.org/10.3390/rs13071263.

Zhou, D., Zhao, S., Liu, S., Zhang, L., & Zhu, C. (2014). Surface urban heat island in China's 32 major cities: Spatial patterns and drivers. *Remote Sensing of Environment, 152*, 51–61. https://doi.org/10.1016/j.rse.2014.05.017.

CHAPTER TWELVE

Future directions in urban plant science and management

Veer Singh[1], Nidhi Singh[2], and Santosh Kumar Singh[2]
[1]Department of Biosciences, School of Physical and Biological Sciences, FoSTA, Manipal University Jaipur, Jaipur, Rajasthan, India,
[2]Centre of Experimental Medicine & Surgery, Institute of Medical Sciences, Banaras Hindu University, Varanasi, Uttar Pradesh, India

12.1 Introduction

The concept of urban air pollution has become one of the most problematic environmental issues in contemporary cities due to the high rate of urbanization, traffic carbon, industrial practices, construction dust, and energy consumption. There are still many urban areas where fine particulate matter, nitrogen oxides, sulfur dioxide, ozone, and volatile organic compounds (VOCs) are above the recommended levels, exposing human populations and vegetation to chronic stress (Das et al., 2026; Singh & Mishra, 2020). Simultaneously, urban green spaces gain importance because plants not only enhance the quality of the landscape and thermal comfort but also trap airborne particles, capture gaseous pollutants, and provide additional regulation to the whole ecosystem (Antenozio et al., 2024). Recently, it has been highlighted that phytoremediation and plant-based air-quality control are becoming viable, low-energy, and nature-based methods of alleviating the pollution load in urban areas (Das et al., 2026; Duan et al., 2023).

Plant adaptation to polluted atmospheres is complicated by a combination of physiological, biochemical, and molecular adaptations. The oxidants or ozone can inhibit photosynthesis, damage membranes, change the operation of stomata, and trigger the formation of reactive oxygen species (ROS), whereas the deposition of particulate matter can decrease the light uptake and block the leaves (Singh & Mishra, 2022; Wedow et al., 2021). These stress reactions influence growth, reproduction, and survival, particularly in delicate species and in highly contaminated corridors, e.g. roadside vegetations and industrial areas (Amaral et al., 2025).

Urban Air Pollution and Plant Responses
DOI: https://doi.org/10.1016/B978-0-443-40320-0.00001-2

Simultaneously, urban vegetation may perform the role of a biological filter by trapping particulates on the leaf surfaces, absorbing some gases via stomata, and reacting with rhizosphere microorganisms that help to transform pollutants (Antenozio et al., 2024). This two-pronged nature of plants as both susceptible perishers and effective mitigating agents makes the study of urban vegetation the key to future air-quality solutions (Amaral et al., 2025; Antenozio et al., 2024).

The increasing rate of interest in the plant responses to pollution has gone beyond the classical physiological studies (Hassan et al., 2025). Recent studies have indicated that omics technologies are useful in the discovery of molecular endpoints of tolerance, such as an alteration in antioxidant mechanisms, stress-signaling networks, and gene expression patterns in relation to exposure to pollutants. These tools enable the comparative study of tolerant and sensitive genotypes, enhance biomonitoring processes, and better select species to be used in urban landscapes (Ahmed et al., 2023; Hassan et al., 2025). This change is significant since most of the conventional planting programs favor ornamental quality or shade but do not carefully consider the tolerance to pollutants, maintenance care, and ecological sustainability of the plant to endure combined pressures like drought, heat, salinity, and air pollution (Raza et al., 2026).

Morphology and cuticle properties at the leaf level determine the deposition efficiency. At the whole-plant level under stress, performance is regulated by growth, transpiration, and carbon assimilation. On a landscape level, the pollutant interception and microclimate control depend on the tree species composition, canopy structure, and spatial distribution (Davi et al., 2007; Yadav et al., 2019). Urban forestry, green infrastructure, and phytoremediation are being progressively incorporated into the planning of sustainable cities at the policy level due to their ability to not only aid in air purification but also provide co-benefits, including enhancement of biodiversity, alleviation of heat islands, and psychological well-being (Das et al., 2026). Recent sources also indicate that plant-based strategies need to be viewed as supplements rather than substitutes for the process of controlling emissions. The use of mechanical and regulatory measures is still necessary, though vegetation can lower exposure and local hotspots, and enhance resilience in areas where the elimination of pollution has not been achieved yet (Antenozio et al., 2024; Kisvarga et al., 2025; Singh et al., 2021).

The chapter is based on the existing evidence of plant tolerance, pollutant-removing capacity, molecular responses, and design principles of urban greening. It encompasses species-specific functionality, seasonal fluctuations, and mixed-pollutant impacts, as well as the scalability of phytoremediation in urbanized metropolitan areas. The chapter synthesizes new developments, hence offering new information that can be utilized in choosing, managing, and breeding plants that can be useful for healthier and more resilient urban landscapes.

12.2 Urban air completions: sources and major pollutants

Transport, industrial activity, power generation, construction, road dust, waste burning, and domestic fuel consumption are factors that generate urban air pollution. These sources in urban areas combine with local meteorology and high-density construction, which enable the accumulation of pollutants in the proximity of the ground and a long-lasting exposure of people and plants (Singh, Singh, Singh, & Verma, et al., 2023). According to the World Health Organization (WHO), the problem of air pollution in the atmosphere is one of the primary health threats in the world, and the aspects that should be considered the most relevant include the presence of fine particulate matter and associated combustion pollutants.

Particulate matter (PM2.5 and PM10), nitrogen oxides (NOx), sulfur dioxide (SO_2), carbon monoxide (CO), ground-level ozone (O_3), VOCs, black carbon, and combustion-related and dust-related toxic metals are the key urban air pollutants. The particular significance of PM2.5 is that it can enter deep into the respiratory system and can also settle on plant surfaces, inhibiting light uptake and gaseous exchange. The most prominent precursors of photochemical smog and ozone are NOx and VOCs, and SO_2 is linked to the burning of fossil fuels and industrial pollutants (Erickson et al., 2020). Table 12.1 provides the summary of the major sources of urban pollution and the pollutants they cause.

According to source distribution studies, transport and industry have still been dominant in most cities, but the pattern is also specific by region and season. New cities have a higher propensity to generate high concentrations of NO_2, CO, and fine particles in their traffic, whereas industrial regions are a greater source of SO_2, metal-containing particulates, and combustion-generated organics. These sources are important to understand, as each of the pollutants has a different impact on urban

Table 12.1 The main urban pollution sources and the pollutants they contribute (Kunak Technologies, 2025; Samet, 2010).

Source	Major pollutants released
Road traffic	NOx, CO, PM2.5, black carbon, VOCs
Industrial facilities	SO2, NOx, PM, metals, toxic organics
Power plants	SO2, NOx, PM, mercury, and other trace metals
Construction activities	PM10, PM2.5, dust, silica particles
Road dust resuspension	PM10, PM2.5, metals, mineral dust
Waste burning	PM, CO, VOCs, dioxins, toxic fumes
Household fuel use	PM2.5, CO, NOx, VOCs
Solvent and fuel evaporation	VOCs, ozone-forming compounds

VOC, volatile organic compound.

plants. The fine particles can be deposited on leaves and close stomata; ozone can cause oxidative stress, and SO_2 or NOx has the power to interfere with photosynthesis and tissue integrity. This source-based approach is required in choosing tolerant plants and planning successful green infrastructure of polluted cities (Erickson et al., 2020).

12.3 The response of the plant to air pollutants

Plants that are exposed to air pollution in urban areas also respond on a morphological, physiological, biochemical, and molecular scale. Such transformations are based on the type of pollutants, the length of exposure, the species of plants, and environmental factors (Banerjee et al., 2022; Singh et al., 2023). Overall, ozone, sulfur dioxide, nitrogen oxides, and particulate matter are some pollutants that may inhibit growth, cause foliage damage, disrupt metabolism, and change stress-signaling pathways (Antenozio et al., 2024).

12.3.1 Morphological responses

The most observable manifestation of air-pollution injury is morphological responses. Characteristic effects are chlorosis, necrosis, leaf curling, premature senescence, loss of leaves, and the suppression of both shoot and root growth (Mehmood et al., 2024; Singh et al., 2024). Contaminated locations are also characterized by shrubbery leaves, epidermal tissue injury, stomatal blockage, and decreased stomatal density or changed stomatal size

(Fadhil et al., 2023). The changes restrict the exchange of gases, and the plants cannot effectively photosynthesize. The degree of sensitivity in species varies quite a bit; thus trees and shrubs are more likely to suffer more foliar damage than others when subjected to the same amount of pollution (Devkota et al., 2024; Fadhil et al., 2023).

12.3.2 Physiological responses

Air pollution disrupts some of the main physiological functions. Stomatal closure is one of the first effects, as it decreases the pollutant entry and, at the same time, limits the CO_2 uptake, thus decreasing the rate of photosynthesis. Pollutants also cause disturbance with the chlorophyll content, pigment stability, transpiration, and membrane integrity (Rani et al., 2023; Singh et al., 2020). Because of this, plants tend to become less efficient in water usage, disturbed in their respiration, and reduced in accumulating biomass (Caporn, 2013). The oxidants, especially ozone, are especially detrimental since they interrupt the process of leaf gas exchange and carbon uptake directly, and the PM can cover the leaf surfaces and decrease the light interception (Caporn, 2013).

12.3.3 Biochemical/molecular responses

Biochemically, the affected plants normally experience the accumulation of ROS, which cause oxidative stress. Plants respond to this by initiating antioxidant defense systems, which include superoxide dismutase, catalase, ascorbate peroxidase, phenolics, flavonoids, and proline. The compounds aid in the free radical scavenging and stabilization of cell structures (Hasanuzzaman et al., 2020). Air pollution can change stress-responsive gene expression and signaling pathways, including jasmonate-linked responses, WRKY, MYB, and AP2/ERF transcription factors molecularly (Hasanuzzaman et al., 2020). The recent studies by omics also indicate that the phenylpropanoid pathway is central to tolerance by promoting the biosynthesis and detoxification of antioxidants. The inability of these defense systems to function in sensitive plants results in membrane lipid peroxidation, loss of pigments, and permanent tissue damage (Antenozio et al., 2024).

12.4 Air pollution plant-based mitigation

Plant-based mitigation also plays a significant role as an addition to emission control due to the ability of vegetation to intercept particles in the atmosphere, absorb gases, and enhance the quality of air locally, as well as provide shade, cooling, and habitat (Li et al., 2023). Recent reviews highlight that the pollutants are removed by plant leaves via adsorption and stomatal absorption, whereas the associated microorganisms in the phyto- and rhizosphere transform and detoxify a part of the compounds (James et al., 2024). This contributes to making vegetation a low-energy, nature-based approach to city settings in which mechanical means of control are sometimes not enough (James et al., 2024; Li et al., 2023).

12.4.1 Phytoremediation

Phytoremediation is the process, commonly used in conjunction with microorganisms, whereby plants are used to eliminate, sequester, convert, or decrease pollutants (Das et al., 2026). It is of particular significance in studies of air pollution of the particulate matter, VOCs, and other airborne pollutants that settle on the leaf surface or enter plant tissues via stomata. Plants can also help in the capturing of particles on the waxy cuticles, trichomes, and rough leaf surfaces, and in the absorption of some gaseous pollutants, which are further metabolized or sequestered (Aryal, 2024). The interactions between the plants and the microbes can contribute a great deal to the effect of the air-pollution remediation in terms of the rate of degradation and detoxification. Due to environmental friendliness and the low-energy requirement, phytoremediation is becoming a feasible solution to the sustainable management of urban air quality (Das et al., 2026; Han et al., 2022; Ravindra & Mor, 2022).

12.4.2 Urban trees and green belts

Large-scale activities that involve urban trees and green cleansing are some of the best strategies that can be used to reduce pollution through plants. Their canopies also serve as biological filters, capturing dust and small particles, particularly in the roadside and industrial environment, where emissions are high (Jeong et al., 2025). Close vegetation and thick cover can also decrease traffic-induced particulate pollution, and typically, the stronger their effectiveness is, the higher the canopy density is and the greater the leaf area (Jeong et al., 2025). Green belts, as well, have the effect of slowing down wind speed and countering the dispersal of pollutants at the ground level, and they forms buffer zones between sources of emissions and residential zones. In addition to air filtration, urban trees cool, sequester carbon, and stabilize the ecological conditions, which reinforces their importance in integrated urban planning (Islam et al., 2012; Jeong et al., 2025).

12.4.3 Vertical gardens and green roofs

Green roofs and vertical gardens are the continuation of vegetation into the built environment where the ground-level planting is restrained. They are used to improve particulate matter, gaseous pollutants, and building microclimates by reducing the surface temperatures and minimizing energy requirements (Kumar et al., 2023). Vertical greenery is particularly beneficial in urban areas where it cannot be cultivated on large plots of land, since it would be utilized to cover more surface area with a relatively small amount of land. Green roofs are also able to control storm water as well as biodiversity and can thus be multifunctional infrastructure instead of single-use pollution filters only (Kumar et al., 2023). These systems, when used with pollutant-resistant species and adequate maintenance, can also play a significant role in the indoor and outdoor air-quality improvement (Han et al., 2022; Kumar et al., 2023).

12.5 More recent and more emergent tools

Recent developments in urban plant science alter the ways in which scholars evaluate the stressors of pollution as well as enhance plant-based mitigation. Rather than using visible damage or simple physiological data, the modern trends involve using a mix of molecular profiling, spatial surveillance, and artificial biological systems to understand the reaction of plants in intricate urban surroundings (Kisvarga et al., 2025). These tools prove to be particularly helpful since urban air pollution frequently does not happen as an individual stressor, but plants are often subjected to combinations of particulate matter, ozone, nitrogen oxides, sulfur dioxide, heat, drought, and heavy metals (Kisvarga et al., 2025).

12.5.1 Omics approaches

Omics methods are genomics, transcriptomics, proteomics, metabolomics, and microbiome analysis. They are collectively used to determine the genes, proteins, metabolites, and microbial partners that are involved in pollutant tolerance. Omics could be used to understand why certain species are more resilient in polluted strips, how the antioxidant defenses are triggered, and what biochemical pathways may be related to detoxification (Jain et al., 2024). They also favor the selection or breeding of pollution-resistant plants for green belts, roadside tree belts, and phytoremediation systems. Due to the magnitude and complexity of the data generated by omics, it is increasingly being used with bioinformatics to find the biomarkers of stress and tolerance. It enhances the selection of plants to be used that is more specific and evidence-based (Naik et al., 2023).

12.5.2 Monitoring by remote sensing and artificial intelligence

The field of remote sensing and artificial intelligence (AI) has increased the size of the scale over which the interaction between plant health and air quality can be studied. The canopy condition, chlorophyll loss, leaf temperature, and patterns of urban pollution can be traced at a city-scale utilizing satellite images, drones, hyperspectral sensors, and ground-based networks (Xu et al., 2025). It is up to AI tools to then process these data to identify stress earlier, forecast hotspots of pollutants, and model the effect of vegetation in reducing exposure under varying urban patterns. The techniques are useful in the urban planning of large urban areas since they connect the performance of plants with real-time environmental variation. They also assist in managing to track the decline of trees, the effectiveness of green covers, and the mitigation of pollution at more cost-effective rates than using field surveys (Sharma & Shivandu, 2024).

12.5.3 Nanotechnology and microbial assistance

Nanotechnology can provide innovative options for making plants more tolerant and eliminating pollutants. Nanoparticles can enhance nutrient transport, enhancing antioxidant protection and, in certain cases, enhancing the capture/conversion of contaminants (Gupta & Saharan, 2026). In other instances, the presence of nano-enabled coating or formulations can assist the leaves to react better with the air quality pollutants. Meanwhile, microbial assistance is also becoming popular as an environmentally friendly solution (Gupta & Saharan, 2026). In certain cases, beneficial rhizobacteria, endophytes, and phyllosphere microbes can enhance plant growth, enhance stress tolerance, degrade pollutants, or minimize oxidative stress. Nanotechnology combined with microbial inoculation could deliver synergy, but their benefits and drawbacks in fields require close observation (Chen et al., 2025).

12.6 Problems and research gaps

Although there is increased attention on the use of plants in solving air pollution in cities, there are still numerous challenges that have restricted their usage. A huge variability of the plant responses among species, cultivars, seasons, and mixtures of pollutants is one of the challenges. A plant that could be tolerant to one urban environment condition might fail to be tolerant to another, particularly ozone, particulate matter, heat stress, drought, and soil contamination. This complicates the process of species selection and makes it site-specific (Khan et al., 2025). The other critical area of concern is the disconnect between laboratory research and actual urban performance. Controlled conditions encompass many experiments, whereas the field environment is far more complex. Pollutant capture and plant health are dependent on the wind patterns, traffic density, the structure of buildings, rainfall, the position of the leaf, and age, as well as the maintenance practices. This can lead to the fact that lab results are not necessarily directly applicable to trustworthy urban results (Barwise & Kumar, 2020).

Most of the studies quantify the short-term alterations in the pollutant concentrations or leaf damage, but not many of them determine the performance of the plants over a period of years (Tomson et al., 2020). There should be long-term observation of the dynamics of canopies, seasonal dynamics, the mortality of plants, and the permanence of the benefits of mitigating pollution. In the absence of this kind of data, sustainable green infrastructure can hardly be designed. Another gap is related to the molecular and ecological processes involved in tolerance (Tomson et al., 2020). The precise mechanisms that dictate resilience remain unclear, even though recent research has found the antioxidant responses, stress-related genes, and interactions among microbiomes. The omics-based and multistress studies should be furthered to elucidate plant adaptation to mixed exposures to pollutants in the actual urban environments. Maintenance and implementation are another matter of concern (Tomson et al., 2020; Zhang et al., 2025). Urban trees, green roofs, and vertical gardens need water, trimming, nutrient care, and replacement at various intervals. These expenses can limit success in the long term in a resource-strained city. Problems like allergen release, invasive propagation, or unreasonable water consumption can also be caused by improper selection of the species (Tomson et al., 2020; Zhang et al., 2025). Future studies on field validation, multipollutant interactions, species-specific tolerance, performance across their lifecycle, and cost-effective management models, thus need to be done. Further association between plant science and urban planning, atmospheric chemistry, and environmental policy is needed to transform plant-based mitigation into a feasible and scalable solution.

12.7 Future perspectives

The future research on urban plants will shift to data-driven, resilient, and multifunctional solutions, which not only solve air pollution but also increase the sustainability of the city. Plant scientists, urban planners, ecologists, and policymakers need to work closely interdisciplinarily to translate the research into practical use. It has several significant points that should be taken into consideration:

- Develop pollution-tolerant plant species through genomics, breeding, and omics-based screening.
- Integrate trees, green roofs, and vertical gardens into urban planning for broader air-quality benefits.
- Use remote sensing, AI, and sensor networks to monitor plant health and pollution reduction in real time.
- Strengthen plant–microbe partnerships to improve pollutant uptake, detoxification, and stress tolerance.
- Test plant-based systems under real urban conditions for long-term effectiveness and scalability.
- Combine green infrastructure with emission control policies for stronger air-quality improvement.
- Focus on multifunctional landscapes that provide cooling, carbon sequestration, biodiversity, and public health benefits

12.8 Conclusion

The study of urban plants is emerging as a constituent aspect of the management of sustainable cities. Plants are not only better for enhancing the beauty of the landscape, but also they help to minimize air pollution and heat absorption and promote biodiversity in cities. Their responses to pollution are highly informative about the state of environmental stress, and new technologies like omics, remote sensors, and AI are making it possible to choose and control tolerant species. Despite the problems with species selection, preservation, and field validation over time, plant-based solutions provide the potential of a low-energy approach to making cities healthier. For future research, scientific, ecological, and policy strategies should be incorporated into future urban planning to ensure that vegetation fully takes part in the purification of air and resilience of the city.

Acknowledgment

Dr. Veer Singh is thankful to Manipal University Jaipur, Jaipur, for providing research and academic support. Dr. Nidhi Singh and Dr. Santosh Kumar Singh are thankful to the Banaras Hindu University, Varanasi, for the necessary research support.

References

Ahmed, G., Jamal, F., Tiwari, R. K., Singh, V., Rai, S. N., Chaturvedi, S. K., Pandey, K., Singh, S. K., Kumar, A., Narayan, S., & Vamanu, E. (2023). Arsenic exposure to mouse visceral leishmaniasis model through their drinking water linked to the disease exacerbation via modulation in host protective immunity: A preclinical study. *Scientific Reports, 13*(1). https://doi.org/10.1038/s41598-023-48642-z.

Amaral, E. A., de Alencar Neves, T., Albuquerque, T. T. D. A., Santos, I. B., Massonetto, M. M., Kumar, P., & Maria, M. C. (2025). The role of plants as biological monitors of air pollution: State of the art on methods and challenges with a focus on Lolium sp. *Science of the Total Environment, 1005*. https://doi.org/10.1016/j.scitotenv.2025.180816.

Antenozio, M. L., Caissutti, C., Caporusso, F. M., Marzi, D., & Brunetti, P. (2024). Urban air pollution and plant tolerance: Omics responses to ozone, nitrogen oxides, and particulate matter. *Plants, 13*(15). https://doi.org/10.3390/plants13152027.

Aryal, M. (2024). Phytoremediation strategies for mitigating environmental toxicants. *Heliyon, 10*(19), e38683. https://doi.org/10.1016/j.heliyon.2024.e38683.

Banerjee, S., Banerjee, A., & Palit, D. (2022). Morphological and biochemical study of plant species- a quick tool for assessing the impact of air pollution. *Journal of Cleaner Production, 339*, 130647. https://doi.org/10.1016/j.jclepro.2022.130647.

Barwise, Y., & Kumar, P. (2020). Designing vegetation barriers for urban air pollution abatement: A practical review for appropriate plant species selection. *Npj Climate and Atmospheric Science, 3*(1). https://doi.org/10.1038/s41612-020-0115-3.

Caporn, S. J. M. (2013). Ecophysiological responses of plants to air pollution. In *Encyclopedia of life sciences*. Wiley. 10.1002/9780470015902.a0003206.pub2.

Chen, X., Wang, S., Lai, H., Deng, L., Zhong, Q., Okoye, C. O., Niu, Q., Jing, Y., Wang, J., & Jiang, J. (2025). Nano-biotechnology in soil remediation: Use of nanomaterials to promote plant growth and stress tolerance. *Nanomaterials, 15*(22), 1743. https://doi.org/10.3390/nano15221743.

Das, S., Saha, A., Saha, D., Roy, K., & Nazir, M. (2026). Phytoremediation for air quality: A sustainable solution for urban resilience and post-covid green recovery. *Current World Environment, 3*(20), 971. https://doi.org/10.12944/cwe.20.3.2.

Davi, H., Barbaroux, C., Dufrêne, E., François, C., Montpied, P., Bréda, N., & Badeck, F. (2007). Modelling leaf mass per area in forest canopy as affected by prevailing radiation conditions. *Ecological Modelling, 211*(3–4), 339–349. https://doi.org/10.1016/j.ecolmodel.2007.09.012.

Devkota, A., Shrestha, S., & Jha, P. K. (2024). Effect of air pollution on the leaf morphology and anatomy of common plant species of Kathmandu Valley. *European Journal of Ecology, 10*(1). https://doi.org/10.17161/eurojecol.v10i1.19487.

Duan, X., Gu, H., Lam, S. S., Sonne, C., Lu, W., Li, H., Chen, X., & Peng, W. (2023). Recent progress on phytoremediation of urban air pollution. *Chemosphere, 349*, 140821. https://doi.org/10.1016/j.chemosphere.2023.140821.

Erickson, L. E., Newmark, G. L., Higgins, M. J., & Wang, Z. (2020). Nitrogen oxides and ozone in urban air: A review of 50 plus years of progress. *Environmental Progress and Sustainable Energy, 39*(6). https://doi.org/10.1002/ep.13484.

Fadhil, A. A., Mohammed, S. J., & Al-Abboodi, A. (2023). Morphological responses of plants to air pollutants: A comparative study on leaf changes in five species. *Iranian Journal of Ichthyology, 10*(Special Issue 1), 286–293.

Gupta, A., & Saharan, B. S. (2026). Nanotech innovations in agriculture: Enhancing plant growth and navigating microbial interactions. *Next Bioengineering, 2*, 100012. https://doi.org/10.1016/j.nxbio.2026.100012.

Han, Y., Lee, J., Haiping, G., Kim, K.-H., Wanxi, P., Bhardwaj, N., Oh, J.-M., & Brown, R. J. C. (2022). Plant-based remediation of air pollution: A review. *Journal of Environmental Management, 301*, 113860. https://doi.org/10.1016/j.jenvman.2021.113860.

Hasanuzzaman, M., Bhuyan, M. H. M., Zulfiqar, F., Raza, A., Mohsin, S., Mahmud, J., Fujita, M., & Fotopoulos, V. (2020). Reactive oxygen species and antioxidant defense in plants under abiotic stress: Revisiting the crucial role of a universal defense regulator. *Antioxidants, 9*(8), 681. https://doi.org/10.3390/antiox9080681.

Hassan, S. H., Simiele, M., Scippa, G. S., Morabito, D., & Trupiano, D. (2025). Omics advancements towards exploring arsenic toxicity and tolerance in plants: A review. *Planta, 261*(4). https://doi.org/10.1007/s00425-025-04646-9.

Islam, M. N., Rahman, K. S., Bahar, M. M., Habib, M. A., Ando, K., & Hattori, N. (2012). Pollution attenuation by roadside greenbelt in and around urban areas. *Urban Forestry and Urban Greening, 11*(4), 460–464. https://doi.org/10.1016/j.ufug.2012.06.004.

Jain, A., Sarsaiya, S., Singh, R., Gong, Q., Wu, Q., & Shi, J. (2024). Omics approaches in understanding the benefits of plant-microbe interactions. *Frontiers in Microbiology, 15*. https://doi.org/10.3389/fmicb.2024.1391059.

James, A., Rene, E. R., Bilyaminu, A. M., & Chellam, P. V. (2024). Advances in amelioration of air pollution using plants and associated microbes: An outlook on phytoremediation and other plant-based technologies. *Chemosphere, 358*. https://doi.org/10.1016/j.chemosphere.2024.142182.

Jeong, J., Kim, C., Choi, S., Sou, H.-D., & Park, C.-R. (2025). Long-term greenness effects of urban forests to reduce PM10 concentration: Does the impact benefit the population vulnerable to asthma? *International Journal of Environmental Research and Public Health, 22*(2), 167. https://doi.org/10.3390/ijerph22020167.

Khan, N. B., Chatterjee, S., Basak, G., Sarkar, P., Singh, V. K., Adit, A., & Barman, C. (2025). Impact of air pollution on reproductive biology of plants: Mechanisms and consequences. *Plant Stress, 18*. https://doi.org/10.1016/j.stress.2025.101119.

Kisvarga, S., Horotán, K., Neményi, A., Jana, T., Istvánfi, Z., & Orlóci, L. (2025). The ecology of urban vegetation: Trade-offs, stressors, and adaptive strategies. *Environmental and Sustainability Indicators, 28*, 100887. https://doi.org/10.1016/j.indic.2025.100887.

Kumar, R., Verma, V., Thakur, M., Singh, G., & Bhargava, B. (2023). A systematic review on mitigation of common indoor air pollutants using plant-based methods: A phytoremediation approach. *Air Quality, Atmosphere & Health, 16*(8), 1501–1527. https://doi.org/10.1007/s11869-023-01326-z.

Kunak Technologies, S. L. (2025). *Urban air pollution: Causes, impacts and solutions*. Kunak June 3.

Li, Y., Chen, X., Sonne, C., Lam, S. S., Yang, Y., Ma, N. L., & Peng, W. (2023). Reduction and control of air pollution: Based on plant-microbe interactions. *Environmental Pollutants and Bioavailability, 35*(1). https://doi.org/10.1080/26395940.2023.2173657.

Mehmood, Z., Yang, H.-H., Awan, M., Ahmed, U., Hasnain, A., Luqman, M., Muhammad, S., Sardar, A., Chan, T.-Y., & Sharjeel, A. (2024). Effects of air pollution on morphological, biochemical, DNA, and tolerance ability of roadside plant species. *Sustainability, 16*(8), 3427. https://doi.org/10.3390/su16083427.

Naik, B., Kumar, V., Rizwanuddin, S., Chauhan, M., Choudhary, M., Gupta, A. K., Kumar, P., Kumar, V., Saris, P. E. J., Rather, M. A., Bhuyan, S., Neog, P. R., Mishra, S., & Rustagi, S. (2023). Genomics, proteomics, and metabolomics approaches to improve abiotic stress tolerance in tomato plant. *International Journal of Molecular Sciences, 24*(3). https://doi.org/10.3390/ijms24033025.

Rani, A., Rajlaxmi, A., & Kumar, M. (2023). Effect of air pollutants on the carbon sequestration rate and other physiological processes in different vegetation types: A review. *MAPAN, 38*(4), 1053–1065. https://doi.org/10.1007/s12647-023-00643-z.

Ravindra, K., & Mor, S. (2022). Phytoremediation potential of indoor plants in reducing air pollutants. *Frontiers in Sustainable Cities, 4*. https://doi.org/10.3389/frsc.2022.1039710.

Raza, A., Li, Y., Charagh, S., Guo, C., Zhao, M., & Hu, Z. (2026). Melatonin-enabled omics: Understanding plant responses to single and combined abiotic stresses for climate-smart agriculture. *GM Crops & Food, 17*(1). https://doi.org/10.1080/21645698.2026.2614130.

Samet, J. M. (2010). Urban air quality. In D. Vlahov, J. I. Boufford, C. E. Pearson, & L. Norris (Eds.). *Urban health: Global perspectives* (pp. 317–338). Hoboken (NJ): John Wiley & Sons.

Sharma, K., & Shivandu, S. K. (2024). Integrating artificial intelligence and Internet of Things (IoT) for enhanced crop monitoring and management in precision agriculture. *Sensors International, 5*, 100292. https://doi.org/10.1016/j.sintl.2024.100292.

Singh, N., Singh, V., & Singh, M. P. (2023). Microbial degradation of lignocellulosic biomass for bioenergy production: A metagenomic-based approach. *Biocatalysis and Biotransformation, 41*(1), 15–25. https://doi.org/10.1080/10242422.2022.2056451.

Singh, V., & Mishra, V. (2020). Coronavirus Disease 2019 (COVID-19): Current situation and therapeutic options. *Coronaviruses, 2*(4), 481–491. https://doi.org/10.2174/2666796701999201005211854.

Singh, V., & Mishra, V. (2022). Sustainable reduction of Cr (VI) and its elemental mapping on chitosan coated citrus limetta peels biomass in synthetic wastewater. *Separation Science and Technology, 57*(10), 1609–1626. https://doi.org/10.1080/01496395.2021.1993921.

Singh, V., Singh, N., Tabassum, N., & Mishra, V. (2020). Microbial System: An emerging application in the bioenergy production. In N. Srivastava, M. Srivastava, P. K. Mishra, & V. K. Gupta (Eds.). *Microbial Strategies for Techno-economic Biofuel Production. Clean Energy Production Technologies* (pp. 249–264). Singapore: Springer. https://doi.org/10.1007/978-981-15-7190-9_9.

Singh, V., Tiwari, R., Chaturvedi, V. K., Singh, N., & Mishra, V. (2021). Microbiological aspects of bioenergy production: Recent update and future directions. In M. Srivastava, N. Srivastava, & R. Singh (Eds.). *Bioenergy Research: Revisiting Latest Development. Clean Energy Production Technologies* (pp. 29–52). Singapore: Springer. https://doi.org/10.1007/978-981-33-4615-4_2.

Singh, V., Singh, J., Singh, N., Verma, M. K., Praveena, S. M., & Mishra, V. (2023). Sustainable removal of Pb (II) from aqueous medium by using chitosan functionalized Citrus limetta peels biomass. *Bioremediation Journal, 28*(3), 245–265. https://doi.org/10.1080/10889868.2023.2269188.

Singh, V., Singh, N., Rai, S. N., Chaturvedi, V. K., Singh, S. K., Kumar, A., Vamanu, E., & Mishra, V. (2024). Chitosan functionalized recyclable and eco-friendly nanoadsorbent for Pb(II) adsorption from water. *Nanocomposites, 10*(1), 201–213. https://doi.org/10.1080/20550324.2024.2347804.

Tomson, M., Kumar, P., Barwise, Y., Perez, P., Forehead, H., French, K., Morawska, L., & Watts, J. F. (2020). Green infrastructure for air quality improvement in street canyons. *Environment International, 146*, 106288. https://doi.org/10.1016/j.envint.2020.106288.

Wedow, J. M., Ainsworth, E. A., & Li, S. (2021). Plant biochemistry influences tropospheric ozone formation, destruction, deposition, and response. *Trends in Biochemical Sciences, 46*(12), 992–1002. https://doi.org/10.1016/j.tibs.2021.06.007.

Xu, L., Chen, L., Luo, Q., Zhao, S., Huang, J., Wang, K., Yang, Z., Weng, X., Fang, K., & Feng, H. (2025). Leveraging UAV hyperspectral imaging for crop physiology and biochemistry: A comprehensive review of feature extraction and selection methods. *Plant Phenomics, 8*(1), 100141. https://doi.org/10.1016/j.plaphe.2025.100141.

Yadav, V. K., Singh, V., & Mishra, V. (2019). Alkaline protease: A tool to manage solid waste and its utility in detergent industry. In *Microbial genomics in sustainable agroecosystems: Volume 2*. Singapore: Springer, 231–254. 10.1007/978-981-32-9860-6_14.

Zhang, B., Cao, Q., Zhang, Q., Shen, H., Zhang, Y., Lu, Z., Guo, X., Yu, Y., & Jia, X. (2025). Bidirectional effect modification of greenness exposure and air pollution on ischemic stroke mortality across different temporal scales. *BMC Public Health, 25*(1). https://doi.org/10.1186/s12889-025-25066-0.

Index

Note: Page numbers followed by "*f*" and "*t*" refer to figures and tables, respectively.

R

S

Printed and bound by CPI Group (UK) Ltd, Croydon, CR0 4YY
17/09/2026
02218551-0001